The Physics of Structural Phase Transitions

Springer
New York
Berlin
Heidelberg
Barcelona
Budapest
Hong Kong
London
Milan
Paris
Santa Clara
Singapore
Tokyo

Minoru Fujimoto

The Physics of Structural Phase Transitions

With 90 Figures

Springer

Minoru Fujimoto
Department of Physics
University of Guelph
Ontario
Canada N1G 2W1

Library of Congress Cataloging-in-Publication Data
Fujimoto, Minoru.
The physics of structural phase transitions / Minoru Fujimoto.
p. cm.
Includes bibliographical references and index.
ISBN 0-387-94856-2 (alk. paper)
1. Phase transformations (Statistical physics) 2. Crystals.
3. Lattice dynamics. I. Title.
QC175.16.P5F86 1997
530.4′14—dc20 96-33166

Printed on acid-free paper.

Production coordinated by Brian Howe and managed by Bill Imbornoni; manufacturing supervised by Johanna Tschebull.
Typeset by Asco Trade Typesetting Ltd., Hong Kong.
Printed and bound by Maple Vail Book Manufacturing Group, York, PA.
Printed in the United States of America.

9 8 7 6 5 4 3 2 1

ISBN 0-387-94856-2 Springer-Verlag New York Berlin Heidelberg SPIN 10547020

To the memory of Professor M. Takéwaki
who inspired me with fantasy in
thermodynamics

Preface

Structural phase transitions constitute a fascinating subject in solid state physics, where the problem related to lattice stability is difficult but challenging to statistical principles for equilibrium thermodynamics. Guided by the Landau theory and the soft-mode concept, many experimental studies have been performed on a variety of crystalline systems, while theoretical concepts, acquired mainly from isotropic systems, are imposed on structural changes in crystals. In any case, since the mean-field approximation is known to be inadequate for critical regions, theories need to be modified to deal with local inhomogeneity and incommensurate aspects, and which are discussed with the renormalization group theory in recent work. In contrast, there are many experimental results that are left unexplained, some of which are even necessary to be evaluated for their relevance to intrinsic occurrence. Under these circumstances, I felt that the basic concepts introduced early on need to be reviewed for the better understanding of structural changes in crystals.

Phase transitions in crystals should, in principle, be an interplay between order variables and phonons. While it has not been seriously discussed so far, I have found that an idea similar to charge-density-wave condensates is significant for ordering phenomena in solids. I was therefore motivated to write this monograph, where basic concepts for structural phase transitions are reviewed in light of the Peierls idea. I have written this book for readers who have a basic knowledge of solid state physics at the level of *Introduction to Solid State Physics* by C. A. Kittel. In this monograph, basic physics for continuous-phase transitions is discussed, referring to experimental evidence, without being biased by existing theoretical models. Since many excellent review articles are available, this monograph is not another comprehensive review of previous experimental results. Although basic concepts are discussed with emphasis, this monograph is by no means theoretical, and can be used as a textbook, or reference material for extended discussions in solid state physics.

The contents of this monograph are divided into two parts for convenience. In Part One, I discuss the basic elements for continuous structural changes to introduce the model of pseudospin condensates, and in Part Two various methods of investigation are discussed, thereby the properties of condensates may be revealed. In Chapter 10, work on representative systems is summarized to conclude this monograph, where the results are interpreted in light of the condensate description of critical fluctuations.

I am enormously indebted to many of my colleagues who helped me in writing this book. I owe a great deal to S. Jerzak, J. Grindley, G. Leibrandt, D. E. Sullivan, H.-G. Unruh, G. Schaack, J. Stankowsky, W. Windsch, A. Janner, and E. de Boer for many constructive criticisms and encouragement. Among them, Professor Windsch took time to read through an early version of the manuscript and gave me valuable comments and advice; Professor Unruh kindly provided me with photographs of discommensuration patterns in K_2ZnCl_4 systems; and Dr. Jerzak helped me to obtain information regarding $(NH_4)_2SO_4$ and $RbH_3(SeO_3)_2$, and to whom I express my special gratitude. Finally, I thank my wife, Haruko, for her continuous encouragement during my writing, without which this book could not have been completed.

"It was like a huge wall!," said a blind man.
"Oh, no! It was like a big tree," said another blind man.
"You are both wrong! It was like a large fan!," said another.
After listening to the blind people, the Lord said, "Alas! None of you have seen the elephant!"

From East-Indian folklore.

A Remark on Bracket Notations

Somewhat unconventional bracket notations are used in this monograph. Here, while the notation $\langle Q \rangle$ generally signifies the spatial average of a distributed quantity Q, $\langle Q \rangle_t$ indicates the temporal average over the timescale t_0 of observation.

In Chapters 8 and 9, the *bra* and *ket* of a vector quantity v, i.e., $\langle v|$ and $|v\rangle$, respectively, are used to express the corresponding row and column matrices in three-dimensional space to facilitate matrix calculations. Although confusing at a glance, I do not think that such bracket notations would cause inconvenience to distinguish them from averages.

Guelph, Ontario Minoru Fujimoto
April 1996

Contents

PART ONE

BASIC CONCEPTS

A structural transformation in a crystal can take place when some distortion is developed in *active* constituent groups either individually or collectively, and which is macroscopically characterized by a change in lattice symmetry. Landau defined the order parameter in terms of irreducible representations of the symmetry element signifying a structural change, while the origin for the phase transition can be attributed to a change in active groups. Being considered as ordering phenomena in crystals, structural transitions should be closely related to the lattice in principle. We consider that an interplay between active groups and their hosting lattice is responsible for a structural change. On the other hand, Cochran introduced early on the concept of soft phonons to deal with lattice instability, which is nevertheless caused by two competing dielectric interactions in ionic crystals.

Although generally acceptable, there is still some confusion about these concepts when applied to structural problems as originally defined. Therefore, I have reconsidered their physical implications for practical crystals, so that critical anomalies observed in various experiments can be interpreted in terms of the interplay between the two aspects of phase transitions. Such anomalies were observed, in fact, depending on the timescale of experiments, although it is generally assumed as infinity in most statistical arguments based on the ergodic hypothesis. In Part One, we pay serious attention to the timescale in reviewing existing concepts, since the observing timescale has to be compared with the characteristic time for critical fluctuations. In Chapter 1, thermodynamical principles are discussed for structural phase transitions, and in Chapter 2 statistical concepts for ordering phenomena are reviewed. In Chapter 3, use of pseudospins is proposed for binary systems, and their correlations are discussed for the singular behavior at transition temperatures. The role played by *soft phonons* is discussed in Chapter 4, where the concept of *condensates* is introduced for the critical region, representing complexes of pseudospins and soft phonons. In Chapter 5, dynamics of condensates and their nonlinear character are discussed in relation to the ordering process in practical crystals.

CHAPTER 1

Thermodynamical Principles and the Landau Theory of Phase Transitions

1.1. Introduction

Phase transitions in condensed matter can basically be interpreted within the scope of thermodynamical principles, while for critical regions precise knowledge of transition mechanisms is essential. In most textbooks on thermodynamics [1], [2], [3], phase equilibria in isotropic media are discussed at some length as simple examples, while structural phase transitions in crystals are complex and described only in a sketchy manner [4], [5]. In nature, there are various types of phase transitions, which Ehrenfest [6] classified in terms of a derivative of the thermodynamical potential exhibiting a discontinuous change at T_c. Among others, the second-order phase transition characterized as a continuous change of the Gibbs potential has attracted many investigations, since the problem is closely related to a fundamental subject of lattice stability. In this chapter, the continuous-phase transition is discussed in the light of thermodynamical principles, but it is significant that critical anomalies and subsequent *domain* formation in the low-temperature phase cannot be properly elucidated, hence pertaining to an area beyond the limit of classical thermodynamics.

Landau [7] formulated a theory of continuous-phase transitions in *binary* systems, which is sketched in Section 1.5. In his theory, a single thermodynamical variable called the *order parameter* emerges at T_c, signifying the ordered phase below T_c by its nonzero value. He proposed that the variation of the Gibbs potential near T_c is expressed by an infinite power series of the order parameter, implying that ordering is essentially a *nonlinear* process. While the order parameter is a well-accepted concept in a uniformly ordered phase, it is well known that the Landau theory is inadequate to explain critical anomalies. The failure can be attributed to the fact that the theory is not dealing with spontaneous inhomogeneity due to distributed critical strains in otherwise uniform crystals. Landau recognized such shortcomings in his abstract theory, and suggested including spatial derivatives of the order parameter in the Gibbs potential for an improved description of phase transitions. In such a revised Landau expansion, an additional term, called the Lifshitz term that is composed of such derivatives, for example, is known to be responsible for a modulated structure in crystals. However, it is still not clear in such a revised theory if critical anomalies can be attributed to a dynamical behavior of the order parameter.

Needless to say, phase transitions are phenomena in macroscopic scale. In a noncritical phase away from T_c, if sufficiently uniform, thermodynamic properties can be described by the *ergodic* average of distributed microscopic variables that correspond to the order parameter. In contrast, the critical region is dictated by short-range correlations among these variables in slow motion at a specific wavevector, for which thermal averages are obviously inadequate. In a modified theory, known as the Landau–Ginzburg theory [8], derivatives of the order parameter are considered to deal with spatial inhomogeneity, although the nature of critical fluctuations is not described properly. Under these circumstances, the traditional thermodynamical approach still provides an approximate pathway to the problem of structural changes. While a reliable model for the transition mechanism has not yet been established, for experimental studies it is a prerequisite to identify the order parameter in terms of constituent ions and molecules. This chapter is devoted to reviewing thermodynamical principles relevant to structural phase transitions in anisotropic crystals, to which the Landau theory has only limited access. In view of the existence of many articles and books on liquid and magnetic systems, particularly an excellent monograph by Stanley [9], our discussion on isotropic systems must be limited to minimum necessity.

1.2. Phase Equilibria in Isotropic Systems

Thermodynamical properties of an isotropic, chemically pure and homogeneous substance can be specified by the Gibbs potential $G(p, T)$, where the pressure p and the temperature T represent the surroundings in equilibrium with it. On the other hand, as evident from an example of liquid–vapor equilibrium, the substance may not necessarily be homogeneous under a given p-T condition. Liquid and vapor phases are homogeneous parts of a condensing system, and hence characterized by the separate Gibbs potentials G_1 and G_2 representing properties of these phases individually. Under a given external condition, the coexisting phases can be in stable equilibrium that is maintained by heat and mass exchanges. Accordingly, these partial potentials should be specified by internal variables related to the numbers of constituent molecules N_1 and N_2 to distinguish one phase from the other, while p and T remain as external variables representing the common surroundings. Although very different from crystalline systems, knowledge of isotropic phase equilibria provides a useful guideline to study structural phase transitions in crystals.

The equilibrium of a substance consisting of two phases under a given p-T condition is determined by minimizing the total Gibbs potential $G = G_1 + G_2$, namely,

$$dG = 0 \qquad \text{where} \quad G_1 = G_1(N_1, p, T) \qquad \text{and} \qquad G_2 = G_2(N_2, p, T)$$

and

$$N_1 + N_2 = N = \text{const.}$$

The phase equilibrium is therefore specified by

$$\left(\frac{\partial G_1}{\partial N}\right)_{p,T} = \left(\frac{\partial G_2}{\partial N_2}\right)_{p,T}.$$

Here the quantity $\mu = (\partial G/\partial N)_{p,T}$ is called the *chemical potential*, being identical to the Gibbs potential per molecule. Using chemical potentials defined for the two phases, the equilibrium condition can be expressed as

$$\mu_1(p, T) = \mu_2(p, T), \tag{1.1}$$

which indicates that the pressure p and the temperature T for equilibrium are not independent. Graphically, in the p-T diagram, the two phases are represented by areas separated by a curve given by (1.1), along which phase equilibria are maintained at all points (p, T).

When such equilibria are compared at two different temperatures T and $T + \delta T$ on the equilibrium line, where $\delta T \ll T$, there should be a change in the equilibrium pressure $\delta p = (dp/dT)\, \delta T$. The rate dp/dT along the line can be obtained from arbitrary variations in equilibrium, i.e.,

$$\delta\mu_1 = \delta\mu_2, \tag{1.2}$$

where

$$\delta\mu_1 = \left\{\left(\frac{\partial \mu_1}{\partial T}\right)_p + \left(\frac{\partial \mu_1}{\partial p}\right)_T \left(\frac{dp}{dT}\right)\right\} \delta T = \left\{-s_1 + v_1\left(\frac{dp}{dT}\right)\right\} \delta T$$

and

$$\delta\mu_2 = \left\{-s_2 + v_2\left(\frac{dp}{dT}\right)\right\} \delta T.$$

Here s_i and v_i (i = 1, 2) are specific entropies and volumes of phases 1 and 2, respectively. Therefore, from (1.2) we can derive the relation

$$\frac{dp}{dT} = \frac{s_1 - s_2}{v_1 - v_2} = \frac{\Delta s}{\Delta v}. \tag{1.3}$$

Here, differences Δs and Δv signify a structural change between the two phases: a finite entropy difference corresponds to the latent heat per molecule $L = T\,\Delta s$, and a finite Δv indicates a packing difference between the two phases. Due to discontinuous first derivatives of the chemical potential, i.e., $s = -(\partial\mu/\partial T)_p$ and $v = (\partial\mu/\partial p)_T$, such phase equilibria are first-order transitions, according to the Ehrenfest classification. Equation (1.3), known as the Clausius–Clapeyron equation, determines the rate at which the equilibrium pressure varies with the equilibrium temperature in the p-T diagram. As an example, the reciprocal derivative $\mathrm{d}T/\mathrm{d}p$ determines a variation of the transition temperature with pressure, which is positive for liquid–vapor transitions, because $v_{\text{vapor}} \gg v_{\text{liquid}}$, where the latent heat is always absorbed by vapor. We also note that the boiling point of liquid always rises with increasing pressure, whereas, when applying (1.3) to a liquid–solid transition, the freezing temperature can either rise or fall, depending on the sign of Δv during solidification.

A useful expression for the vapor pressure of a liquid can be derived by integrating (1.3). Ignoring very small v_{liquid} as compared with much larger v_{vapor}, the vapor pressure p_1 can be determined from the equation

$$\frac{\mathrm{d}p_1}{\mathrm{d}T} = \frac{L}{Tv_1} = \frac{Lp_1}{\mathrm{k_B}T^2},$$

where the vapor is assumed to obey the ideal gas law, i.e., $v_1 = \mathrm{k_B}T/p_1$, and $\mathrm{k_B}$ is the Boltzmann constant. Assuming further that the latent heat L is a constant independent of temperature, the above differential equation can be easily integrated, and the vapor pressure is expressed by

$$p_1 = p_0 \exp\left(-\frac{L}{\mathrm{k_B}T}\right),$$

where p_0 is the constant of integration, corresponding to the pressure in an ideal gas, or $L = 0$. It is clear from this result that for a nonzero L the vapor pressure p_1 remains constant during an isothermal condensation process. As will see in Section 1.4, this conclusion supplements a correct thermodynamical interpretation to a mathematical conjecture involved in the van der Waals isotherms.

1.3. Chemical Potentials in First-Order Phase Transitions and Metastable States

With the aid of a *phase diagram*, it is instructive to see how chemical potentials of two coexisting phases behave in the vicinity of their equilibrium. For a uniform substance, the chemical potential μ is a continuous function of p and T, which can be illustrated by a smooth mathematical surface in the three-dimensional μ-p-T space (Fig. 1.1). In such a space for a liquid–vapor

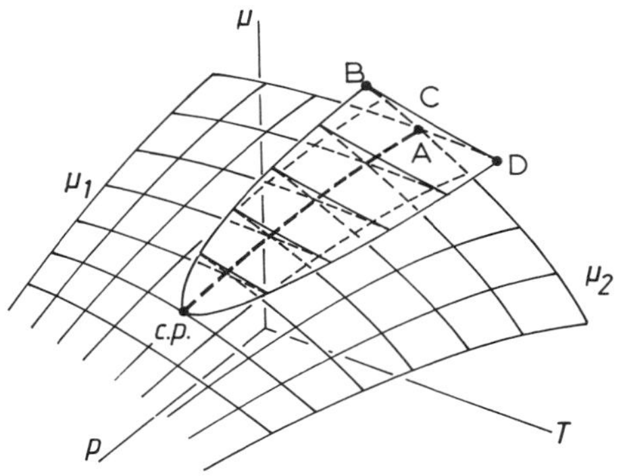

Figure 1.1. Chemical potential surfaces $\mu_1(p, T)$ and $\mu_2(p, T)$ for two phases in equilibrium. The equilibrium is represented by the intersection A...c.p. shown by the thick broken line, where A is in equilibrium and c.p. is the critical point. Points B and D represent possible metastable states. Thin lines on these surfaces are curves at constant T and at constant p, respectively.

equilibrium, surfaces for the two phases that can be in equilibrium intersect in a curve, along which the two chemical potentials take an equal value. Two phases exist in equilibrium on such a curve, but when comparing two points away from the intersecting curve, only one of these on the lower μ surface can be thermally stable.

Consider a simple isotropic substance that exhibits three different phases; solid, liquid, and vapor. Generally, three surfaces representing these phases may intersect in pairs to give three equilibrium curves. If a point lying on all three surfaces (or three intersecting curves) can be found, all three phases can be in equilibrium at this point, called the *triple point*. A phase diagram can be drawn in two dimensions, e.g., with p and T at a constant μ, which is obtained as a projection of the equilibrium curves onto the p-T plane. Similar projections can be made on the μ-p and μ-T planes, providing us with useful diagrams at constant T and constant p conditions, respectively.

Although equilibrium curves in such phase diagrams represent primarily thermally stable states, it is emphasized here that in practical systems there are always *metastable* states, as indicated by a point x on the extension of a constant μ line in Fig. 1.2. While thermodynamically unstable, such a metastable state is often observed, as if stable in practical systems. For example, it is known that a vapor can usually be compressed to a pressure beyond its vapor pressure, if there are no appreciable *nuclei* for initiating condensation. Here, the word "nuclei" is rather vaguely defined, but expressing the presence of unavoidable impurities and foreign particles playing a significant role as condensing nuclei. Such metastable vapor is said to be *supersaturated*. Metastable vapor is mechanically unstable against disturbance like external shock, resulting in sudden condensation.

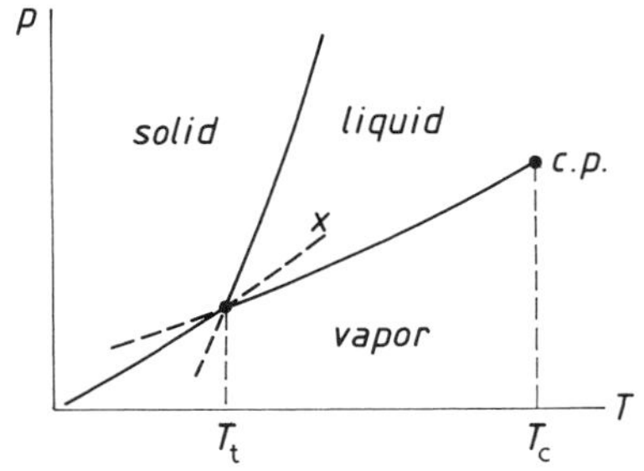

Figure 1.2. The triple point T_t and the critical point T_c for equilibria in a system having three phases. The point x on the extension of the solid–vapor equilibrium line represents a supersaturated state.

We can examine the nature of a metastable state in a phase diagram. Figure 1.3(b) shows two μ curves in a μ-T diagram that are crossing at temperature T_x, while p is kept constant. In this case, the stable phase of the substance is specified by the lower μ curve, and a state y on the μ curve above the other is thermodynamically unstable, but may be observed as metastable in practice. Similar to supersaturated vapor, a liquid can be heated to a temperature above boiling point, and is called *superheated*. While stability of a metastable state depends on the nature of rather ill-defined nuclei, it is significant that metastable states play an essential role in phase transitions of practical systems.

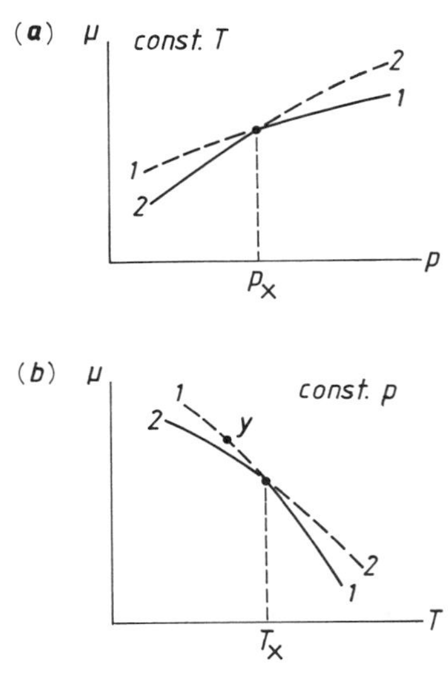

Figure 1.3. Two-phase equilibrium. (a) At a constant T, where p_x is an equilibrium pressure, (b) At a constant p, where T_x is an equilibrium temperature.

The slope of a μ curve at constant T is given by the derivative $(\partial\mu/\partial p)_T = v$, representing the specific volume that is always positive (Fig. 1.3(a)). In a solid–liquid equilibrium, either phase can be thermodynamically stable, depending on which phase is characterized by a smaller v. In contrast, in vapor–liquid and vapor–solid transitions, the vapor phase is always stable because of much larger v_{vapor} than v_{liquid} and v_{solid}. In a μ-T diagram at a constant p, on the other hand, the slope given by $(\partial\mu/\partial T)_p = -s$ is always negative (Fig. 1.3(b)). It is noted that the stable phase has a larger entropy, and hence the stable phase with lower μ_1 on one side of T_x is transformed at a constant p to the phase with lower μ_2 on the other side, when the temperature is varied through T_x. In this case, the temperature T_x is the transition temperature between the two phases at the pressure p.

In a liquid–vapor equilibrium, the specific volume of the liquid v_{liquid} can be increased up to the value of v_{vapor} by heating at a constant p. Or, v_{vapor} can be reduced to v_{liquid} by increasing the pressure under a constant T. At the mathematical limit of $v_{\text{vapor}} = v_{\text{liquid}}$ the two phases become indistinguishable, and so the equilibrium curve must be terminated at this point (p_c, T_c), where the rate $\mathrm{d}p/\mathrm{d}T$ cannot be determined from the Clausius–Clapeyron equation. Such a state is called *critical*, where p_c and T_c are referred to as the critical pressure and the critical temperature, respectively. Beyond the critical point, the two phases are identical, so that the substance exhibits only one homogeneous phase.

Microscopically however, v_{vapor} and v_{liquid} may not be exactly equal in the critical limit, if molecular clusters in finite size are required for initiating condensation. Although attractive molecular interactions are essential for clustering, it is known empirically that the presence of nuclei is responsible for initial condensation in practical vapor. In this case, the volume difference $\Delta v = v_{\text{vapor}} - v_{\text{liquid}}$ cannot be zero in the critical limit, despite the unknown mechanism of nucleation. It is also conceivable that a finite latent-heat energy $T_c\,\Delta s$ is involved in forming molecular clusters, and hence Δs is nonzero. In this context, the transition cannot be continuous, since Δv and Δs cannot converge to zero toward the critical point.

Such a molecular cluster at the condensation threshold is regarded as a small droplet of liquid consisting of a certain number of molecules around a nucleus, which is distinct from normal droplets in a larger size below T_c in terms of the surface-to-bulk ratio. Formation of minimum droplets near T_c were evidenced by opalescent light from some condensing liquids, occurring when the droplet size is comparable to the wavelength of incident light. Only qualitative though, such opalescence signifies a discontinuous change in density at the threshold of condensation. For further details, interested readers are referred to [9].

As remarked, the transition temperature T_x is not a unique parameter for isotropic substances, depending on the pressure p. For a phase transition in solids, on the other hand, the transition is normally observed at a unique transition temperature T_c under ambient atmospheric pressure $p = 1$ atm, which is therefore regarded as a parameter significant for the crystal.

1.4. The van der Waals Equation of State

Thermodynamical properties of a practical gas are adequately described by the van der Waals equation of state, which is used to explain a significant feature of condensation. Although only approximate, it is instructive to see how the van der Waals theory can deal with the condensation phenomenon as a first-order phase transition. It is important, however, that the theory contains some mathematical conjectures, which need to be delineated in the light of thermodynamical principles.

In the van der Waals theory, such essential features as attractive molecular interactions and a nonzero molecular volume are taken into account for the equation of state of a nonideal gas. Attractive molecular forces should reduce the vapor pressure from the corresponding value of p in idealized gas by ignoring interactions, and the effective gas volume for moving molecules should be equal to the container volume minus the total molecular volume. For such a reduced pressure in a practical gas, van der Waals postulated that actual attractive interactions be replaced by their *long-range average*. It is realized that such averaged interaction is correct only in the mean-field approximation.

The van der Waals equation of state is written for one mole of a gas as

$$\left(p + \frac{a}{V^2}\right)(V - b) = RT, \tag{1.4}$$

where a and b represent the averaged interaction potential and the effective molar volume, respectively, and R is the gas constant ($R = 8.314\,\text{j}\cdot\text{deg.}^{-1}$ per mole). Values of these constants a and b are tabulated for representative gases in many reference books. (See, e.g., [2].)

To discuss the general features of the van der Waals isotherms, (1.4) is rewritten in the form

$$V^3 - \left(b + \frac{RT}{p}\right)V^2 + \left(\frac{a}{p}\right)V - \frac{ab}{p} = 0. \tag{1.4a}$$

This cubic equation has either one or three real roots for a given set of external parameters p and T. Among various p-V curves for isothermal changes shown in Fig. 1.4, it is noticed that there is an isotherm at a specific temperature T_c, at which the point of inflection is characterized by a horizontal tangent. Mathematically, for all isotherms above T_c, only one real root corresponds to a given p, whereas below T_c each curve intersects with a horizontal line $p = p_0$ at three roots V_1, V_2, and V_3 at the states A, C, and E, respectively, in Fig. 1.4. The former case represents a uniform state of the gas that can be uniquely specified by given values of p and T, while the latter is considered for a condensing state comprising liquid and vapor. However, it is noted that there are some mathematical implications that conflict with physical realities, as discussed in the following.

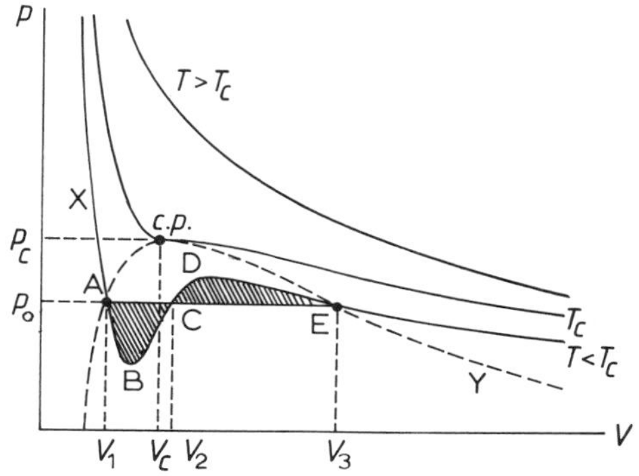

Figure 1.4. The van der Waals isotherms for vapor–liquid equilibrium in a pV diagram. At T_c the two phases are in critical equilibrium. The c.p. state is specified by the critical pressure p_c and temperature T_c. At temperatures $T_< < T_c$, the liquid and vapor phases are represented by chemical potentials $\mu(\text{A})$ and $\mu(\text{E})$, respectively, and states on the horizontal line AE at the constant p_0 consist of two coexisting phases. The value of p_0 can be determined from area(ABC) = area(CDE).

First, we notice that the isotherm changes through T_c continuously from one of the above categories to the other, implying that the condensation is continuous at T_c. In fact, the three roots of (1.4a) for $T < T_c$ become equal to a single value V_c, when T_c is approached from below. In this limit, (1.4a) can be written as

$$(V - V_c)^3 = 0.$$

Therefore, we have relations

$$3V_c = b + \frac{RT_c}{p_c}, \qquad 3V_c^2 = \frac{a}{p_c} \qquad \text{and} \qquad V_c^3 = \frac{ab}{p_c},$$

indicating that all the critical variables p_c, V_c, and T_c can be expressed in terms of the molecular constants a and b. Namely,

$$p_c = \frac{a}{27b^2}, \qquad V_c = 3b, \qquad \text{and} \qquad T_c = \frac{8a}{27b}. \tag{1.5}$$

Using these results, we can confirm that

$$\left(\frac{\partial p}{\partial V}\right)_{T_c} = 0 \qquad \text{and} \qquad \left(\frac{\partial^2 p}{\partial V^2}\right)_{T_c} = 0,$$

which are the conditions for the isotherm at T_c to have a point of inflection at (p_c, V_c). It is realized that the continuity of the van der Waals transitions at T_c originates from the mean-field assumption for molecular interactions, thereby the whole system is regarded as homogeneous. This assumption is in obvious contradiction to the fact that liquid droplets can coexist with vapor at the threshold of condensation. In this context, the van der Waals theory is incorrect in the critical region of liquid–vapor transitions.

Second, the thermodynamical inequality relation

$$\left(\frac{\partial p}{\partial V}\right)_T < 0 \tag{1.6}$$

must be held for any practical substance, suggesting that the pressure should always decrease with its increasing volume. Contrary to this inequality, below T_c a part of the van der Waals isotherms that is marked BD in Fig. 1.4 has a positive slope, and such a contradiction to the physical reality of (1.6) is considered as a mathematical conjecture involved in the van der Waals equation. In fact, in Section 1.2 it was shown that the vapor pressure should remain constant at p_0 during isothermal condensation. Consequently, all real isothermal states between B and D should be located on the straight horizontal line $p = p_0$. We can then interpret that states E and A represent the beginning and final states of a condensation process when the volume is decreased from V_3 to V_1, and at states between E and A, liquid and vapor can coexist in equilibrium. Negative slopes on the van der Waals isotherm outside region EA indicate that supersaturated metastable vapor and liquid can exist in practical systems.

When the vapor is compressed from state Y along an isotherm below T_c, condensation begins to form liquid droplets at state E. On further decreasing volume toward state A, more liquid is formed while the vapor pressure is unchanged. In region ECA where liquid and vapor are mixed inhomogeneously, the Gibbs potential at an arbitrary state P between E and A can be expressed as

$$G_E(\text{vapor}) = G_P(\text{vapor}) + G_P(\text{liquid}) - p_0(V_3 - V_P),$$

and in the process of $P \to A$ at a constant T

$$G_P(\text{vapor}) \to 0 \qquad \text{and} \qquad G_P(\text{liquid}) \to G_A(\text{liquid})$$

to achieve total condensation. In this limit the relation

$$G_A(\text{liquid}) = G_E(\text{vapor}) + p_0(V_3 - V_1) \tag{1.7}$$

can be obtained. Assuming that the van der Waals equation is physically meaningful over the whole region of the isotherm at $T < T_c$, we can express the Gibbs potential at the state P as

$$G_P(p, T) = G(p_0, T) - \int_{v_1}^{v} \left(\frac{\partial G}{\partial V}\right)_T \mathrm{d}V = G(p_0, T) - \int_A^P p\,\mathrm{d}V \tag{1.8}$$

for integration from A to P, and we write $G_A(\text{liquid}) = G(p_0, T)$. As compared with (1.7), it is clear that

$$\int_A^E p\,\mathrm{d}V = -p_0(V_3 - V_1)$$

representing the area under curve ABCDE. Therefore, the straight line ACE at $p = p_0$ must be drawn in such a way that the areas under the curves ABC and CDE are equal in magnitude but opposite in sign. The above procedure for correcting the van der Waals equation to physical reality is known as the Maxwell equal-area construction. Because of the discontinuity expressed by (1.7) at temperatures below T_c, the van der Waals equilibrium is discontinuous and first-order, although the equilibrium at T_c is continuous.

We have considered in the above argument that all states are thermodynamically stable, but the states B and D, for example, are metastable in practical systems, so that phase diagrams for condensing processes should be as shown in Fig. 1.5. In this diagram, it is shown schematically that the vapor pressure p_0 at the states A and E converges to p_c at the virtual critical point (c.p.) in a practical system.

When the temperature is lowered through T_c, the low-temperature phase consists of liquid droplets in the system, which is characterized by a decreasing value of $\Delta v = v_{\text{vapor}} - v_{\text{liquid}}$. Such a quantity as Δv can specify the degree of ordering in condensing gas, however as the order parameter we should refer to a quantity that changes from zero to a nonzero value. In this case, it is practical to use the density difference between the two phases, i.e.,

$$\eta = \Delta\rho = \rho_{\text{liquid}} - \rho_{\text{vapor}} \approx \rho_{\text{liquid}}. \tag{1.9}$$

Similar to (1.9), the order parameter for a binary transition is defined as the density difference between the ordered and disordered phases, i.e.,

$$\eta = \Delta\rho = \rho_{\text{order}} - \rho_{\text{disorder}}. \tag{1.9a}$$

Order parameters defined in Chapters 2 and 3 for practical binary systems are seemingly different from (1.9a), however all are attributed essentially to the density difference.

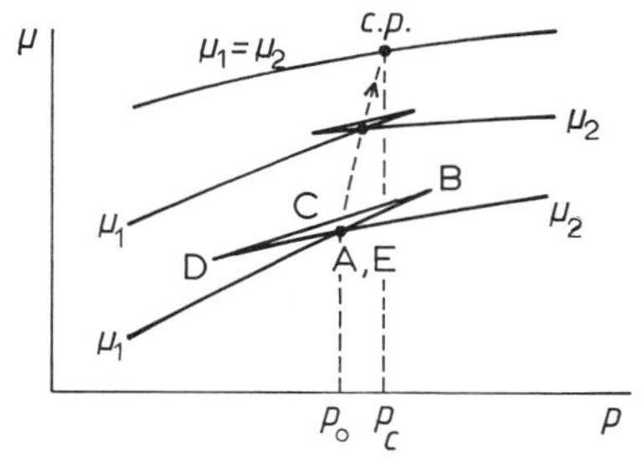

Figure 1.5. Representative vapor–liquid equilibria in the μp plane, showing changes of an intersecting point between two $\mu(p)$ curves when approaching c.p., which corresponds to the van der Waals isotherm at T_c. Compare with the three-dimensional drawing in Fig. 1.1.

1.5. The Ehrenfest Definition of the Second-Order Phase Transition and the Landau Theory for a Binary Transition

In Sections 1.3 and 1.4, general features of first-order phase equilibria were illustrated in phase diagrams, with which the van der Waals theory of liquid–vapor transitions was discussed as an example. Representing different phases under a constant p condition, two chemical-potential curves $\mu_1(T)$ and $\mu_2(T)$ may generally intersect for a transition between them at a temperature T_x, at which the slopes are not equal, i.e., $(\partial\mu_1/\partial T)_p \neq (\partial\mu_2/\partial T)_p$. This indicates that at T_x there is a discontinuity in specific entropy $\Delta s = s_1 - s_2$, hence a finite-latent heat $T_x\,\Delta s$ characterizes the transition of first order. In a μ-p phase diagram at a constant T, the discontinuous change may occur at the crossing point p_x, where a finite change in specific volume $\Delta v = v_1 - v_2$ or a discontinuous amount of work $-p_x\,\Delta v$ specifies a mass exchange $\mu\,\Delta N$ between the two phases.

According to Ehrenfest, even if first-order derivatives of the chemical potentials are equal, two distinct phases could be in equilibrium, provided that the second- or higher-order derivatives of μ_1 and μ_2 are discontinuous at their crossing point. In Fig. 1.6(a) such a case is illustrated in the μ-T diagram, where curves $\mu_1(T)$ and $\mu_2(T)$ are just touching at a single point T_c with a common tangent, and hence $-\Delta s = (\partial\,\Delta\mu/\partial T)_{T=T_c} = 0$. However, in this case, one of these curves is lower than the other on both sides of T_c, hence representing no transition between the two phases.

On the other hand, we notice that the Ehrenfest condition for a second-order phase transition can be fulfilled mathematically by considering only a single potential μ, instead of μ_1 and μ_2, if a discontinuous change of the curvature, i.e., $(\partial^2\,\Delta\mu/\partial T^2)_{T=T_c} \neq 0$, occurs at T_c. In this case, we must consider a substance consisting of a single phase, where the chemical potential represents a continuous change as a whole from one phase to another. Now that we consider a single thermodynamic phase where the number of constituents N is constant, it is more convenient to use the Gibbs function G than the chemical potential μ. Thus, such a second-order phase transition can be characterized by $\Delta(\partial G/\partial T)_p = 0$ and $\Delta(\partial^2 G/\partial T^2)_p \neq 0$ at T_c.

Under a constant p condition, we consider that the Gibbs potential is $G_0(T)$ for $T > T_c$, which changes continuously to $G(T)$ for $T < T_c$ when the temperature is changed through T_c where $G_0(T_c) = G(T_c)$, as illustrated in Fig. 1.6(b). Writing for $T < T_c$ that

$$G(T) = G(T_c) + \Delta(T),$$

the continuous deviation $\Delta(T)$ from $G(T_c)$ can be expressed in the vicinity of T_c as

$$\Delta G(T) = G(T) - G(T_c) = -\frac{1}{2}\left(\frac{\partial^2\,\Delta G}{\partial T^2}\right)_{T=T_c}(T_c - T)^2 + \cdots. \tag{1.10}$$

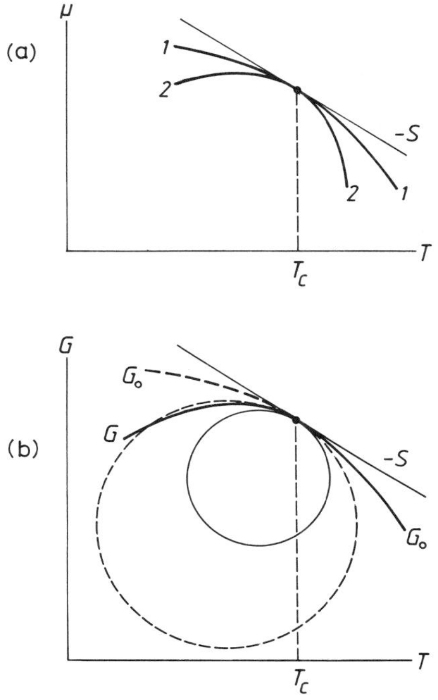

Figure 1.6. The Ehrenfest second-order phase transition illustrated in the μ-T diagram. (a) Interpreted as two phases in equilibrium. (b) Interpreted as a continuous change in entropy S and discontinuous in the curvature at T_c. Here two circles of different radii of curvature are shown.

Here, in the series of (1.10) $(\partial\, \Delta G/\partial T)_{T=T_c} = -\Delta S = 0$, and the first-order term of $T_c - T$ is absent. Further, the coefficient $(\partial^2\, \Delta G/\partial^2 T)_{T=T_c} = -(\Delta C_p)_{T=T_c}$ represents a discontinuous change in the heat capacity C_p at T_c. In (1.10), we assumed that the second-order derivative is nonzero, but ΔG could be dominated by a term higher than the second. In this case, such a continuous change may be called a higher-order phase transition. However, lacking significant examples in practical systems, such a higher-order transition may remain just as a mathematical consequence.

Although isotropic substances are considered so far, structural changes take place normally in anisotropic crystals. Nevertheless, whenever a structural change is continuous, being characterized by the absence of a *hysteresis*, the transition is regarded as second-order, and described by a single Gibbs thermodynamical potential. We find typical examples of second-order phase transitions among so-called *order–disorder* phenomena, for which the Gibbs potential G is regarded as a function of a variable η called the order parameter. Landau [7] formulated the thermodynamical problem of a continuous phase transition in *binary* systems, where the Gibbs potential $G(\eta)$ is invariant under *inversion* of the order parameter, i.e., $\eta \leftrightarrow -\eta$. The Landau theory is abstract, where η is unspecified, but invariance of $G(\eta)$ under inversion is physically significant for a binary system. Namely,

$$G(\eta) = G(-\eta). \tag{1.11}$$

It is realized that inversion symmetry assumed in the Landau theory may be considered as *reflection* on a mirror plane which is often significant in anisotropic crystals, where the order parameter is a vector. Also significant is that the order parameter can be responsible to an applied field or stress, in which case the Gibbs potential is modified accordingly. Here, we discuss the second-order phase transition that occurs spontaneously, leaving the external interaction to later discussions.

Landau proposed that the Gibbs potential in the absence of an external field or stress can be expressed as an infinite series of η, i.e.,

$$G(\eta) = G_0 + \tfrac{1}{2}A\eta^2 + \tfrac{1}{4}B\eta^4 + \tfrac{1}{6}C\eta^6 + \cdots, \tag{1.12}$$

where $G_0 = G(0) = G(T_c)$. The coefficients in (1.12), A, B, C, ... are normally smooth functions of temperature. It is noted that in the expansion of $G(\eta)$ there is no term in odd power, owing to the inversion symmetry expressed by (1.11).

At temperatures close to T_c, the magntide of η is sufficiently small, so that the expansion in (1.12) can be trimmed at the quartic term $\frac{1}{4}B\eta^4$. In this case, the order parameter can be easily determined by minimizing the truncated Gibbs potential

$$G(\eta) = G_0 + \tfrac{1}{2}A\eta^2 + \tfrac{1}{4}B\eta^4. \tag{1.13}$$

The value of η in thermal equilibrium can then be obtained from the equation

$$\frac{\partial G}{\partial \eta} = A\eta + B\eta^3 = \eta(A + B\eta^2) = 0.$$

Therefore the solutions can be either

$$\eta = 0 \tag{1.14a}$$

or

$$\eta = \pm\left(-\frac{A}{B}\right)^{1/2}. \tag{1.14b}$$

Supposing that $A > 0$ and $B > 0$, (1.14a) is the only real solution, since (1.14b) is imaginary, and hence the solution $\eta = 0$ represents the disordered state above T_c. On the other hand, if $A < 0$ and $B > 0$, (1.14b) gives a real solution of a nonzero value, hence representing the ordered phase below T_c. In this context, the phase transition is signified by changing the sign of the coefficient A. Landau wrote that

$$A = A'(T - T_c) \qquad \text{where} \quad A' > 0. \tag{1.15}$$

The two solutions, (1.14a) and (1.14b), must be consistent at $T = T_c$ for a continuous transition, for which it is sufficient to consider $A \geq 0$ for $T \geq T_c$. The coefficient B can be regarded as absent from $G(\eta)$ at and above T_c, whereas it signifies the presence of a positive *quartic potential* $\frac{1}{4}B\eta^4$ below T_c. In the vicinity of T_c, the corresponding order parameter in the low-temperature phase can therefore be expressed by

$$\eta = \pm\left[\left(\frac{A'}{B}\right)(T_c - T)\right]^{1/2}. \tag{1.16}$$

Considering η as a continuous variable, thermodynamical states of the substance can be specified at minima of the potential curve $G(\eta)$, as shown schematically in Fig. 1.7. Here, assuming that $G(0) = 0$, the parabolic $G(\eta) = \frac{1}{2}A\eta^2$ $(A > 0)$ above T_c has the minimum at $\eta = 0$, whereas in a double-well potential $G(\eta) = \frac{1}{2}A\eta^2 + \frac{1}{4}B\eta^4$ $(A < 0$ and $B > 0)$ below T_c there are two minima at $\pm\eta_0 = \pm(-A/B)^{1/2}$ related by inversion. These two minima emerge as the temperature is lowered through T_c, shifting their positions symmetrically away from $\eta = 0$. As seen from (1.16), the order parameter η exhibits a *parabolic* temperature-dependence in the vicinity of T_c. In a later discussion, the *quartic* term $\frac{1}{4}B\eta^4$ can be attributed to correlations among pseudospins emerging at T_c, and the parabolic η given by (1.16) is found as a consequence of the *mean-field approximation.*

Ordering processes in practical systems cannot be correctly described by the simple Landau theory, while the parabolic temperature-dependence of η can be examined experimentally for ferroelectric and ferromagnetic systems. Some discrepancies from the parabolic dependence were always indicative in experimental results, signifying inadequacy of the mean-field approximation particularly at temperatures close to T_c. Furthermore, from a parabolic η, the transition temperature may be determined as the intersect T_0 of a linear extrapolation of η^2-T plots with the T axis, but such a T_0 was always found higher than the transition temperature T_c determined as the ordering threshold, and the difference $T_c - T_0$ is often quoted for the abnormality in the

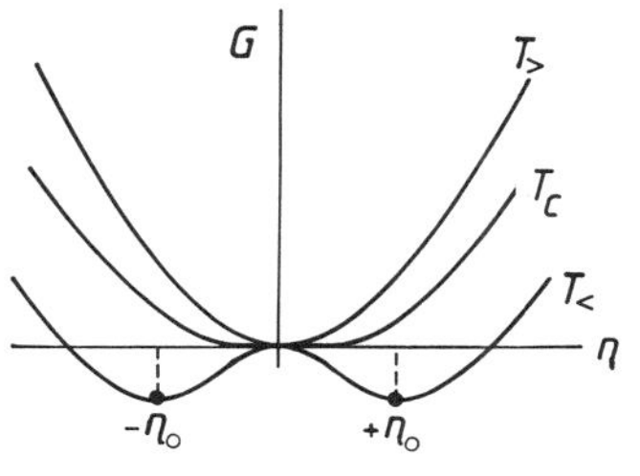

Figure 1.7. The behavior of a Gibbs potential $G(\eta)$ in the vicinity of T_c. A typical change from a parabolic G above T_c to a double-well potential below T_c is shown, as given by the truncated Gibbs function, (1.12). The equilibrium below T_c is signified by fluctuations between $\pm\eta_0$.

critical region. In practice, the temperature-dependence is analyzed by means of an empirical expression $(T_c - T)^\beta$, where the exponent β showed values considerably different from the mean-field value 0.5. According to the contemporary theory of critical exponents [10], the value of β depends on the dimensionality of η, which was not considered in the Landau theory.

Following the Landau theory, the expressions for entropies and heat capacities can be obtained at and below T_c for a qualitative test of the theory. Entropies are calculated from the relation $S = -(\partial G/\partial T)_p$, which are

$$S(T) = S(T_c) + \left(\frac{A'^2}{2B}\right)(T_c - T) \qquad \text{for} \quad T \leq T_c.$$

The heat capacity can be expressed by the relation $C_p = -T(\partial S/\partial T)_p$, but the value at T_c is twofold. Namely,

$$(C_p)_{T>T_c} = -T_c\left\{\frac{\partial S(T_c)}{\partial T}\right\}_{p,T>T_c} = C_0,$$

if T_c is approached from above, whereas

$$(C_p)_{T<T_c} = C_0 - \frac{A'^2 T_c}{2B},$$

when approached from below. Therefore, there is a discontinuous change in the heat capacity

$$\Delta C_p = \frac{A'^2 T_c}{2B} \qquad \text{at} \quad T = T_c,$$

as consistent with (1.10). However, as illustrated in Fig. 1.8(a) schematically, such a discontinuous change in C_p has never been observed from continuous phase transitions in any practical systems, while observed C_p shows a sharp rise to virtual infinity at T_c followed by a gradual decrease. Such a curve resembles the Greek letter λ as in Fig. 1.9, where shown is a typical example of so-called λ anomaly in the order–disorder transition of β brass [11]. Indicating the failure of the mean-field theory, many systems undergoing continuous-phase transitions appear to exhibit λ anomalies of the heat capacity in the critical regions. Although unexplained thermodynamically, the λ anomaly can be regarded as an essential feature of the second-order phase transitions.

In practice, such an anomalous heat capacity measured as a function of $T - T_c$ is expressed by means of critical exponents α and α'. Namely, anomalies can be analyzed by assuming that

$$C_p \propto (T - T_c)^{-\alpha} \qquad \text{and} \qquad C_p \propto (T_c - T)^{-\alpha'}$$

for $T > T_c$ and $T < T_c$, respectively. Although calculated with the Landau theory only at T_c, such exponents are considered to represent deviations from the mean-field values in the critical region.

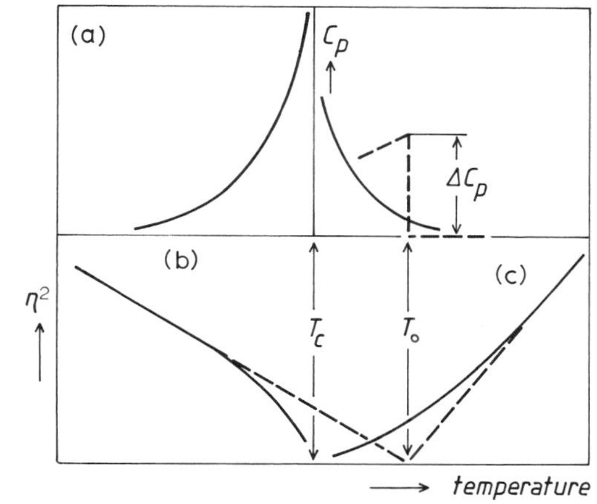

Figure 1.8. Critical anomalies in a second-order phase transition as observed in: (a) the specific heat C_p (λ anomaly); (b) the squared order parameter η^2 (a deviation from the parabolic η); and (c) the susceptibility χ (the Curie–Weiss anomaly). The broken lines indicate predictions by the mean-field theory.

In the above, we discussed the second-order phase transition on the basis of the Landau theory. However, it is noted that the Gibbs potential expressed as a power series of the order parameter is not necessarily for continuous transitions. Blinc and Zeks [12] showed that the Gibbs expansion including the term $C\eta^6$ in (1.12) may lead to a first-order phase transition, if $B < 0$ and $C > 0$, whereas the transition is second order if $B > 0$. Apart from the physical implication of these constants, their argument is not inconsistent to ours,

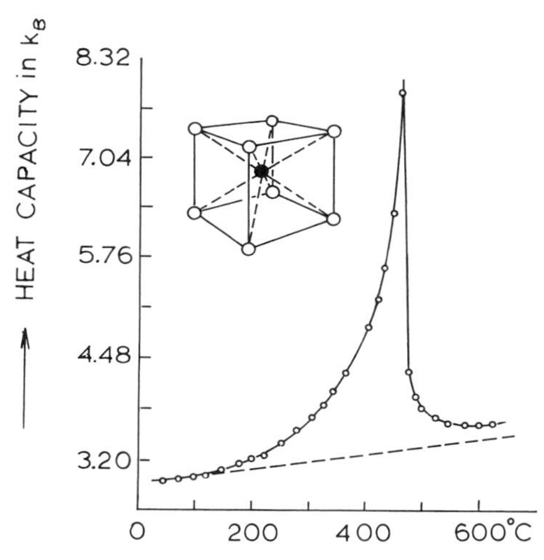

Figure 1.9. The specific curve of β brass, showing a typical λ anomaly. (From F. C. Nix and W. Shockley, *Rev. Mod. Phys.* **10**, 1 (1938).)

if we consider the fact that in the first-order transition the order parameter may emerge with a finite amplitude at T_c, so that higher-order terms like $C\eta^6$ may not be quite negligible even at temperatures close to T_c. In contrast, in this region of a second-order transition, the amplitude of η is infinitesimal, so that the expansion can be truncated at the quartic term $B\eta^4$.

1.6. Order-Parameter Susceptibility, the Curie–Weiss Law, and Domains

Depending on the physical nature, the order parameter η can respond to an applied field or stress F, and the Gibbs potential should include the interaction energy $-\alpha\eta F$, where α is a constant. Therefore, it is noted that in the presence of a nonzero F the Gibbs potential can no longer be invariant under inversion (or reflection) $\eta \leftrightarrow -\eta$. In this case, at temperatures close to T_c, the Gibbs potential can be written as

$$G(\eta) = G_0 + \tfrac{1}{2}A\eta^2 + \tfrac{1}{4}B\eta^4 - \alpha\eta F, \tag{1.17}$$

where the expansion is truncated in the vicinity of T_c, and F is assumed to be weak so that $\alpha\eta F$ is smaller than the ordering energy expressed by the power terms of η.

On applying an external F, the system is forced to be ordered to some extent even at temperatures above T_c, and therefore no clear-cut transition temperature is expected, resulting in a *diffuse* transition. Nevertheless, the thermal equilibrium can be specified by minimizing G with respect to η, i.e., by solving the equation

$$\frac{\partial G}{\partial \eta} = A\eta + B\eta^3 - \alpha F = 0.$$

Clearly, as long as $F \neq 0$, this equation has no solution $\eta = 0$ for complete disorder, and the state is signified by a small but finite η at temperatures above T_c. Ignoring the term $B\eta^3$ for such a small η, the response of η to the applied F is expressed by the *static susceptibility*

$$\chi = \frac{\alpha\eta}{F} = \frac{\alpha^2}{A} = \frac{\alpha^2/A'}{T - T_c}, \tag{1.18}$$

which is known as the Curie–Weiss law. In (1.18) the susceptibility is characterized as $\chi \to \infty$ when T_c is approached from above. Although in the presence of F the phase transition appears diffuse, the temperature T_c can be determined with reasonable accuracy from the susceptibility in the limit of $F \to 0$. In practice, the reciprocal susceptibility χ^{-1} is plotted against T to see a deviation from a straight line predicted by the mean-field approximation. Usually, a substantial discrepancy is observed at temperatures close to T_c, as illustrated schematically for $T > T_c$ in Fig. 1.8(c), and the temperature-dependences for $T > T_c$ and $T < T_c$ are normally described by using another set of critical exponents as

$$\chi \propto (T - T_c)^{-\gamma} \qquad \text{and} \qquad \chi \propto (T_c - T)^{-\gamma'},$$

respectively. Here, as will be discussed in Chapter 4, below T_c the order parameter η is often different from the polar mode responsible for χ that behaves as obeying the Curie–Weiss law, but is related to χ in the critical region where the exponent γ' is significantly deviated from the mean-field value. Mean-field values of these exponents are given by $\gamma = \gamma' = 1$.

When observed with an oscillating low-frequency field, the response χ above T_c can be shown to be related to a relaxation time described by $\tau \propto (T - T_c)^{-1}$ that shows a temperature-dependence identical to the static susceptibility (see Chapter 7). In this context, such a divergent susceptibility is regarded as representing *critical slowing down*, signifying the onset of critical behavior as T_c is approached from above.

In the following, we discuss the concept of a *molecular field*, which was first introduced by Weiss for magnetic systems below T_c. (See, e.g., Chapter 15 in [8].) For a ferromagnet, Weiss considered that the transition temperature T_c manifests an internal field, which was interpreted microscopically as due to emerging correlations among microscopic magnetic spins. Although not deducible by thermodynamic principles, he postulated the presence of an internal magnetic field B_i in the magnetized phase, and wrote it as related to the magnetization M, i.e.,

$$B_i = \lambda M, \tag{i}$$

where λ is constant. As already remarked, in an applied field B_0 the crystal is magnetized to some extent even in the paraelectric phase, which becomes appreciable when the temperature is close to T_c. Under the circumstances, we can consider that the induced magetization M by the combined field $B_i + B_0$ is expressed as

$$M = \chi_0(B_i + B_0), \tag{ii}$$

where χ_0 is the paramagnetic susceptibility that obeys the Curie law

$$\chi_0 = \frac{C}{T} \quad (C\text{: Curie's constant}). \tag{iii}$$

Combining these relations (i), (ii), and (iii) together, the magnetic susceptibility can be expressed by

$$\chi = \frac{M}{B_0} = \frac{C}{T - C\lambda},$$

which is the Curie–Weiss law, where the constant $C\lambda$ written as T_0 is the transition temperature. Here, the relation $T_0 = C\lambda$ suggests a close relation between the transition temperature and the Weiss molecular field. Therefore in the mean-field approximation, we can state that the phase transition is initiated by such an ordered spin cluster that is responsible for the Weiss field in minimal strength. For an isotropic paramagnet, using the standard expressions for C and λ we can express this relation as

$$k_B T_0 = \frac{2zJ}{3}, \tag{iv}$$

where J is the exchange integral, and z the number of nearest neighbors. (For further details, see, e.g., G. H. Wannier, *Statistical Physics*, Chapter 15, Wiley, New York, 1966). The relation (iv) suggests that the spin cluster of z ordered spins produces the minimum internal field. Assuming a limited range for correlations, such a model can be generalized for estimating the cluster size at a transition threshold, provided that a reliable value of J is available. Similar to critical droplets of condensing liquid, such clusters of order variables appear to play an essential role for the magnetic response that is *singular* at the transition temperature.

The Weiss postulate of a molecular field can properly be applied to a single domain crystal. According to the Landau theory of binary systems, two domains are characterized by two values of the order parameter $\pm\eta$, or by magnetization $\pm M$ in a uniaxial magnet. Although thermodynamically equivalent, such domains as magnetized in opposite directions are magnetically different, being characterized by two distinct magnetizations $\pm M$, or opposite the Weiss fields $\pm B_i$. Hence, the internal magnetization energy expressed by $(\pm M)(\pm B_i)$ is unchanged by reflection $M \leftrightarrow -M$, whereas the external magnetic energy $-MB_0$ changes sign.

A significant feature of the domain structure is the nature of so-called *domain walls*, that are boundary regions between oppositely magnetized domains. The domain walls can easily be displaced in a "soft" magnet by an externally applied field B_0 with a modest strength. By a soft magnet, we mean generally a magnetic crystal that is relatively free from lattice defects, where a domain wall can be described theoretically by a *spin soliton* that is mobile in perfect crystals. Bloch (see, e.g., Chapter 15 in [8]) depicted such a domain wall as a region where spins are spirally arranged to reverse the direction from one domain to the other, although experimentally the wall structure has not yet been revealed in detail. In practice, such an empirical feature of domain walls is taken for granted.

Owing to the high mobility of domain walls in a soft magnet, the volume ratio between opposite domains can easily be changed by an applied field. Assuming that virtually no energy dissipation is involved in domain-wall displacements, we can write the Gibbs potential of a soft magnet below T_c as

$$G = N_+\mu_+ + N_-\mu_-,$$

where

$$\mu_\pm = \mu_0 - (\pm M)(\pm B_i + B_0)$$

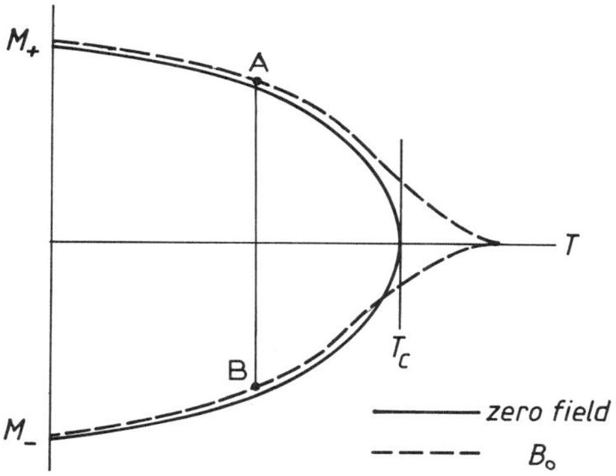

Figure 1.10. Magnetization curves of a soft ferromagnet with no external field, and when a weak field B_0 is applied. In the former case, the transition is sharp at T_c, whereas the latter case is characterized by a diffuse transition. Magnetized domains A and B behave as if they were two phases in first-order equilibrium, when B_0 is varied. In the presence of B_0, the magnetization curves are generally asymmetrical.

are the chemical potentials, and $N_\pm$ are the numbers of spins of the two domains, respectively, and $N_+ + N_- = N$ is constant. The chemical potential μ_0 represents thermal properties of the whole crystal when $B_0 = 0$. It is noted that the difference

$$\Delta\mu = \mu_+ - \mu_- = -2MB_0$$

signifies the transition between two domains, as if they are in first-order equilibrium, shifting from one domain to the other by varying the external field B_0. By increasing B_0, either a process for $(N_+ \to N, N_- \to 0)$ or $(N_+ \to 0, N_- \to N)$ can take place, ending up as a single domain crystal with either magnetization M or $-M$. Figure 1.10 illustrates schematically such equilibrium between two opposite domains as influenced by a weak applied field B_0, where the domain conversion occurs analogously to liquid–vapor transitions in the van der Waals theory. It is realized that such a conversion depends on the softness of a magnet due primarily to the content of lattice imperfections that may immobilize domain walls. In this context, such imperfections appear to play a similar role to nuclei in a condensing vapor. Although introduced originally for magnetic systems, the concept of the Weiss field has a significant implication applicable to other ordering systems as well, and is considered in general as a valid concept in the mean-field accuracy.

1.7. Beyond Classical Thermodynamics, the Nature of Critical Fluctuations

As summarized in Fig. 1.8, significant deviations from thermodynamical predictions prevail in the vicinity of T_c of second-order phase transitions, indicating the failure of the mean-field theory. Microscopically, as Onsager's theoretical work [13] has indicated, critical anomalies are due to fluctuations in the system of microscopic-order variables. Although the problem as such is beyond the scope of classical thermodynamics, it is worth considering the nature of fluctuations, prior to the following chapters where the origin is discussed.

First, we pay attention to a phase transition of first order, particularly the transition threshold under a constant p and T condition. Denoting two phases as 1 and 2, the equilibrium process can be described by

$$\Delta(G_1 + G_2) = (\mu_1 + \mu_2)\langle n \rangle_t = 0,$$

where n, if positive, represents the number of molecules being transferred from phase 1 to 2, but if negative, the molecular transfer is in the opposite direction. The balance can be specified by the time average $\langle n \rangle_t$ that vanishes under a normal equilibrium condition. In fact, the transfer rate dn/dt is so fast that during the time interval between t and $t + t_0$ the molecular transfer

$$\langle n \rangle_t = \left\langle \int \frac{dn}{dt} dt \right\rangle_t = t_0^{-1} \int_0^{t_0} \frac{dn}{dt} dt$$

is averaged out when observed at a timescale t_0 longer than the characteristic time τ. On the other hand, if τ is longer than t_0, the average $\langle n \rangle_t$ may not vanish, resulting in nonzero $\Delta(G_1 + G_2)$. Such averaging in a slow process [14] can, in principle, be expressed as

$$\langle n \rangle_t = 0 \quad \text{if} \quad t_0 \gg \tau, \quad \text{or practically} \quad t_0 \sim \infty. \tag{1.19}$$

Although very unusual for first-order equilibria, such a slow process as $\tau \gg t_0$ may occur in structural phase transitions of second order, as indicated by recent low-frequency experiments in the critical region, where fluctuations are in the range of microwave frequencies lower than 10^{-10} Hz. (See Chapter 4 for the experimental detail.) Therefore, such a slow process is not entirely academic, but representing a realistic process in the critical region of second-order phase transitions.

Despite the presence of appreciable transition anomalies, the second-order phase transition is continuous theoretically. Being governed by the Gibbs function of a single-order parameter, the Landau theory assumes $\Delta G = 0$ at a continuous transition, however, there is no intrinsic mechanism for such fluctuations as those considered for first-order equilibria. Therefore, we must consider the presence of a hidden mechanism, which is primarily unrelated to G but indirectly responsible for a nonzero ΔG in the critical region.

For ordering in crystals, it is logical to consider an influence of lattice vibration under specific circumstances in the critical region. As evidenced from resulting symmetry change, there should be a close relation with the lattice. According to Born and Huang [15], a crystal cannot be uniformly stressed under an equilibrium condition. In this context, it is conceivable that at the threshold of a structural change the lattice should be strained at least at the lowest level of an acoustic excitation. We can then consider that the collective mode of order variables couples with the inhomogeneously strained lattice in the critical region. Under the circumstances, we can assume that correlated-order variables grouped in a small volume V_i at a position i are described by a local Gibbs function $G_i(p_1, T_1)$ that is different from the $G_{0i}(p, T)$ in the corresponding uncorrelated crystal. Here each group is considered as sufficiently homogeneous inside the volume V_i, so that it is represented by the homogeneous Gibbs function. In equilibrium, the difference

$$\Delta G_i = G_i(p_1, T_1) - G_{0i}(p, T)$$

can be balanced with the lattice counterpart $-(\Delta G_{\text{lattice}})_i$ representing distributed strains in the lattice. Following Landau and Lifshitz [7], such a local thermodynamical excitation ΔG_i of order variables can be determined in the inequality relation

$$\Delta G_i \geq -(T_1 - T)\,\Delta S_i + (p_1 - p)\,\Delta V_i = -(\Delta G_{\text{lattice}})_i, \tag{1.20}$$

where ΔS_i is the corresponding change in entropy in the small part V_i in the system. Here the unequal sign represents an irreversibility for such a local process, hence implying diverse ΔG_i for distributed clusters. Anomalies at second-order phase transitions should therefore be due to these ΔG_i, whose time average $\langle \Delta G_i \rangle_t$ is measurable as nonzero fluctuations, when sampled by probes with a short timescale t_0 of observation. (For the experimental detail, see Chapter 9)

On the other hand, as related to the sum $\sum_i \langle \Delta G_i + (\Delta G_{\text{lattice}})_i \rangle_t$ over the distribution, the macroscopic fluctuation $\Delta G + \Delta G_{\text{lattice}}$ is responsible for anomalies exhibited by thermodynamical quantities in the critical region. Whereas $\Delta G_{\text{lattice}}$ due to lattice distortion occurs only in the critical region, $\Delta G_{\text{lattice}} = 0$ and $\Delta G = 0$ at all noncritical temperatures T. At $T = T_c$, the equilibrium condition is given by $dG = 0$ for infinitesimal variations dS and dV, while at temperatures T away from T_c we have a quadratic relation

$$\Delta G = \tfrac{1}{2}\left[\left(\frac{\partial^2 U}{\partial S^2}\right)(\Delta S)^2 + 2\left(\frac{\partial^2 U}{\partial S\,\partial V}\right)\Delta S\,\Delta V + \left(\frac{\partial^2 U}{\partial V^2}\right)(\Delta V)^2\right]$$
$$= \tfrac{1}{2}[\Delta T\,\Delta S - \Delta p\,\Delta V]$$

for finite deviations ΔS and ΔV, where $U = U(\Delta S, \Delta V)$ is the internal energy, and the thermodynamical relations

$$\left(\frac{\partial U}{\partial S}\right)_V = \Delta T \qquad \text{and} \qquad \left(\frac{\partial U}{\partial V}\right)_S = -\Delta p$$

were used for deriving the last expression. Since $\Delta G_{\text{lattice}} = 0$ at noncritical temperatures, $-\Delta G/T$ represents an entropy increase in the total system, for which the probability w can expressed from the Boltzmann relation $-\Delta G = \mathrm{k_B} T \ln w$. Namely,

$$w = \exp\left[\frac{-\Delta T\,\Delta S + \Delta p\,\Delta V}{2\mathrm{k_B} T}\right]. \tag{1.21}$$

Considering ΔS and Δp as functions of ΔT and ΔV, for example,

$$\Delta S = \left(\frac{\partial S}{\partial T}\right)_V \Delta T + \left(\frac{\partial S}{\partial V}\right)_T \Delta V \quad \text{and} \quad \Delta p = \left(\frac{\partial p}{\partial T}\right)_V \Delta T + \left(\frac{\partial p}{\partial V}\right)_T \Delta V, \tag{1.22}$$

and so the exponent in (1.21) is related to squared fluctuations $(\Delta T)^2$ and $(\Delta V)^2$. Notice here, the cross-term $(\Delta T)(\Delta V)$ is zero because of the relation $(\partial S/\partial V)_T = (\partial p/\partial T)_V$, signifying that fluctuations ΔT and ΔV are independent. Equation (1.21) gives a *Gaussian* distribution in homogeneous crystals, but for finite ΔT and ΔV the probability w is very small, although depending on the coefficients in (1.22). In any case, critical fluctuations in continuous-phase transitions should be originated from a mechanism $\Delta G_{\mathrm{lattice}}$ that is unrelated to order variables, whereas the thermodynamical fluctuation is generally negligible at all noncritical temperatures.

1.8. Remarks on Critical Exponents

Thermodynamically critical anomalies are due to the sum $\langle \sum_{\mathrm{i}} \Delta G_{\mathrm{i}} \rangle_{\mathrm{t}} = \langle \Delta G \rangle_{\mathrm{t}}$ averaged over the timescale of measurement, which are observed as a deviation from the Gibbs potential $G(T_{\mathrm{c}})$. Experimentally, such a deviation is expressed in term of a critical exponent, as discussed in this chapter for those thermodynamical quantities like the order parameter η, its susceptibility χ, and the specific heat C_p observed as a function of $\Delta T = T - T_{\mathrm{c}}$. Namely,

$$\eta = \eta_0(-\Delta T)^{\beta}, \qquad \chi = \chi_0(\Delta T)^{-\gamma},$$

and

$$C_{p+} \propto (\Delta T)^{-\alpha} \quad \text{and} \quad C_{p-} \propto (-\Delta T)^{-\alpha'}.$$

These expressions are obviously hypothetical, for which no rigorous mathematical proof can be given. However, by this hypothesis we mean that a leading mechanism prevails for phase transitions, which may be described by such simple exponents. Although difficult to deduce their physical implications, some theoretical relations can be found among these exponents, constituting the foundation for the *scaling theory* [9].

In the scaling theory, thermodynamic functions are considered to be homogeneous, and hence logically applied to isotropic liquid and magnetic systems. On the other hand, in our discussion on anisotropic crystals, fluctuations should be inhomogeneous under critical conditions, as verified with microscopic probes in various measurements. In this context, the underlying hypothesis of scaling theory is not always acceptable for structural phase transitions in crystals. Furthermore, deviations from mean-field averages arise generally from long-range order, and here the basic problem is how the temperature variation and nonlinear propagation of ordering can be analyzed, as will be discussed in later chapters.

CHAPTER 2

Order Variables and Their Correlations, and the Mean-Field Approximation

2.1. Order Variables and Their Mean-Field Average

For a phase transition in a crystal the order parameter η is a thermodynamical variable signifying the ordered phase at temperatures below the critical region. Originating from microscopic variables σ_m attached to ions or molecules that are *active* at sites m for a structural change, their *ensemble average* is considered to represent the macroscopic-order parameter η. Such a variable σ_m is a function of space–time coordinates at the site m, so that the time variation as well as the spatial distribution are significant for the averaging of distributed variables. In the disordered phase above T_c, those variables σ_m are usually in fast random motion so that the time average $\langle\sigma_m\rangle_t$ vanishes at each lattice point, hence independent of the site m. In contrast, at temperatures below T_c, they are correlated in slow motion, so that the ordered phase is dominated by their spatial distribution. In a *binary* system, as illustrated schematically in Fig. 2.1, the "ordered" phase below T_c are topologically inhomogeneous, comprising either domains or sublattices [16]. The ensemble average in such a state is meaningful, only if calculated for one of these subsystems instead of the whole crystal.

While it is essential to identify the active group in a given system, in practice the variable σ_m is not always self-explanatory from the chemical formula or the unit-cell structure of a given crystal, except in a few simple cases. Despite pending identification of σ_m, as is often the case with practical systems, it is essential to deal with their dynamical behavior during structural changes. The critical region cannot be described in terms of thermodynamical principles, but is known to be dominated by the slow dynamical behavior of order variables in collective motion. We shall hereafter call σ_m the *order*

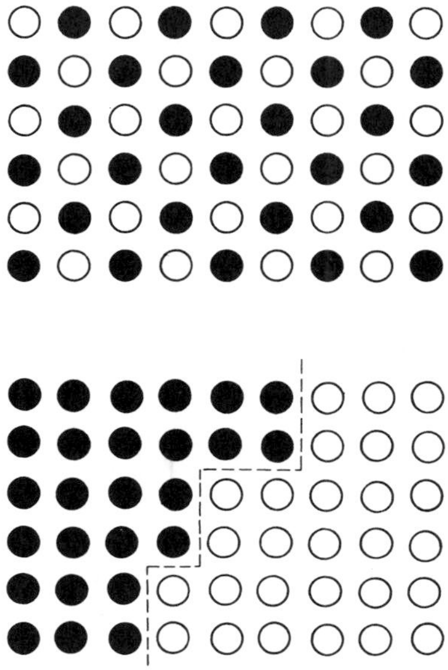

Figure 2.1. Schematic ordered phases in a two-dimensional binary system, where two states of the pseudospin are indicated by open and filled circles. (a) Ordered phase consisting of intermingling sublattices. (b) Ordered phase of two "opposite" domains. Domain walls are shown by broken lines.

variable to distinguish the microscopic mechanism from the corresponding macroscopic order parameter η.

Order variables σ_m are related to the order parameter η as expressed by

$$\eta = \sum_m \frac{\langle \sigma_m \rangle_t}{N}, \tag{2.1a}$$

if these σ_m are not strongly correlated with each other. In this case, time averages $\langle \sigma_m \rangle_t$ are first calculated at all sites, and then averaged over the subsystem consisting of **N** variables. On the other hand, if they are correlated in collective motion, the average should first be calculated for such a group of σ_m. Namely

$$\eta = \langle \eta_i \rangle \qquad \text{where} \quad \eta_i = \left\langle \sum_m \sigma_m \right\rangle_t. \tag{2.1b}$$

Perfectly ordered and disordered crystals are macroscopically homogeneous, where order variables in time average $\langle \sigma_m \rangle_t$ are all identical and independent of the site m. For an ordering process, in contrast, (2.1b) suggests that the order parameter η can be obtained as the spatial average of distributed variables η_i, which should be first calculated. It is clear that such *mean-field averages*, as expressed by spatial averages given by (2.1a) and (2.1b), are meaningful, only when the crystal can be regarded as macroscopically uniform at least approximately. Accordingly, the order parameter does not represent the critical region where the crystal is spontaneously inhomogeneous.

Ordering in crystalline systems is considerably more complex than in isotropic systems, in that strains in the host lattice may play a hidden role in the process [17]. Moreover, owing to σ_m in collective motion, the structural change takes place at a slow rate, for which the timescale t_0 of observation is significant, particularly when the characteristic time τ of the collective motion is comparable to or longer than t_0. Normally, observing conditions for slow solid-state processes are not quite so serious as in the unique critical situation during structural changes. Nevertheless, by observing with the timescale $t_0 \le \tau$, significant information can be obtained for the slow mechanism, as will be discussed in later chapters.

In this Chapter 2, we review existing statistical theories of binary systems in the light of a slow variation of order variables near T_c. It is noticed that in most theories possible values of $\langle\sigma_m\rangle_t$ are dealt with in terms of probabilities for individual $\langle\sigma_m\rangle_t$, which are generally a valid concept for slow processes. Setting aside the significance of t_0 until we discuss practical methods of observation, here we formulate the *pseudospin* model to represent probabilities for ordering in binary systems.

2.2. Probabilities, Short-Range Correlations, and the Order Parameter

In binary alloys AB such as Cu–Zn (β brass), atomic ordering takes place below the transition temperature T_c, arising from a diffusive rearrangement of atoms among lattice sites. Since such a process is slow and often quasistatic, the rearrangement can be described by a variable σ_m defined by the relations

$$\sigma_m = p_m(A) - p_m(B), \tag{2.2}$$

where

$$p_m(A) + p_m(B) = 1 \tag{2.3}$$

at a representative site m in the lattice. Here $p_m(A)$ and $p_m(B)$ are probabilities for the site m to be occupied by an atom A and by an atom B, respectively.

In a disordered phase where atoms at lattice sites are totally uncorrelated, these probabilities can take only two values, either 1 for an occupied site or 0 for a vacant site. On the contrary, in a partly-ordered phase, atoms are correlated, where the site occupation depends on the instantaneous atomic distribution in immediate surroundings. In this case, these probabilities can take a variety of values that are virtually continuous in the range between 1 and 0. Furthermore, lattice sites are not all identical during the ordering process, where the probabilities can be regarded as continuous functions of the site position and time. Generally, a classical order variable is signified by such continuous probabilities, whereas such quantum mechanical variables as for spin states $\pm\frac{1}{2}$, are characterized by discrete probabilities, if uncorrelated. It is noted however, as will be discussed in Section 2.5, that even in a quantum system, the order variable behaves as if classical, if correlated heavily with surrounding variables, and is specified by continuous probabilities.

Defined by (2.2), the order variable σ_m is virtually time-independent in a slow process in a binary system, representing short-range order at the site m. On the other hand, the macroscopic order parameter η is given by the spatial average of those variables, as calculated with (2.1a), i.e.,

$$\eta = \langle \sigma_m \rangle = \frac{\sum_m \sigma_m}{N},$$

where the summation is taken over the whole subsystem. Needless to say, such an average is meaningful only if the spatial variance of distribution is sufficiently small. The validity for a mean-field average η is evaluated by the *binary correlation function* for these variables σ_m, which is defined as

$$\Gamma(r_{mn}) = \langle (\sigma_m - \eta)(\sigma_n - \eta) \rangle = \langle \sigma_m \sigma_n \rangle - \eta^2 \delta_{mn}, \tag{2.4}$$

where r_{mn} is the distance between σ_m and σ_n, and the last expression was obtained by using the relations $\langle \sigma_m \rangle = \langle \sigma_n \rangle = \eta$ that signify the site-independence of the mean-field average. Here δ_{mn} is Kronecker's delta, whose value is 1 for m = n, and = 0 for m $\neq$ n.

For a statistically meaningful η, the function $\Gamma(r_{mn})$ should be negligibly small for all m $\neq$ n. A correlated system is characterized by nonzero $\Gamma(r_{mn})$ for all unequal pairs, for which correlations expressed by the products $\sigma_m \sigma_n$ are significant, while $\langle {\sigma_m}^2 \rangle^{1/2} = \eta$. On the other hand, for a uncorrelated system $\Gamma(r_{mn}) = 0$, signifying that $\sigma_m \sigma_n = 0$ for all unequal pairs m $\neq$ n, and $\eta = 0$. It is therefore logical to consider that the correlation energy in a correlated system is generally expressed by the Hamiltonian

$$\mathscr{H}_{m \neq n} = -J_{mn} \sigma_m \sigma_n, \tag{2.5}$$

where J_{mn} is a parameter for the magnitude of the correlation between σ_m and σ_n, and the negative sign is attached for convenience.

In the following, we show that (2.5) can be derived from a physical description of short-range interactions in crystals. While (2.5) expresses correlations J_{mn} between two lattice points m and n in general, for many applications only interactions at short distances r_{mn} are sufficient, and we only consider interactions with nearest neighbors, or including up to the next nearest neighbors in some cases. As an example, we consider a binary system AB, where A and B represent the binary states (or an atom A or an atom B) at a site m in a simple lattice. The short-range correlation energy E_m arising from those interactions between the site m and the neighboring sites n can be expressed in terms of local probabilities, $p_m(A)$, $p_m(B)$, $p_n(A)$, and $p_n(B)$. Namely,

$$E_m = \sum_n [p_m(A)p_n(A)\varepsilon_{AA}(m, n) + p_m(B)p_n(B)\varepsilon_{BB}(m, n)$$
$$+ p_m(A)p_n(B)\varepsilon_{AB}(m, n) + p_m(B)p_n(A)\varepsilon_{BA}(m, n)], \quad (2.6)$$

where the energy parameter $\varepsilon_{AB}(m, n)$, for instance, represents the interaction between A and B located at sites m and n, respectively. This expression can be further simplified for a cubic lattice by taking only the nearest neighbors where distances r_{mn} are all equal, in which case no site specifications (m, n) are needed. In this case, from (2.2) and (2.3),

$$p_m(A) = \tfrac{1}{2}(1 + \sigma_m) \quad \text{and} \quad p_m(B) = \tfrac{1}{2}(1 - \sigma_m).$$

Substituting for the probabilities in (2.6), the energy E_m can be expressed in terms of order variables σ_m and σ_n. Namely,

$$E_m = \sum_n E_{mn},$$

where

$$E_{mn} = \tfrac{1}{2}(2\varepsilon_{AB} + \varepsilon_{AA} + \varepsilon_{BB}) + \tfrac{1}{4}(\varepsilon_{AA} - \varepsilon_{BB})(\sigma_m + \sigma_n)$$
$$+ \tfrac{1}{4}(2\varepsilon_{AB} - \varepsilon_{AA} - \varepsilon_{BB})\sigma_m\sigma_n.$$

Or

$$E_{mn} = \text{const.} - K(\sigma_m + \sigma_n) - J\sigma_m\sigma_n, \quad (2.7)$$

where

$$K = \tfrac{1}{4}(\varepsilon_{BB} - \varepsilon_{AA}) \quad \text{and} \quad J = \tfrac{1}{4}[(\varepsilon_{AA} + \varepsilon_{BB}) - 2\varepsilon_{AB}].$$

The parameter J is for binary correlations with nearest neighbors, and essentially the same as J_{mn} in (2.5), whereas K is zero for most binary states $\varepsilon_{AA} = \varepsilon_{BB}$, or negligibly small in binary alloys where $\varepsilon_{AA} \approx \varepsilon_{BB}$. The first constant term in (2.7) is independent of order variables, and so insignificant for ordering processes. Therefore (2.5) or (2.7) is considered as an expression for pseudospin interactions that is generally acceptable for correlation energies.

Considering z identical nearest neighbors $n = 1, 2, \ldots, z$ in the immediate vicinity of σ_m, and assuming $K = 0$, the short-range energy E_m is given by

$$E_m = \text{const.} - J\sigma_m \sum_n \sigma_n. \quad (2.8)$$

In this expression, the quantity $J\sum_n \sigma_n$ may be interpreted as the local field F_m at the site m due to the nearest group of σ_n. Representing local order, the average $\langle \sum_n \sigma_n \rangle$ taken over the group of z neighbors may be replaced by $z\eta$ in the mean-field approximation applied to the whole subsystem, and hence $F = \langle F_m \rangle = Jz\eta$, which is analogous to the Weiss field in a ferromagnetic domain.

As remarked, the ordered phase of a binary system consists of two subsystems characterized by $\pm\eta$, which are however thermodynamically indistinguishable because of the invariant Gibbs potential under inversion $\eta \to -\eta$. Although implicit in thermal properties, the Weiss field is a significant quantity in each of the subsystems when distinguished by applying an external field.

In the mean-field approach, using local probabilities averaged over all lattice sites in the subsystems, i.e.,

$$p(\mathrm{A}) = \langle p_{\mathrm{m}}(\mathrm{A})\rangle \qquad \text{and} \qquad p(\mathrm{B}) = \langle p_{\mathrm{m}}(\mathrm{B})\rangle,$$

where

$$p(\mathrm{A}) + p(\mathrm{B}) = 1,$$

the order parameter can be defined as

$$\eta_1 = \eta = p(\mathrm{A}) - p(\mathrm{B}) \quad \text{for subsystem 1}$$

and

$$\eta_2 = -\eta = p(\mathrm{B}) - p(\mathrm{A}) \quad \text{for subsystem 2.}$$

Here the average probabilities $p(\mathrm{A})$ and $p(\mathrm{B})$ can take values in the continuous range between 1 and 0, whereas the order parameters η_1 and η_2 are continuous in the ranges (1, 0) and (0, −1), respectively. For complete disorder, $\eta_1 = \eta_2 = 0$, and hence $p(\mathrm{A}) = p(\mathrm{B}) = \frac{1}{2}$. On the other hand, ordered states $\eta_1, \eta_2 = 1$ correspond to $p(\mathrm{A}) = 1$, $p(\mathrm{B}) = 0$ and $p(\mathrm{A}) = 0$, $p(\mathrm{B}) = 1$, respectively.

From (2.8), we can obtain the expression for the mean-field average of local interaction energies

$$E_1, E_2 = \langle E_{\mathrm{m}}\rangle = \text{const.} - (\tfrac{1}{2}\mathbf{N})Jz\langle\sigma_{\mathrm{m}}\rangle\eta = \text{const.} - \tfrac{1}{2}\mathbf{N}Jz\eta^2, \tag{2.9}$$

where the factor $\frac{1}{2}$ is included for correct counting of the number of interacting pairs. Equation (2.9) can be interpreted in a simple manner as related to counting the number of "unlike" pairs that should diminish toward perfect order. The average number of variables in state A in subsystem 1 is given by $\mathbf{N}p(\mathrm{A})$. These variables in state A are surrounded by variables in unlike state B in z nearest-neighbor sites, whose number is $zp(\mathrm{B})$. Hence, the average number of interacting AB pairs is $\mathbf{N}zp(\mathrm{A})p(\mathrm{B})$. Similarly, the average number of BA pairs in the same subsystem is $\mathbf{N}zp(\mathrm{B})p(\mathrm{A})$, and hence the total number of unlike pairs is given by $2\mathbf{N}zp(\mathrm{A})p(\mathrm{B})$. We can also count the number of unlike pairs in subsystem 2, resulting in exactly the same number as in subsystem 1. Here the average probabilities can be expressed in terms of η as

$$p(\mathrm{A}) = \tfrac{1}{2}(1+\eta) \qquad \text{and} \qquad p(\mathrm{B}) = \tfrac{1}{2}(1-\eta).$$

Therefore, the total number of those unlike pairs in the whole system is

$$2 \times 2\mathbf{N}zp(\mathrm{A})p(\mathrm{B}) = 4(\tfrac{1}{2}Nz) \times \tfrac{1}{4}(1-\eta^2) = \tfrac{1}{2}Nz(1-\eta^2),$$

which becomes zero in the limit of $\eta \to 1$. In this context, the last term of (2.9) represents energies of unlike pairs with respect to like pairs at the nearest-neighbor distance. As indicated in (2.9), $E_1 = E_2$, but thermodynamically only the total energy $E = E_1 + E_2$ is significant. Namely,

$$E(\eta) = E_1 + E_2 = \text{const.} - \tfrac{1}{2}NzJ\eta^2. \tag{2.10}$$

2.3. Mean-Field Theory of Order–Disorder Statistics

Bragg and Williams [18] introduced in their statistical theory the concept of short- and long-range order for ordering in binary alloys. They assumed that thermal properties of an ordered alloy can be specified by the long-range order parameter η, which then determines the mean-field average of the short-range correlation energy $E(\eta)$. As remarked in Section 2.2, the short-range energy is due to interacting unlike pairs AB in the subsystem, where there are a great many combinations for choosing variables in different states A and B from $\mathbf{N}$ variables, giving the degeneracy $g(\eta)$ for $E(\eta)$. The number of such combinations is

$$g(\eta) = \binom{\mathbf{N}}{\mathbf{N}p(\mathrm{A})}\binom{\mathbf{N}}{\mathbf{N}p(\mathrm{B})} = \binom{\mathbf{N}}{\frac{1}{2}\mathbf{N}(1+\eta)}\binom{\mathbf{N}}{\frac{1}{2}\mathbf{N}(1-\eta)}.$$

We can therefore write the following partition function for a partially ordered system,

$$Z(\eta) = Z(0)g(\eta)\exp\left(\frac{\frac{1}{2}NzJ\eta^2}{\mathrm{k_B}T}\right),$$

where $Z(0)$ is the partition function for variables other than η in the system. Assuming that the lattice is unchanged during the ordering process, the thermal equilibrium under a constant volume can be determined by minimizing the Helmholtz free energy, $F = -\mathrm{k_B}T\ln Z(\eta)$, with respect to η, i.e.,

$$\frac{\partial}{\partial\eta}\left\{\ln Z(0) + \ln g(\eta) + \left(\frac{\frac{1}{2}NzJ}{\mathrm{k_B}T}\right)\eta^2\right\} = 0.$$

For large N, the term $\ln g(\eta)$ can be evaluated by the Stirling formula, i.e.,

$$\frac{\partial \ln g(\eta)}{\partial\eta} = -\tfrac{1}{2}N\ln\frac{1+\eta}{1-\eta}.$$

Therefore, the order parameter for equilibrium can be given as a solution of the equation

$$\left(\frac{zJ}{\mathrm{k_B}T}\right)\eta = \ln\frac{1+\eta}{1-\eta}$$

or

$$\eta = \tanh\frac{zJ}{2\mathrm{k_B}T}\eta. \tag{2.11}$$

The solution of (2.11) can be obtained graphically from the intersection between the straight line

$$y = \left(\frac{zJ}{2k_B T}\right)\eta$$

and the hyperbolic curve

$$\eta = \tanh y,$$

as illustrated in Fig. 2.2(a). It is noticed that for $2k_B T/zJ \geq 1$ the intersection is only at $\eta = 0$, whereas for $2k_B T/zJ < 1$ there is another intersection at which the nonzero η represents a partially ordered state. The transition is therefore described as $0 \leftrightarrow \eta \neq 0$, and the transition temperature is given by

$$T_c = \frac{zJ}{2k_B}. \tag{2.12a}$$

Writing $y = (T/T_c)\eta$, from (2.11)

$$\left(\frac{T}{T_c}\right)\eta = \tanh^{-1}\eta \approx \eta + \frac{\eta^3}{3}$$

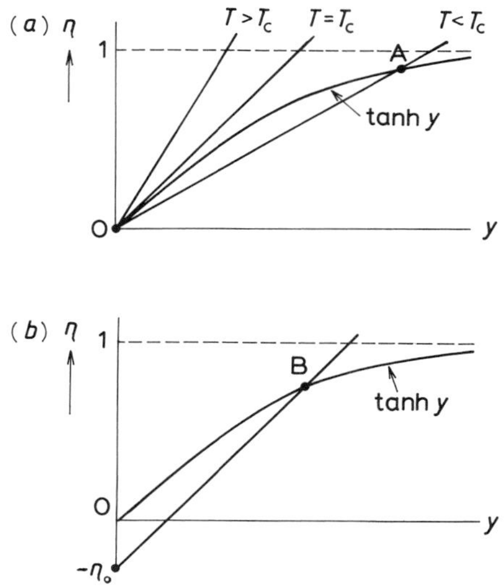

Figure 2.2. (a) Graphical solutions for the order parameter. For $T < T_c$, the straight line $y = (zJ/2k_B T)\eta$ intersects with the curve $y = \tanh \eta$ at A, whereas for $T > T_c$ the only real solution is $\eta = 0$, signifying disorder. (b) Graphical solutions for ferromagnetic order. The intersection B between the straight line $y = (T/T_c)\eta - \eta_0$ and the curve $y = \tanh \eta$ can be uniquely determined below T_c, while there is no solution above T_c.

for a small value of η. Hence we obtain

$$\eta^2 \approx \frac{3(T_c - T)}{T_c}, \tag{2.12b}$$

indicating a parabolic temperature-dependence of η in the vicinity of T_c. It is noted that such a parabolic-order parameter is a consequence of the mean-field approximation, which is obtained either from a spherical (isotropic) or cubic system. According to contemporary theories, the critical exponent β in the empirical expression $\eta \propto (T_c - T)^\beta$ depends on the dimensionality of η, while in the mean-field theory $\beta = \frac{1}{2}$ for the present case of a scalar parameter.

As discussed in Section 1.5, a discontinuity in the heat capacity ΔC_V is expected at T_c, which is calculated from the mean-field average of the odering energy $E(\eta)$. Namely,

$$C_V = \left\{\frac{\partial E(\eta)}{\partial T}\right\}_V = \left(\frac{\mathrm{d}E}{\mathrm{d}\eta}\right)\left(\frac{\mathrm{d}\eta}{\mathrm{d}T}\right) = (-NzJ\eta)(-\tfrac{3}{2}/T_c\eta) = 3N\mathrm{k_B}$$

for $T < T_c$, whereas $C_V = 0$ at $T = T_c$. Hence, $\Delta C_V = 3N\mathrm{k_B}$, which is a typical result of the mean-field approximation for second-order phase transitions.

2.4. The Ising Model for Spin–Spin Correlations

Quantum-mechanical spin–spin correlations in magnetic crystals are expressed by the Heisenberg exchange interaction

$$\mathscr{H}_{\mathrm{mn}} = -2J_{\mathrm{mn}}\boldsymbol{s}_{\mathrm{m}} \cdot \boldsymbol{s}_{\mathrm{n}} \tag{2.13}$$

between spins $\boldsymbol{s}_{\mathrm{m}}$ and $\boldsymbol{s}_{\mathrm{n}}$, where the parameter J_{mn} is the exchange integral between unpaired electrons of magnetic ions at lattice sites m and n. However, when these spins are correlated with many other neighboring spins in the crystal, a spin $\boldsymbol{s}_{\mathrm{m}}$ may behave like a classical vector, and their interactions can be expressed by a simplified form known as the Ising model,

$$\mathscr{H}_{\mathrm{mn}} = -2J_{\mathrm{mn}}s_{\mathrm{mz}}s_{\mathrm{nz}}, \tag{2.14}$$

which includes only z components of the spin vectors. The Ising interactions for classical spins are often used as a simplified model for problems of magnetic ordering. Here, the presence of a unique z direction is not specified for the model, while the origin is related to a magnetic anisotropy in a practical magnetic crystal.

In the previous section, we showed that short-range interactions in binary systems can be described by (2.5), which is similar to the Ising Hamiltonian (2.14). By analogy, the z component of a magnetic spin can be interpreted as probabilities for spin states, when the spin is considered as a classical vector. It is noted that the factor 2 in (2.14) originates from convenience for the spin quantum number $\frac{1}{2}$, while its absence in (2.5) is due to the order variable σ_{m} representing probabilities for binary states.

Such a uniaxial magnet consisting of microscopic magnetic moments of spin $\frac{1}{2}$ provides an instructive example of binary order. With respect to the axis denoted by z, all spins s_m are quantized, being characterized by eigenvalues $\pm\frac{1}{2}$ of its component s_{mz}. In the classical analogue, the spin s_m is precessing around the z axis, and the precessing components s_x and s_y are time-dependent and expressed by off-diagonal matrices. In the disordered phase, these spins are in independent motion and uncorrelated, and hence two individual spin states $\pm\frac{1}{2}$ are signified by an equal probability. On the other hand, when correlated in the ordered phase, where these probabilities at each site m are unequal for the $\pm\frac{1}{2}$ states, depending on the spin distribution in the immediate surroundings. To validate such a statistical description, we assume that the time average of perpendicular components vanishes, i.e., $\langle s_{mx}s_{nx} + s_{my}s_{ny}\rangle_t = 0$ in (2.13), so that the Ising formula (2.14) can be used instead. In terms of precessing spins, this assumption is equivalent to the *random-phase* approximation.

In this assumption, the correlated spin state at a site m is given by

$$|m\rangle = a_m|+\rangle + b_m|-\rangle, \tag{2.15a}$$

where $|\pm\rangle$ represent uncorrelated spin states $|\pm\frac{1}{2}\rangle$, and for the coefficients a_m and b_m we have the normalization relation

$$a_m{}^2 + b_m{}^2 = 1. \tag{2.15b}$$

For spins in the disordered phase $a_m = b_m$, whereas in the ordered phase $a_m \neq b_m$, by which we can define the order variable as

$$\sigma_m = a_m{}^2 - b_m{}^2. \tag{2.15c}$$

Writing $a_m = \cos\theta_m$ and $b_m = \sin\theta_m$ for convenience, the order variable σ_m can be regarded as a vector whose direction is specified by the angle $2\theta_m$ from the z axis. Corresponding to $-1 \le a_m, b_m \le 1$, the range for the angle is $0 \le \theta_m \le \pi$. As remarked, the above statistical description is valid for a slow spin rearrangement process among lattice sites.

Assuming nearest-neighbor interactions only, the short-range energy is expressed by

$$\begin{aligned} E_m &= \sum_n \langle m, n|\mathcal{H}_{mn}|n, m\rangle \\ &= -2J\sum_n \{a_m{}^2a_n{}^2\langle ++|s_{mn}s_{nz}|++\rangle + b_m{}^2b_n{}^2\langle --|s_{mz}s_{nz}|--\rangle \\ &\quad + a_m{}^2b_n{}^2\langle +-|s_{mz}s_{nz}|+-\rangle + b_m{}^2a_n{}^2\langle -+|s_{mz}s_{nz}|-+\rangle\}, \end{aligned}$$

for which $z = 8$ and $J_{mn} = J$ are assumed for a cubic lattice. For spins $\frac{1}{2}$, these matrix elements are

$$\langle ++|s_{mz}s_{nz}|+\rangle = \langle --|s_{mz}s_{nz}|--\rangle = \tfrac{1}{4} \quad \text{and}$$

$$\langle +-|s_{mz}s_{nz}|+-\rangle = \langle -+|s_{mz}s_{nz}|-+\rangle = -\tfrac{1}{4},$$

and

$$E_m = 2J \sum_n [-\tfrac{1}{4}(a_m{}^2 a_n{}^2 + b_m{}^2 b_n{}^2) + \tfrac{1}{4}(a_m{}^2 b_n{}^2 + b_m{}^2 a_n{}^2)\}$$
$$= -2J \sum_n [(a_m{}^2 - a_m{}^2)(b_m{}^2 - b_n{}^2)]$$
$$= -\tfrac{1}{2} J\sigma_m \sum_n \sigma_n. \tag{2.16}$$

Replacing the factor $\frac{1}{2}J$ by $2J_{mn}$, we obtain equivalently $E_{mn} = -J_{mn}\sigma_m\sigma_n$ as given by (2.5). Thus, the Ising model for order variables gives their correlation energy in an identical form to interacting spin components $s_{mz}s_{nz}$.

The sum $\sum_n \sigma_n$ over the nearest neighbors of σ_m can be considered as representing the local field F_m at the site m, which may be written as $F_m = \frac{1}{2}Jz\langle\sigma_n\rangle_m$. Therefore, the long-range average of $\langle E_m\rangle_1$ in domain 1 is given as the mean-field average of "local Zeeman energies"

$$E_1 = \langle -\sigma_m F_m\rangle = -\tfrac{1}{4}NzJ\eta^2,$$

where $\eta = \langle\sigma_m\rangle = \langle a_m{}^2\rangle - \langle b_m{}^2\rangle$ is the order parameter. The total short-range energy given by the above is exactly the same as (2.10). Beyond this point, the order–disorder statistics in the mean-field approximation is completely analogous to that for an alloy described in Section 2.3.

The order variable σ_m can be expressed in terms of a classical spin component as $s_{mz} = \frac{1}{2}\sigma_m$, that is, related to the magnetic moment $\mu_{mz} = g\beta s_{mz} = \frac{1}{2}g\beta\sigma_m$. Here, g is the Landé factor, and β is the Bohr magneton. The short-range energy can therefore be expressed as $-\sigma_m F_m = -\mu_{mz}B_m$, where $B_m = (2/g\beta)F_m$ is the local magnetic field at site m. It is noted that the average internal field is expressed either by $\langle F_m\rangle = F$, or $\langle B_m\rangle = B_i$, that is the Weiss field discussed in Section 1.6.

In the previous section we found that the order–disorder transition is determined by (2.11) in the mean-field accuracy, which can be reexpressed as

$$\eta = \langle\sigma_m\rangle = \tanh\left\{\left(\frac{zJ}{2k_B T}\right)\langle\sigma_m\rangle\right\} = \tanh\left(\frac{F}{k_B T}\right)$$
$$= \frac{\exp\{(+1)F/k_B T\} - \exp\{(-1)F/k_B T\}}{\exp\{(+1)F/k_B T\} + \exp\{(-1)F/k_B T\}}.$$

Here, the maximum and minimum values of η are denoted by $+1$ and -1, for which the probabilities are

$$p(+1) = Z^{-1}\exp\frac{(+1)F}{k_B T} \quad \text{and} \quad p(-1) = Z^{-1}\exp\frac{(-1)F}{k_B T}, \tag{2.17}$$

where

$$Z = \exp\frac{(+1)F}{k_B T} + \exp\frac{(-1)F}{k_B T}$$

is the partition function of the system of effective magnetic moments, $+1$ and

-1, in the Weiss field F. We can then see the probabilities

$$p(+1) = \langle p_{\mathrm{m}}(+)\rangle \qquad \text{and} \qquad p(-1) = \langle p_{\mathrm{m}}(-)\rangle$$

are identical to the Boltzmann probabilities. Therefore, (2.11) can be interpreted as originated from the magnetic spin moment with components ± 1 in the Weiss magnetic field.

In a ferromagnet, the magnetized phase is characterized by the internal Weiss field B_{i}, which is therefore proportional to magnetization M in the mean-field approximation. In the presence of an external field B_0 along the z direction, the internal energy associated with η can be written as

$$E(\eta) = -\tfrac{1}{2}\mathbf{N}Jz\eta^2 - \mathbf{N}(\tfrac{1}{2}g\beta\eta)B_0 = -\mathbf{N}(\tfrac{1}{2}g\beta\eta)(B_{\mathrm{i}} + B_0).$$

Here the Weiss field should be given by $B_{\mathrm{i}} = Jz/g\beta$, but can be derived from the following argument.

Now that we know the order parameter can be specified by thermal probabilities $p(\pm 1)$ for $\eta = \pm 1$, the corresponding energies per spin can be given by

$$\varepsilon(+1) = -\tfrac{1}{2}zJp(+1) - g\beta(+1)B_0 \qquad \text{and}$$
$$\varepsilon(-1) = -\tfrac{1}{2}zJp(-1) - g\beta(-1)B_0,$$

where

$$p(+1) = Z^{-1}\exp\frac{-\varepsilon(+1)}{\mathrm{k_g}T} \qquad \text{and} \qquad p(-1) = Z^{-1}\exp\frac{-\varepsilon(-1)}{\mathrm{k_B}T}.$$

Here Z is the partition function $\sum_{\pm}\exp\{-\varepsilon(\pm 1)/\mathrm{k_B}T\}$. Therefore,

$$\frac{p(+1)}{p(-1)} = \exp\frac{zJ\eta + 2g\beta B_0}{\mathrm{k_B}T} = \frac{1+\eta}{1-\eta},$$

and

$$\eta = \tanh\frac{\tfrac{1}{2}zJ\eta + g\beta B_0}{\mathrm{k_B}T}, \tag{2.18}$$

which is similar to (2.11). Graphically, in exactly the same way as for (2.11), we can solve (2.18) by finding an intersection between the straight line

$$y = \left(\frac{zJ}{2\mathrm{k_B}T}\right)\eta + \frac{g\beta B_0}{\mathrm{k_B}T} \tag{2.19}$$

and

$$y = \tanh^{-1}\eta.$$

Writing $\mathrm{k_B}T_{\mathrm{c}} = \tfrac{1}{2}zJ$ as in (2.12), these equations can be reexpressed as

$$\eta = \left(\frac{T}{T_{\mathrm{c}}}\right)y - \frac{g\beta B_0}{\mathrm{k_B}T_{\mathrm{c}}} \qquad \text{and} \qquad \eta = \tanh y. \tag{2.20}$$

As sketched in Fig. 2.2(b), the first straight line intercepts the η axis at $\eta_0 = -g\beta B_0/k_B T_c$, which is however very small in practical cases. For $\beta = 1$ Bohr magneton, $B_0 = 3\ \mathrm{w/m^2}$ and $T_c \approx 10^3$ K in a typical case of a ferromagnet, η_0 is only of the order of 10^{-2}. It is noticed that in the presence of B_0 there is always a real solution for η at all temperatures, and no critical point while T_c is the singular point at the Weiss field B_i.

As pointed out in Section 1.6, even at temperatures above T_c, the system is forced to be ordered to an extent by an applied field B_0. Hence, the above argument for short-range interactions is valid for $T \geq T_c$ as in the range $T \leq T_c$. In the former case if the temperature is close to T_c, $y \cong \eta$, and hence from (2.20) we obtain

$$\eta\left(1 - \frac{T}{T_c}\right) = -\frac{g\beta B_0}{k_B T_c}$$

or

$$\eta = \frac{g\beta B_0/k_B}{T - T_c} \qquad \text{for} \quad T > T_c,$$

which is the Curie–Weiss law as expressed in the form

$$\chi = \frac{N\frac{1}{2}g\beta\eta}{B_0} = \frac{C}{T - T_c}$$

where $C = Ng^2\beta^2/2k_B$ is the Curie constant.

CHAPTER 3

Pseudospins and Their Collective Modes in Displacive Crystals

3.1. Pseudospins in Displacive Crystals

Among crystalline systems undergoing phase transitions, there is a group of crystals where the structural change is signified by active ions or molecules involved in continuous displacements. At the threshold of such transitions, these active groups begin to exhibit linear or rotational displacements in some cases and distortion in others, as evidenced by X-ray diffraction [19] and magnetic resonance results [20], [21]. Arising from continuous displacements, these so-called *displacive* phase transitions are second order according to the Ehrenfest classification, and characterized by a classical displacement vector signifying symmetry change at the transition temperatures.

Structural changes in perovskite crystals provide typical examples of displacive phase transitions. Given chemical formula ABO_3, the unit cell in the normal phase consists of an octahedral complex $BO_6{}^{2-}$ that is surrounded by eight A^{2+} ions at corners of the cubic cell, as shown in Fig. 3.1(a). Crystals of the perovskite family are rich in types of structural changes, exhibiting a variety of displacement schemes. For instance, in the ferroelectric phase transition of $BaTiO_3$ at 405 K the structural change is signified by an off-center displacement of the central Ti^{4+} ion along a C_4 axis parallel to one of the cubic axes (Fig. 3.1(b)). For the structural transition in $SrTiO_3$ at 105 K, on the other hand, a rotational displacement of $TiO_6{}^{2-}$ around a C_4 axis is responsible (Fig. 3.1(c)).

Structural transitions in these perovskites are known as binary, where the low-temperature phase is characterized by two directions of the vector-order parameter that are related by inversion or reflection. Accordingly, there should be two directions for each of the microscopic displacements $\boldsymbol{\sigma}_m$, which are also vector variables. While a continuous-phase transition described by

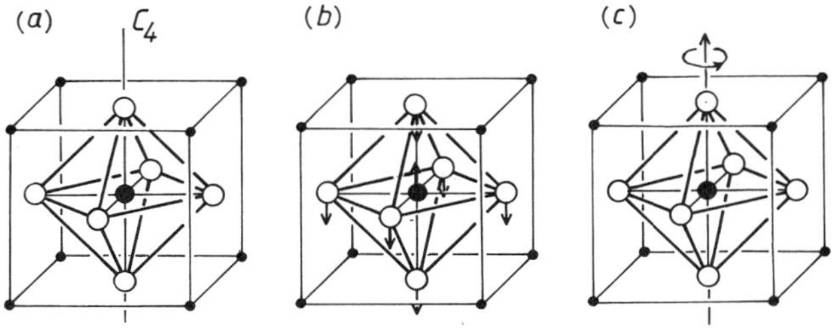

Figure 3.1. Unit cells in the perovskite structure: (a) the normal phase; (b) linear ionic displacements along the C_4 axis in $BaTiO_3$; and (c) rigid-body rotation of a TiO_6 octahedron around the C_4 axis of $SrTiO_3$ crystals.

the Landau theory appears to be one dimensional as specified by a scalar-order parameter, it can be two- or three-dimensional as signified by vector-order variables $\boldsymbol{\sigma}_m$. For such a displacement variable $\boldsymbol{\sigma}_m$, we can consider a local potential $V(\boldsymbol{\sigma}_m)$ that is related to the Gibbs potential as the mean-field average, i.e., $\Delta G = \langle V(\boldsymbol{\sigma}_m)\rangle$. For example, taking the C_4 axis of an octahedral complex $BO_6{}^{2-}$ along the symmetry direction axis as the z direction, the following potential can be considered for the vector variable $\boldsymbol{\sigma}_m = (\sigma_{mx}, \sigma_{my}, \sigma_{mz})$ for $T > T_c$. Namely,

$$V_>(\boldsymbol{\sigma}_m) = \tfrac{1}{2}a_x\sigma_{mx}{}^2 + \tfrac{1}{2}a_y\sigma_{my}{}^2 + \tfrac{1}{2}a_z\sigma_{mz}{}^2, \qquad (3.1a)$$

where the coefficients a_x, a_y, and a_z are positive and independent of m. Figure 3.2(a) shows such a potential in two dimensions, where the displacement $\boldsymbol{\sigma}_m$ is harmonic, and the time average $\langle\boldsymbol{\sigma}_m\rangle_t = 0$ for $T > T_c$, signifying that the crystal is stable when $\langle\boldsymbol{\sigma}_m\rangle_t = 0$. Statistically, it is clear that $\langle V_>(\boldsymbol{\sigma}_m)\rangle_t = 0$ in thermal equilibrium, where the average correlations are zero, i.e., $\langle\boldsymbol{\sigma}_m\boldsymbol{\sigma}_n\rangle_t = 0$.

If, on the other hand, correlations $\sigma_{mz}\sigma_{nz}$ in the z direction become significant in the critical region, the potential of (3.1a) should change to

$$V_<(\boldsymbol{\sigma}_m) = \tfrac{1}{2}a_x\sigma_{mx}{}^2 + \tfrac{1}{2}a_y\sigma_{my}{}^2 + \tfrac{1}{2}a_z\sigma_{mz}{}^2 + \tfrac{1}{4}b_z\sigma_{mz}{}^4 \qquad (3.1b)$$

for $T < T_c$, where $a_z < 0$ and $b_z > 0$ while a_x and a_y remain positive and unchanged from (3.1a). Figure 3.2(b) sketched the potential $V_<(\boldsymbol{\sigma}_m)$ in two dimensions. In the potential $V_<(\boldsymbol{\sigma}_m)$ the vector $\boldsymbol{\sigma}_m$ can move between $\pm\sigma_0 = +(-a_z/b_z)^{1/2}$ over the saddle point at the origin in either the zx or zy plane, depending on which of the coefficients a_x and a_y is larger.

In Fig. 3.2(b), we have assumed for brevity that $a_y = 0$, thereby the vector $\boldsymbol{\sigma}_m$ is confined to the zx plane. In this case, owing to correlations among these order variables for $T < T_c$, we can define probabilities for the z component σ_{mz} to be in the two positions $\pm\sigma_0$, as in the Ising pseudospin, i.e.,

$$\sigma_{mz} = \sigma_0\{p_m(+\sigma_0) - p_m(-\sigma_0)\}. \qquad (3.2)$$

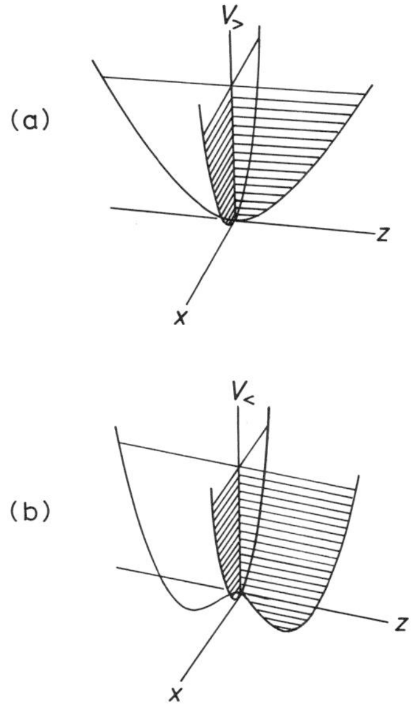

Figure 3.2. Local crystalline potentials in a quasi-two-dimensional lattice: (a) paraboloidal $V_>$ above T_c; and (b) double-well paraboloidal $V_<$ below T_c.

Writing $\sigma_{mz} = \sigma_0 \cos \theta_m$, the other component of σ_m can be expressed as

$$\sigma_{mx} = \sigma_0 \sin \theta_m = \sigma_0 \{2p_m(+\sigma_0)p_m(-\sigma_0)\}^{1/2},$$

giving a complete description of the vector $\boldsymbol{\sigma}_m$.

Here, we realize that if potentials $V_<(\boldsymbol{\sigma}_m)$ are uncorrelated, $p_m(\sigma_0) = p_m(-\sigma_0)$ hence $\sigma_{mz} = 0$, and the potential $V_<(\boldsymbol{\sigma}_m)$ becomes the same as $V_>(\boldsymbol{\sigma}_m)$ of (3.1a). However, signified by a nonzero $\langle \boldsymbol{\sigma}_m \rangle_t$ observed with a very short timescale t_0, the quartic potential $\frac{1}{4}b_z \sigma_{mz}{}^4$ in $V_<(\boldsymbol{\sigma}_m)$ should be associated with the presence of pseudospin correlations below T_c.

The interaction energy between two vector-order variables $\boldsymbol{\sigma}_m$ and $\boldsymbol{\sigma}_n$ at different sites m and n can be assumed as generally given as proportional to the scalar product $\boldsymbol{\sigma}_m \cdot \boldsymbol{\sigma}_n$. Although similar in form to the Heisenberg coupling, the interaction is temperature-dependent, and the parameter J_{mn} is not so well defined as in the exchange coupling due partly to the presence of intervening ions or molecules in-between. Nevertheless, as in magnetic cases, the sign of J_{mn} is essential for the pseudospin order: either parallel or antiparallel. Furthermore, while generally in a short-range, it is significant that such correlations, as extended beyond nearest neighbors, may violate translational symmetry of the lattice. The order variable is basically a part of the active group, so that the lattice is strained by ordering. However, in this chapter we assume for the time being that the lattice structure remains unchanged, leaving the problem arising from lattice distortion to Chapter 4.

By analogy to a magnetic spin, we shall refer to such a vector variable $\boldsymbol{\sigma}_m$ in a binary system as a *pseudospin*, and postulate the following Hamiltonian for correlations among pseudospins at lattice points, i.e.,

$$\mathscr{H}_{mn} = -J_{mn}\boldsymbol{\sigma}_m \cdot \boldsymbol{\sigma}_n. \tag{3.3}$$

Similar to a classical spin, as so often assumed in theories of magnetism, the pseudospin is a classical vector written as

$$\boldsymbol{\sigma}_m = \sigma_0 \boldsymbol{e}_m, \tag{3.4}$$

where σ_0 is the amplitude, and $\boldsymbol{e}_m$ is the unit vector along the direction of $\boldsymbol{\sigma}_m$. Near the threshold of a continuous-phase transition, the infinitesimal amplitude σ_0 can virtually be site-independent, and only its direction $\boldsymbol{e}_m$ is considered as significant. In this case, (3.3) can be expressed by

$$\mathscr{H}_{mn} = -\sigma_0{}^2 J_{mn}\boldsymbol{e}_m \cdot \boldsymbol{e}_n. \tag{3.3a}$$

For displacive ordering, pseudospin correlations expressed by (3.3) and (3.3a) give an adequate description of short-range interactions as shown by examples in Section 3.5, although the parameter J_{mn} cannot be defined from first principles.

As remarked, finite displacements $\boldsymbol{\sigma}_m$ in the active groups may violate local lattice symmetry, resulting in distributed strains in the lattice. Such a strain effect is significant in the critical region, for which a specific coupling is responsible between pseudospins and displacements of active groups from their original sites, while these are basically independent at all noncritical temperatures.

3.2. The Landau Criterion for Classical Fluctuations

In Section 3.1, the classical pseudospin was defined to express a binary displacement. In a double-well potential, the displacing motion of a particle between two minima is quantum-mechanical tunneling in principle, when the central potential barrier is sufficiently high. On the other hand, if the kinetic energy is higher than the height, the motion between the two potential minima is regarded as a classical jumping over the barrier. Landau has given a general criterion for *classical fluctuations* that are expressed in terms of continuous probabilities for a binary case.

We consider that a small but finite displacement $\boldsymbol{\sigma}_m$ from the center of the potential at a site m is involved in a long-wave excitation at a small energy ε or at a low frequency $\omega = \varepsilon/\hbar$ in the crystal. While not quite identifiable in the lattice spectra, such a low-frequency excitation can be attributed to an acoustic vibrational mode at long wavelengths. Suppose the local distortion at a lattice point x in a continuum crystal is described by a function $\sigma(x)$, thermodynamically there should be a negative entropy change $\Delta S < 0$, for which a continuous probability $p(x)\,dx$ can be defined for $\sigma(x)$ to be found in a range between x and $x + dx$. Using the Boltzmann formula, the relation between $\Delta S(x)$ and $p(x)$ can be written as

$$\Delta S(x) = k_B \ln \frac{p(x)}{p_0} = k_B \ln \frac{1 - \Delta p(x)}{p_0},$$

where $p_0 = \frac{1}{2}$ is the probability for complete disorder between $\pm\sigma_0$, and $\Delta p = p_0 - p$. Assuming that $|\Delta p| \ll p_0$, the value of $|\Delta S(x)|$ is substantially smaller than the Boltzmann constant k_B at any arbitrary point x, and hence the inequality

$$\text{(i)} \qquad \frac{|\Delta S(x)|}{k_B} \ll 1$$

can be considered for an entropy change for classical fluctuations. At a temperature T the entropy decreases by $\Delta S(x)$, corresponding to a local excitation energy $\Delta\varepsilon(x)$, as in the relation

$$\text{(ii)} \qquad -\Delta\varepsilon(x) = T\,\Delta S(x),$$

which causes a deviation from the minimum of the Gibbs potential, but returning normally to the stable-ordered state as a thermal relaxation. If such an excitation is quantum-mechanical, in contrast, $\Delta\varepsilon$ has to be subjected to the uncertainty principle

$$\text{(iii)} \qquad \Delta\varepsilon\,\Delta t \approx \hbar,$$

where Δt may be regarded as the relaxation time τ for the excitation to decay to thermal equilibrium. In any case, such an excitation energy $\Delta\varepsilon$ will decrease the entropy of the system by ΔS, but relaxing back to thermal equilibrium. Combining the relations (ii) and (iii), $|\Delta S| = \Delta\varepsilon/T = \hbar/(\tau T)$, which is then incorporated into the inequality (i) to obtain the criterion for classical fluctuations in terms of τ and T, i.e.,

$$\tau T \gg \hbar k_B \sim 10^{-11}\ \text{s.K.} \tag{3.5}$$

An example of this Landau criterion is that, if the characteristic time τ is longer than 0.5×10^{-13} s in a phase transition at $T_c = 200$ K, the inequality of (3.5) is fulfilled, and so the critical fluctuations are classical. For a typical displacive phase transition, empirically τ is of the order of 10^{-11} s, so that observed critical anomalies should be due to classical fluctuations. On the other hand, proton tunneling in hydrogen-bonding crystals is quantum-mechanical at their structural transitions. In this case, the characteristic time is of the order of $\tau \sim 10^{-13}$ s and T_c is typically about 50 K in a hydrogen-bonding crystal, and so $\tau T_c < 10^{-13}$, which definitely violates the condition given by (3.5). It is known that, in general, order–disorder phase transitions due to proton rearrangements are discontinuous, owing to an inevitably discrete energy change at T_c.

3.3. Quantum-Mechanical Pseudospins and Their Correlations in Crystals

When tunneling motion prevails, the displacement in a double-well potential cannot be continuous. On the other hand, we suggested in Section 3.1, for a classical displacement below T_c, that the low-temperature phase is characterized by a quartic potential $\frac{1}{4}B\eta^4$ due to emerging correlations. Nevertheless, such a potential due to correlations should emerge below T_c in whatever types of structural ordering. In contrast to the statistical argument in Section 3.1, we can see from a quantum mechanical argument how pseudospin correlations are associated with a quartic potential. A quantum theory of tunneling particles and their correlations in a periodic lattice are sketched in this section, following Blinc and Zeks [12].

A particle tunneling through the barrier of a double-well potential in (3.1b) is given by an equal quantum state ε_0 in the two potential minima, but perturbed by the potential V in-between. The doubly degenerate unperturbed energy ε_0 is the ground state of the Schrödinger equation

$$\mathscr{H}\varphi_m = \varepsilon_0\varphi_m$$

at the site m. The degenerate eigenstate ε_0 is invariant by inversion

$$\sigma_z = \begin{pmatrix} 1 & 0 \\ 0 & -1 \end{pmatrix}$$

operated on $\mathscr{H}$, where the suffix z is referred to the direction of tunneling motion. Therefore, the unperturbed Hamiltonian $\mathscr{H}$ and inversion σ_z are commutable, i.e.,

$$[\mathscr{H}, \sigma_z] = 0,$$

and diagonalized simultaneously. It is noted that the inversion operator σ_z is qualified as the binary-order variable with eigenvalues ± 1.

As a consequence of the perturbing potential barrier V, the degenerate energy ε_0 of the particle for $\sigma_z = \pm 1$ is split into two levels $\varepsilon_\pm = \varepsilon_0 \pm \frac{1}{2}V$, which are characterized by symmetric and antisymmetric combinations of wavefunctions φ_{mL} and φ_{mR}, i.e.,

$$\psi_{m+} = 2^{-1/2}(\varphi_{mL} + \varphi_{mR}) \quad \text{and} \quad \varphi_{m-} = 2^{-1/2}(\varphi_{mL} - \varphi_{mR}). \tag{3.6a}$$

Here the indexes L and R are referred to left and right minima of the potential $V_<(\sigma_m)$. In the second-quantization scheme, the functions φ_{mL}, φ_{mR} and the corresponding $\varphi_{mL}{}^\dagger$, $\varphi_{mR}{}^\dagger$ represent annihilation and creation operators of a particle at these positions, respectively, at the site m. For these operators, the normalization relations are expressed as

$$\psi_{m+}{}^\dagger\psi_{m+} + \psi_{m-}{}^\dagger\psi_{m-} = \varphi_{mL}{}^\dagger\varphi_{mL} + \varphi_{mR}{}^\dagger\varphi_{mR} = 1. \tag{3.6b}$$

Since $\varphi_{mL}{}^\dagger\varphi_{mL}$ and $\varphi_{mR}{}^\dagger\varphi_{mR}$ represent densities at the left and right minima, we can define the order variable as

$$\sigma_{mz} = \psi_{m+}{}^{\dagger}\psi_{m+} - \psi_{m-}{}^{\dagger}\psi_{m-} = \varphi_{mL}{}^{\dagger}\varphi_{mL} - \varphi_{mR}{}^{\dagger}\varphi_{mR}, \tag{3.7a}$$

which is 0 in a uncorrelated case, where the two minima L and R are equivalent. On the other hand, in a correlated system, the short-range average of σ_{mz} may not be zero, serving as the order variable.

The tunneling between two minima on the left and right may be described by the operators defined as

$$\sigma_{mx} = \varphi_{mL}{}^{\dagger}\varphi_{mR} + \varphi_{mR}{}^{\dagger}\varphi_{mL} \tag{3.7b}$$

and

$$\sigma_{my} = \varphi_{mL}{}^{\dagger}\varphi_{mR} - \varphi_{mR}{}^{\dagger}\varphi_{mL}. \tag{3.7c}$$

It is noted that these operators are related by the commutation relations, i.e.,

$$[\sigma_{mx}, \sigma_{my}] = i\sigma_{mz}, \qquad [\sigma_{my}, \sigma_{mz}] = i\sigma_{mx} \qquad \text{and} \qquad [\sigma_{mz}, \sigma_{mx}] = i\sigma_{my}, \tag{3.8}$$

where the three variables σ_{mx}, σ_{my}, and σ_{mz} constitute components of a quantum-mechanical pseudospin vector $\boldsymbol{\sigma}_m$, being given by the Pauli matrices, namely,

$$\sigma_{my} = \begin{pmatrix} 0 & -i \\ i & 0 \end{pmatrix} \qquad \text{and} \qquad \sigma_{mx} = \begin{pmatrix} 0 & 1 \\ 1 & 0 \end{pmatrix}.$$

For the diagonalized σ_{mz}, the matrices σ_{mx} and σ_{my} are off-diagnonal. Using the vector $(\sigma_{mx}, \sigma_{my}, \sigma_{mz})$, the steady-state energies of a uncorrelated particle are given by $\varepsilon_m = \varepsilon_0 \pm \frac{1}{2}V$ for $\sigma_{mz} = \pm 1$, while σ_{mx} and σ_{my} are responsible for transition between L and R.

The correlation energy between σ_m and σ_n is generally expressed as

$$\mathscr{H}'_{m,n} = \sum_{\alpha\beta\gamma\delta} \psi_{m\alpha}{}^{\dagger}\psi_{m\beta}(V^{mn}{}_{\alpha\beta\gamma\delta})\psi_{n,\gamma}{}^{\dagger}\psi_{n,\delta}, \tag{3.9}$$

where V^{mn}, unspecified though, represents generally the interaction potential at distance r_{mn}, but for the short-range correlations between nearest neighbors we omit the superscript ^{mn} and write $v_{\alpha\beta\gamma\delta}$. Here (3.9) indicates that the correlations between σ_m and σ_n are related to components of the density matrices $(\psi^{\dagger}\psi)_m$ and $(\psi^{\dagger}\psi)_n$, which can be reexpressed in terms of the pseudospin notations defined by (3.7a,b,c) as follows, i.e.,

$$\psi_{m+}{}^{\dagger}\psi_{m+} = \tfrac{1}{2}(\varphi_{mL}{}^{\dagger}\varphi_{mL} + \varphi_{mR}{}^{\dagger}\varphi_{mR} + \varphi_{mL}{}^{\dagger}\varphi_{mR} + \varphi_{mR}{}^{\dagger}\varphi_{mL}) = \tfrac{1}{2}(1 + \sigma_{mx}),$$

$$\psi_{m+}{}^{\dagger}\psi_{m-} = \tfrac{1}{2}(\varphi_{mL}{}^{\dagger}\varphi_{mL} - \varphi_{mR}{}^{\dagger}\varphi_{mR} - \varphi_{mL}{}^{\dagger}\varphi_{mR} + \varphi_{mR}{}^{\dagger}\varphi_{mL})$$
$$= \tfrac{1}{2}(\sigma_{mz} - \sigma_{mx}), \text{ etc.}$$

For the interaction potential $v_{\alpha\beta\gamma\delta}$, we can write as

$$v_{++--} = v_{--++}, \qquad v_{+-+-} = v_{-+-+} = v_{-++-},$$

and all other asymmetric elements, such as v_{++-+}, vanish by virtue of symmetry. Applying these results to the pseudospins σ_m and σ_{m+1}, (3.9) for short-range energy can be written as

$$\mathscr{H}'_{m,m+1} = -J_{m,m+1}\sigma_{mz}\sigma_{m+1,z} - K_{m,m+1}\sigma_{mx}\sigma_{m+1,x}, \tag{3.10a}$$

where

$$J_{m,m+1} = 4v_{+-+-} \quad \text{and} \quad K_{m,m+1} = 2v_{++--} - v_{++++} - v_{----}. \tag{3.10b}$$

Therefore, the total Hamiltonian for the two correlated pseudospins is given by

$$\begin{aligned}\mathscr{H} &= \mathscr{H}_m + H_{m+1} + \mathscr{H}'_{m,m+1}\\ &= -\Omega(\sigma_m + \sigma_{m+1}) - J_{m,m+1}\sigma_{mz}\sigma_{m+1,z} - K_{m,m+1}\sigma_{mx}\sigma_{m+1,x},\end{aligned} \tag{3.10c}$$

where

$$\Omega = -V + v_{++++} - v_{----} \approx -V.$$

Equation (3.10c) is known as quantum-mechanical correlation energy between two pseudospins σ_m and σ_{m+1}. For a proton-tunneling system Blinc and Zeks [12] showed that the terms of Ω and $K_{m,m+1}$ in (3.10c) are negligibly small as compared with the term of $J_{m,m+1}$, and so a conventional Ising Hamiltonian of the type (3.3) can be used for the analysis.

As remarked, pseudospins at normal lattice sites are correlated primarily by short-range interactions at temperatures below T_c. Consequently, in the low-temperature phase the pseudospin component σ_{mz} represents no longer their states ± 1 with an equal probability, but with unequal probabilities as in (3.7a). Thus, pseudospins tend to exhibit a classical character with increasing correlations. On the other hand, if the transition threshold is signified by a discontinuous potential V with the characteristic time $\tau \sim V^{-1}$ shorter than 10^{-11} s, critical fluctuations must be interpreted quantum-mechanically, while correlations are insignificant.

Considering that the operator $\psi_{m\pm}$ or $\varphi_{mL,R}$ represents the displacement of an order variable at the lattice point m, the density element $\psi_{m+}{}^\dagger\psi_{m+}$ should give rise in the classical limit to a displacement of the particle in a quadratic potential energy proportional to $\sigma_{m+}{}^2$. Accordingly, such density correlations as $\psi_{m+}{}^\dagger\psi_{m+}v_{++++}\psi_{m+1,+}{}^\dagger\psi_{m+1,+}$ suggest a close relation to a quartic potential proportional to $\sigma_{m+}{}^2\sigma_{m+1,x}{}^2$, $\sigma_{m+}{}^2\sigma_{m+1,y}{}^2$, etc. We can also consider density fluctuations such as $\psi_{m+}{}^\dagger\psi_{m+}v_{++--}\psi_{m-}{}^\dagger\psi_{m-}$, etc., at the site m, which correspond to a potential proportional to $\sigma_{m+}{}^2\sigma_{m-}{}^2$, etc. We therefore assume that in the critical region pseudospin correlations give rise to anharmonic quartic potentials in the lowest order. In this context, for classical pseudospin vectors $\boldsymbol{\sigma}_m$ below T_c we can logically consider an anharmonic potential $V_<(\sigma_m)$ as expressed by (3.1a,b), where the quartic term $\frac{1}{2}B\sigma^4$ is considered to emerge at T_c. Based on the classical model of pseudospins, the quartic potential is attributed to correlations only statistically, while it is clear in the quantum description that such correlations signifying the critical region is represented by a quartic term in the Landau expansion.

3.4. Collective Pseudospin Modes in Displacive Systems

For tunneling protons in a hydrogen-bonding crystal, their motion is involved in energy dissipation through the potential barrier, so that they are virtually independent, and their average positions are not significantly related by the periodicity of the lattice. In contrast, for a continuous structural change in displacive systems, it is significant that the motion of classical pseudospins at regular lattice sites is collective, resulting from their correlations. As evidenced by experimental results, such a collective mode of pseudospins is signified by a low-frequency ϖ and a long wavelength $2\pi/q$, and hence experimentally it is a basic task to determine the dispersion relation $\varpi = \varpi(q)$ for such a specific collective mode of pseudospins that should occur as compatible to the symmetry of a given crystal.

Such a collective mode of pseudospins $\boldsymbol{\sigma}_\mathrm{m}$ can generally be described in the reciprocal lattice by the Fourier transform $\boldsymbol{\sigma}_{\boldsymbol{q}}$ of $\boldsymbol{\sigma}_\mathrm{m}$, i.e., $\boldsymbol{\sigma}_\mathrm{m} = \sum_{\boldsymbol{q},\varpi} \boldsymbol{\sigma}_{\boldsymbol{q}} \exp\{i(\boldsymbol{q}\cdot\boldsymbol{r}_\mathrm{m} - \varpi t)\}$. However, since $\boldsymbol{\sigma}_\mathrm{m}$ is a real vector, i.e., $\boldsymbol{\sigma}_\mathrm{m}{}^* = \boldsymbol{\sigma}_\mathrm{m}$ for a single mode at a frequency ϖ, the real mode should be written as

$$\begin{aligned}\boldsymbol{\sigma}_\mathrm{m} &= \exp(-i\varpi t)\{\boldsymbol{\sigma}_{\boldsymbol{q}} \exp(i\boldsymbol{q}\cdot\boldsymbol{r}_\mathrm{m}) + (\boldsymbol{\sigma}_{\boldsymbol{q}} \exp(i\boldsymbol{q}\cdot\boldsymbol{r}_\mathrm{m}))^*\}\\ &= \exp(-i\varpi t)\{\boldsymbol{\sigma}_{\boldsymbol{q}} \exp(i\boldsymbol{q}\cdot\boldsymbol{r}_\mathrm{m}) + \boldsymbol{\sigma}_{\boldsymbol{q}}{}^* \exp(-i\boldsymbol{q}\cdot\boldsymbol{r}_\mathrm{m})\}.\end{aligned}$$

Therefore, at $t = 0$ the relation

$$\boldsymbol{\sigma}_{\boldsymbol{q}}{}^* = \boldsymbol{\sigma}_{-\boldsymbol{q}} \tag{3.11}$$

is held for the Fourier transform, implying that there should always be two waves characterized by wavevectors $+\boldsymbol{q}$ and $-\boldsymbol{q}$. In the critical region of a phase transition, we therefore consider that the transforms $\boldsymbol{\sigma}_{\pm\boldsymbol{q}}$ represent the collective modes of classical pseudospins, where their direction vectors $\boldsymbol{e}_{\pm\boldsymbol{q}}$ should satisfy the relation similar to (3.11) at $t = 0$, i.e.,

$$\boldsymbol{e}_{\boldsymbol{q}}{}^* = \boldsymbol{e}_{-\boldsymbol{q}}, \tag{3.12}$$

and at an arbitrary time t

$$\boldsymbol{e}_{\pm\boldsymbol{q}} = \sum_\mathrm{m} \boldsymbol{e}_\mathrm{m} \exp\{-i(\pm\boldsymbol{q}\cdot\boldsymbol{r}_\mathrm{m} - \varpi t_\mathrm{m})\}. \tag{3.13}$$

For the short-range energy $\mathscr{H}_\mathrm{m}$ at a site m, we have to deal with the correlations between $\boldsymbol{\sigma}_\mathrm{m}(\boldsymbol{r}_\mathrm{m}, t_\mathrm{m})$ and all other $\boldsymbol{\sigma}_\mathrm{n}(\boldsymbol{r}_\mathrm{n}, t_\mathrm{n})$ in the surroundings, where the time difference $\Delta t = t_\mathrm{m} - t_\mathrm{n}$ plays a significant role in relation to the timescale t_0 of observation.

$$\begin{aligned}\mathscr{H}_\mathrm{m} &= \sum_\mathrm{n} \mathscr{H}_\mathrm{mn}\\ &= -\sigma_0{}^2 \sum_\mathrm{n} J_\mathrm{mn} \exp\{i(\pm\boldsymbol{q})\cdot(\boldsymbol{r}_\mathrm{m} - \boldsymbol{r}_\mathrm{n})\} \exp\{-i\varpi(t_\mathrm{m} - t_\mathrm{n})\}\boldsymbol{e}_{-\boldsymbol{q}}\cdot\boldsymbol{e}_{\boldsymbol{q}}.\end{aligned}$$

The observed correlation energy should correspond to the time average of $\mathscr{H}_m$ over the symmetrical interval between $-t_0$ and t_0 (time-reversal symmetry), i.e.,

$$\langle \mathscr{H}_m \rangle_t = -\sigma_0{}^2 \Gamma_t \sum_n J_{mn} \exp\{i\boldsymbol{q}\cdot(\boldsymbol{r}_m - \boldsymbol{r}_n)\} \boldsymbol{e}_{-\boldsymbol{q}}\cdot\boldsymbol{e}_{\boldsymbol{q}},$$

where

$$\Gamma_t = (2t_0)^{-1}\left\{\int_{-t_0}^{t_0} \exp(-i\varpi\Delta t)\, \mathrm{d}\Delta t\right\}$$

$$= (2t_0)^{-1}\int_0^{t_0} \{\exp(-i\varpi\Delta t) + \exp(i\varpi\Delta t)\}\, \mathrm{d}\Delta t = \frac{\sin(\varpi t_0)}{\varpi t_0} \tag{3.14}$$

is the time correlation factor, which is nearly equal to 1 for such a slow variation as $2\pi/\varpi > t_0$. Under the circumstances, the spatial correlations dominates in the observed $\langle \mathscr{H}_m \rangle_t$, which are otherwise averaged out in a long timescale $t_0 > 2\pi/\varpi$.

Assuming that $\Gamma_t \sim 1$, the observable short-range energy can be expressed by

$$\langle \mathscr{H}_m \rangle_t = -\sigma_0{}^2 \boldsymbol{e}_{-\boldsymbol{q}}\cdot\boldsymbol{e}_{\boldsymbol{q}} J_m(\boldsymbol{q}), \tag{3.15a}$$

where

$$J_m(\boldsymbol{q}) = \sum_n J_{mn} \exp\{i\boldsymbol{q}\cdot(\boldsymbol{r}_m - \boldsymbol{r}_n)\}. \tag{3.15b}$$

Here, the summation in (3.15b) is usually extended to the nearest and the next-nearest neighbors, giving competitive contributions to the effective short-range interaction parameter $J_m(\boldsymbol{q})$. It is significant to realize that only when $\Gamma_t \sim 1$ or $2\pi/\varpi > t_0$ such quasi-static correlations can be explicitly observed.

We consider that in a rigid crystal collective pseudospin modes are in thermal equilibrium in a modulated lattice, if the correlation energy $\langle \mathscr{H}_m \rangle_t$ takes a minimum value, corresponding to a maximum of the parameter $J_m(\boldsymbol{q})$. It is noted that $J_m(\boldsymbol{q}) = 0$ corresponds to zero correlations in the disordered state, and hence the threshold of a phase transition is characterized by a maximum of the positive $J_m(\boldsymbol{q})$. Mathematically, the equation

$$\mathrm{grad}_{\boldsymbol{q}}\, J_m(\boldsymbol{q}) = 0 \tag{3.16}$$

is to be solved for the wavevector $\boldsymbol{q}$, and the solutions $\boldsymbol{q} = +\boldsymbol{k}$ and $\boldsymbol{q}' = -\boldsymbol{k}$ of (3.16) should satisfy the normalization condition

$$N = \sum_{\boldsymbol{q},\boldsymbol{q}'} \boldsymbol{e}_{\boldsymbol{q}}\cdot\boldsymbol{e}_{\boldsymbol{q}'} \exp\{i(\boldsymbol{q} - \boldsymbol{q}')\cdot\boldsymbol{r}_m\}$$

$$= 2\boldsymbol{e}_{\boldsymbol{k}}\cdot\boldsymbol{e}_{-\boldsymbol{k}} + \boldsymbol{e}_{\boldsymbol{k}}{}^2 \exp(2i\boldsymbol{k}\cdot\boldsymbol{r}_m) + \boldsymbol{e}_{-\boldsymbol{k}}{}^2 \exp(-2i\boldsymbol{k}\cdot\boldsymbol{r}_m), \tag{3.17}$$

where N is the total number of pseudospins and independent of m. In order for (3.17) to be independent of $\boldsymbol{r}_m$, the solutions $\boldsymbol{k}$ and $-\boldsymbol{k}$ cannot be arbitrary. It is clear that such specifics $\boldsymbol{k}$ and $-\boldsymbol{k}$, as to satisfy (3.17), should be either one of the following cases:

$$\boldsymbol{k} = 0, \tag{3.18a}$$

$$\boldsymbol{k} = \tfrac{1}{2}\boldsymbol{G} \quad \text{(a half of the reciprocal lattice vector)}, \tag{3.18b}$$

and

$$\boldsymbol{e}_{\boldsymbol{k}}{}^2 = \boldsymbol{e}_{-\boldsymbol{k}}{}^2 = 0. \tag{3.18c}$$

Equations (3.18a,b) give *ferrodistortive* and *antiferrodistortive* arrangements of pseudospins [17], while (3.18c) signifies a specific arrangement of pseudospins given by the equation

$$e_{kx}{}^2 + e_{ky}{}^2 + e_{kz}{}^2 = 0. \tag{3.18d}$$

In fact, (3.18d) can be considered as a general requirement for specific solutions of (3.16), for which any value of the wavevector in the range $0 \le |\boldsymbol{k}| \le \frac{1}{2}G$ can be assigned to the pseudospin mode for calculating the value of the interaction parameter $J(\boldsymbol{k})$. As will be shown in Section 3.5, such a specific wavevector can be *irrational* in the reciprocal lattice unit, while *rational* at 0 and $\frac{1}{2}G$. Such a collective pseudospin mode with an irrational $\boldsymbol{k}$ can be visualized by the vector $\boldsymbol{e}_{\boldsymbol{k}}$, for example, $e_{ky} = 0$ and $e_{kx} = \pm i e_{kz}$, satisfying (3.18d), which can be transformed back to the crystal space to obtain the displacement $\boldsymbol{e}_{\mathrm{m}}$, i.e.,

$$e_{\mathrm{mz}} = \cos\phi, \qquad e_{\mathrm{mx}} = \pm\sin\phi, \qquad \text{and} \qquad e_{\mathrm{my}} = 0,$$

where $\phi = \boldsymbol{k}\cdot\boldsymbol{r}_{\mathrm{m}} + \phi_0$ represents the phase of the collective mode, and $\phi_0 = \boldsymbol{k}\cdot\boldsymbol{r}_0$. Due to a large number of regular lattice points x_{m} on the direction of $\boldsymbol{r}_{\mathrm{m}} - \boldsymbol{r}_0$ and an irrational wavevector $\boldsymbol{k}$, the phase ϕ can be a virtually continuous angle that is distributed in the whole range $0 \le \phi \le 2\pi$ in repetition. Thus, the vector $\boldsymbol{e}_{\mathrm{m}}$ represents a sinusoidal wave propagating along the x_{m} in the xz plane.

In the above argument, it was shown that a collective pseudospin mode characterized by a specific wavevector $\boldsymbol{k}$ can be responsible for initiating a phase transition, corresponding to a minimum correlation energy. For that matter, it is necessary to find a macroscopic quantity of the crystal that exhibits a singular behavior at this particular wavevector $\boldsymbol{k}$. While in Section 3.5 examples of such modes are discussed for representative systems, in Section 3.6 the Weiss field at $\boldsymbol{k}$ in the mean-field approximation is calculated, which exhibits a singularity at the transition temperature T_{c}.

3.5. Examples of Collective Pseudospin Modes

In the theory of magnetism, the method for "minimum correlations" is widely used to classify magnetic orderings, i.e., ferromagnetic, antiferromagnetic, and spiral spin arrangements [22]. In this section, this method is applied to some representative displacive crystals to see how a modulated pseudospin structure can emerge at the transition threshold. Among calculated results, we can find a pseudospin mode consistent with a modulated structure that was observed by magnetic resonance sampling in practical systems, in spite of unknown correlation parameters. It is therefore postulated that phase transitions are initiated by such a collective mode of pseudospins at a minimum energy, although supportive examples are limited to a few representative cases. Phase transitions in perovskite crystals and in organic calcium chloride compound, *tris-sarcosine calcium chloride* (TSCC), are selected as model systems, since rather comprehensive experimental results are available in the literature from crystallographic, soft-mode, and magnetic resonance studies on these crystals.

3.5.1. Strontium Titanate and Related Perovskites

Structural phase transitions in crystals of the perovskite family provide simple examples of displacive systems, to which a prototype model of active groups for classical fluctuations can be adapted for their collective motion. In the ferroelectric phase transition of $BaTiO_3$, the lattice symmetry changes from cubic to tetragonal, when ionic displacements from normal sites take place along one of the cubic axes. Evidenced by diffuse X-ray diffraction patterns, it is known that below the transition temperature T_c Ti^{4+} and six O^- ions in the $TiO_6{}^{2-}$ octahedron displace in opposite directions along the cubic axis, resulting in tetragonal symmetry. As a consequence, three differently oriented tetragonal domains appear in the low-temperature phase. For a single domain crystal where the a axis signifies such tetragonal distortion, the order variable can be defined as

$$\sigma_m \propto (u_+ - 2u_- - 4u_-')_m,$$

where u_+, u_-, and u_-' represent, respectively, displacements of Ti^{4+}, two O^- on the axis, and four O^- ions in the perpendicular symmetry plane of an octahedral $TiO_6{}^{2-}$. Normalizing to ± 1 for two opposite directions of distortion along the a axis, the order variable can be expressed by a vector signified by its axial component $\sigma_{mz} = p_m(+a) - p_m(-a)$.

On the other hand, the structural change in $SrTiO_3$ crystals is due to an axial rotation of $TiO_6{}^{2-}$ octahedra, which behave like a rigid body. Letting $\delta\theta_\pm$ be small angular displacements around the a axis in opposite directions, the order variable can be expressed by the axial component of a pseudospin vector $\boldsymbol{\sigma}_m$, i.e.,

$$\sigma_{mz} \propto (\delta\theta_+ - \delta\theta_-)_m.$$

Figure 3.3 shows the pseudospin arrangement in the bc plane of a perovskite crystal, where significant correlations are conceivable between the pseudospin $\boldsymbol{\sigma}_m$ and six surrounding $\boldsymbol{\sigma}_n$'s at its nearest neighbors, and between $\boldsymbol{\sigma}_m$ and

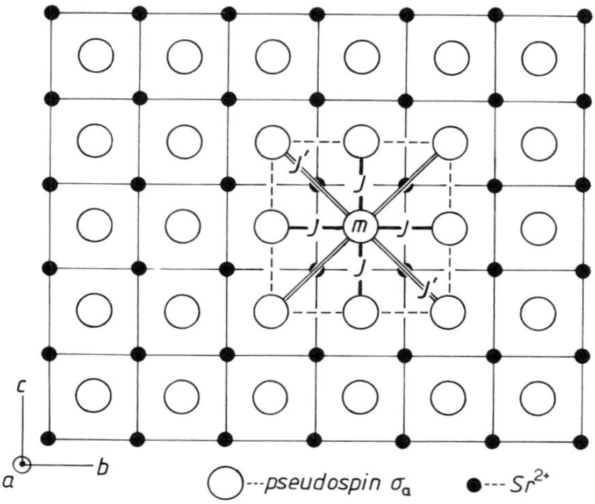

Figure 3.3. The pseudospin arrangement in a perovskite lattice in the *bc* plane. The range for initial correlations are assumed to be a square area shown by the broken lines.

eight next-nearest neighbors. These interactions are denoted by J and J' as indicated in the figure, respectively.

The interaction parameter $J_m(\boldsymbol{q})$ defined by (3.15) can be expressed explicitly for such a group of pseudospins as

$$J_m(\boldsymbol{q}) = 2J\{\cos(q_a a) + \cos(q_b b) + \cos(q_c c)\} + 4J'\{\cos(q_b b)\cos(q_c c) + \cos(q_c c)\cos(q_a a) + \cos(q_a a)\cos(q_b b)\}.$$

Differentiating this $J_m(\boldsymbol{q})$ with respect to q_a, q_b, and q_c, maxima of the short-range energy can be determined from the equations,

$$\sin(q_a a)\{J + 2J'\cos(q_b b) + 2J'\cos(q_c c)\} = 0,$$

$$\sin(q_b b)\{J + 2J'\cos(q_c c) + 2J'\cos(q_a a)\} = 0,$$

and

$$\sin(q_c c)\{J + 2J'\cos(q_a a) + 2J'\cos(q_b b)\} = 0.$$

A variety of solutions can be obtained from these equations as listed below:

(i) $\sin(q_a a) = \sin(q_b b) = \sin(q_c c) = 0,$

(ii.1) $\sin(q_a a) = 0, \qquad \cos(q_b b) = \cos(q_c c) = -\left(1 + \frac{J}{2J'}\right),$

(ii.2) $\sin(q_b b) = 0, \qquad \cos(q_c c) = \cos(q_a a) = -\left(1 + \frac{J}{2J'}\right),$

$$\text{(ii.3)} \quad \sin(q_c c) = 0, \qquad \cos(q_a a) = \cos(q_b b) = -\left(1 + \frac{J}{2J'}\right),$$

$$\text{(iii.1)} \quad \cos(q_a a) = 0, \qquad \cos(q_b b) = \cos(q_c c) = -\frac{J}{2J'},$$

$$\text{(iii.2)} \quad \cos(q_b b) = 0, \qquad \cos(q_c c) = \cos(q_a a) = -\frac{J}{2J'},$$

$$\text{(iii.3)} \quad \cos(q_c c) = 0, \qquad \cos(q_a a) = \cos(q_b b) = -\frac{J}{2J'}.$$

Solution (i) gives the wavevector $\boldsymbol{q} = \boldsymbol{k}_1$ with components

$$k_{1a} = \frac{\pi}{a} l, \qquad k_{1b} = \frac{\pi}{b} m, \qquad \text{and} \qquad k_{1c} = \frac{\pi}{c} n,$$

where l, m, and n are 0 or $\pm$ odd integers. The wavevector $\boldsymbol{k}_1$ is commensurate with the lattice, and the parameter is given by

$$J_m(\boldsymbol{k}_1) = 6J + 12J',$$

indicating that all the pseudospins considered for short-range correlations are arranged in parallel.

The first set of solution (ii) indicates that the a component of the wavevector $\boldsymbol{q} = \boldsymbol{k}_2$ is rational in the reciprocal lattice unit a^*, whereas b and c components are irrational in the units $\boldsymbol{b}^*$ and $\boldsymbol{c}^*$, provided that $|1 + J/2J'| < 1$. Thus, the wavevector $\boldsymbol{k}_2$ gives an incommensurate pseudospin arrangement in the bc plane, and commensurate along the a direction, and the components are usually expressed as

$$k_{2a} = l\frac{2\pi}{a} \quad (l = 0, \text{ or half integers}), \qquad k_{2b} = (\tfrac{1}{2} - \delta_b)b^*, \qquad \text{and}$$

$$k_{2c} = (\tfrac{1}{2} - \delta_c)c^*,$$

where δ_b and δ_c are called incommensurate parameters. The energy parameter for solution (ii) is given by

$$J_m(\boldsymbol{k}_2) = -2J - 4J' - \frac{J^2}{J'},$$

where can be positive to yield a small negative correlation energy, depending on the values of J and J' that are opposite in sign.

Solution (iii) also gives a two-dimensional incommensurate pseudospin arrangement similar to solution (ii). In the first set (iii.1), if $|J/2J'| < 1$, the wavevector $\boldsymbol{q} = \boldsymbol{k}_3$ can have components

$$k_{3a} = l\frac{\pi}{a} \quad (l = \text{integers}), \qquad k_{3b} = (\tfrac{1}{2} - \delta_b)b^*, \qquad \text{and} \qquad k_{3c} = (\tfrac{1}{2} - \delta_c)c^*,$$

corresponding to the energy parameter

$$J_m(\mathbf{k}_3) = -\frac{J^2}{J'}.$$

It is noted that such a commensurate pseudospin arrangement along the a direction is characterized by $k_{3a} = l(\frac{1}{2}a^*)$, being antiferrodistortive along the a axis, in contrast to the ferrodistortive arrangement by k_{2a}. In the bc plane, both cases give incommensurate arrangements but are specified with different sets of parameters δ_b and δ_c.

Despite unknown correlation parameters, the above analysis suggests possible incommensurate structures, as well as uniform and cell-doubling transitions, although modulated structures in perovskite crystals have not been confirmed. It is noted however that Comès et al. [19] observed two-dimensional diffuse X-ray diffraction patterns from $NaNbO_3$ crystals near the transition temperature, and that Müller et al. [20] reported an anomalous lineshape in the EPR spectra of Fe^{3+} ions in doped $SrTiO_3$ at 105 K, which is similar to that in Mn^{2+} spectra in TSCC. These anomalies are due undoubtedly to spatially modulated pseudospins in the critical region, although such interpretation was not given by these authors. Nevertheless, in the above argument the cell-doubling feature of the transition in $SrTiO_3$ is clearly recognized experimentally, as discussed in case (iii) for the correlation parameters satisfying the relation $|J/2J'| < 1$.

3.5.2. Tris-Sarcosine Calcium Chloride and Related Crystals

Tris-sarcosine calcium chloride (TSCC) with the formula unit (sarcosine)$_3$-$CaCl_2$, where the sarcosine is methyl glycine $H_3C—NH_2—CH_2COOH$, crystalizes in a pseudo-orthorhombic structure. At room temperature, TSCC crystals exhibit twinning due presumably to the pseudostructure, where *ferroelastic* domains can be easily identified in twinned crystals by viewing a sample through a pair of crossed polarizers [23]. Single-domain samples of high optical quality can then be cut out of twinned crystals for studies of the ferroelectric phase transition at 120 K to a phase polarized along the b axis. In spite of the chemical complexity, TSCC crystals show a simple structural change characterized by a loss of mirror symmetry on the b plane, offering an example of continuous uniaxial ferroelectric phase transitions.

Figure 3.4(a) illustrates the molecular arrangement in TSCC at room temperature, which was determined by Kakudo et al. [24] from their X-ray study. In the figure, mirror symmetry on the b plane is clearly visible in the quasi-triginal structure. Nakamura et al. [25] showed from their X-ray studies that the active group for the ferroelectric phase transition is the Ca(sarcosine)$_6$ complex where the Ca^{2+} ion is surrounded near-octahedrally by six carbonyl oxygens, as shown in Fig. 3.4(b). It was evidenced by diffuse diffraction spots from the O^- ions located outside the mirror plane. Using

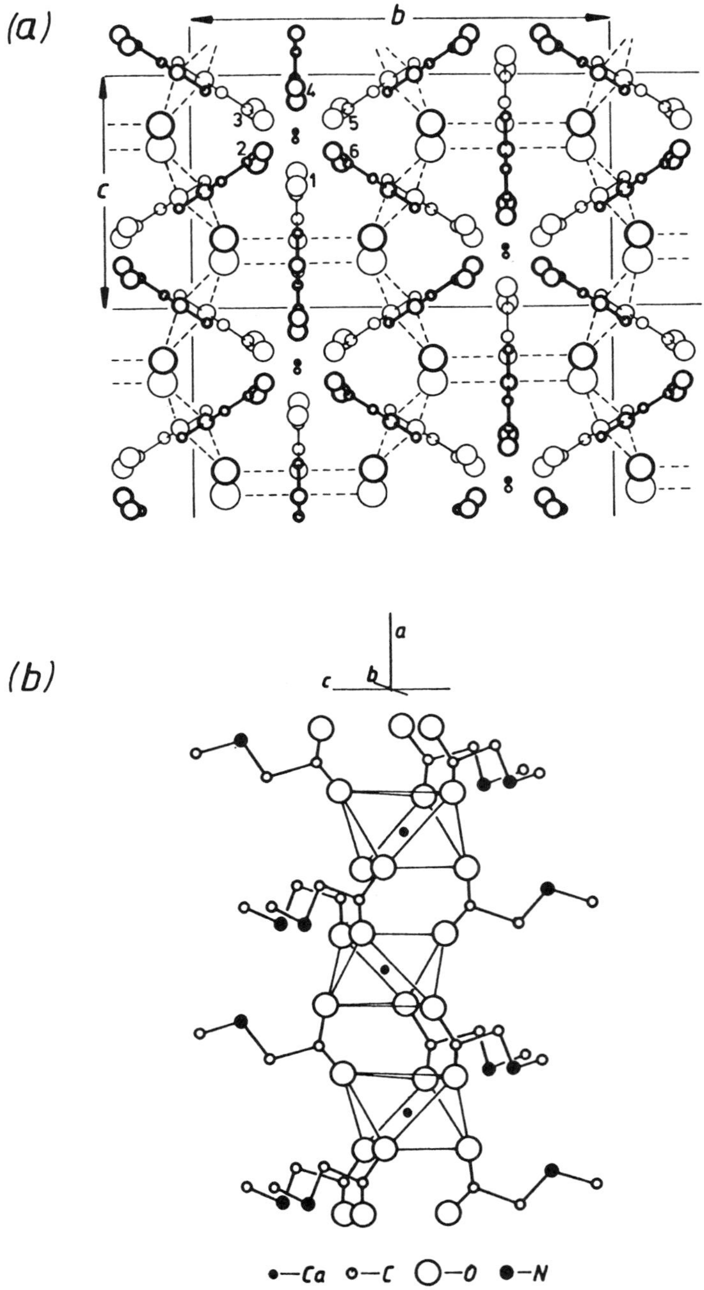

Figure 3.4. (a) The molecular arrangement in the normal (ferroelastic) phase in TSCC crystals. (b) The ligand structure in Ca^{2+}(sarcosine)$_6$ complexes in TSCC.

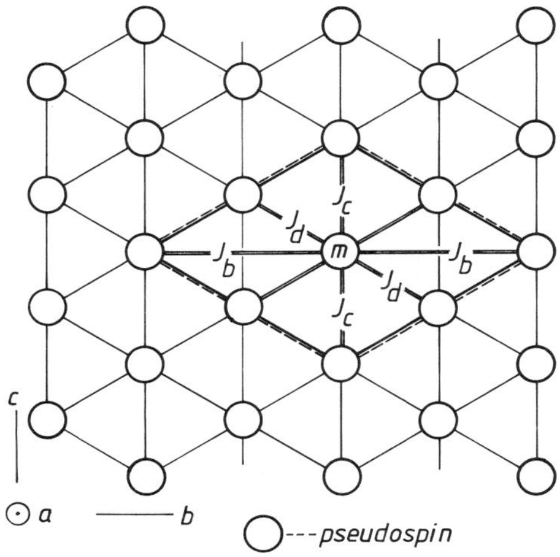

Figure 3.5. The pseudospin lattice in TSCC crystals viewed in the *bc* plane. The range for initial correlations are indicated by the broken lines.

paramagnetic probes substituted for the central Ca^{2+} ion, Jerzak and Fujimoto [26] carried out an EPR study on the ferroelectric phase transition in TSCC, confirming that the active group is $Ca(sarcosine)_6$.

Figure 3.5 shows the lattice of pseudospins representing active groups in TSCC, which is abstracted from the molecular arrangement determined by Kakudo et al. [24]. We can consider that correlations J_a, J_b, J_c, and J_d in the figure are essential for initial clustering. Here, J_a, J_c, and J_d represent the nearest-neighbor interactions, whereas J_b signifies correlations with the next-nearest neighbors in the direction perpendicular to the mirror plane. In the figure for the projection to the *bc* plane, the parameter J_a is not shown.

For such a group of correlated pseudospins, the function $J_m(\boldsymbol{q})$ can be expressed as

$$J_m(\boldsymbol{q}) = 2J_a \cos(q_a a) + 2J_b \cos(q_b b) + 2J_c \cos(q_c c) + 4J_d \cos(\tfrac{1}{2}q_b b)\cos(\tfrac{1}{2}q_c c), \tag{3.19}$$

where the lattice constant b is the largest, and $a \approx c$. Maximizing $J_m(\boldsymbol{q})$ with respect to $\boldsymbol{q}$, the wavevector $\boldsymbol{k}$ to satisfy (3.16) is given by

$$\text{(i)}\quad k_{1a} = \frac{\pi}{a}l, \qquad k_{1b} = \frac{\pi}{b}m, \qquad k_{1c} = \frac{\pi}{c}n,$$

$$\text{(ii)}\quad k_{2a} = \frac{\pi}{a}l, \qquad \cos(\tfrac{1}{2}k_{2b}b) = -\frac{J_d}{2J_b}, \qquad k_{2c} = \frac{2\pi}{c}n,$$

$$\text{(iii)}\quad k_{3a} = \frac{\pi}{a}l, \qquad k_{3b} = \frac{2\pi}{b}, \qquad \cos(\tfrac{1}{2}k_{3c}c) = -\frac{J_d}{2J_b},$$

where l, m, and n are 0 or $\pm$ integers.

Here solution (i) represents a commensurate pseudospin arrangement with the lattice, for which

$$J_{\rm m}(\boldsymbol{k}_1) = 2J_{\rm a} + 2J_{\rm b} + 2J_{\rm c} + 4J_{\rm d},$$

corresponding to complete order. Solutions (ii) and (iii) give incommensurate arrangements along the b and c directions, respectively, provided that $|-J_{\rm d}/2J_{\rm b}| < 1$. For the ferroelectric phase transition in TSCC, we naturally select solution (ii) that breaks the mirror reflection symmetry on the b plane with the correlation energy proportional to

$$J_{\rm m}(\boldsymbol{k}_2) = 2J_{\rm a} + 2J_{\rm b} + 2J_{\rm c} - \frac{2J_{\rm d}^{\,2}}{J_{\rm b}}.$$

It is noted that if $|-J_{\rm d}/2J_{\rm b}| = 1$, $J_{\rm m}(\boldsymbol{k}_1) = J_{\rm m}(\boldsymbol{k}_2)$, for which $k_{2\rm b} = 0$ gives ferroelectric order along the b axis, whereas if $|-J_{\rm d}/2J_{\rm b}| < 1$, $k_{2\rm b}$ is irrational in unit of b^*, representing an incommensurate pseudospin order. As will be discussed in Chapter 9, such a specific wavevector $\boldsymbol{k}$ can be inferred from EPR anomalies described by distributed $\boldsymbol{e}_{\rm m}$ that is the Fourier transform of $\boldsymbol{e}_{\boldsymbol{k}}$. Fujimoto et al. [21] determined the vector $\boldsymbol{e}_{\rm m}$ from EPR anomalies of Mn^{2+} spectra in the critical transition region in TSCC, showing that

$$e_{\rm ma} = \pm\sin\phi, \qquad e_{\rm mb} = \cos\phi, \qquad \text{and} \qquad e_{\rm mc} = 0,$$

where $\phi = k_{2\rm b}y_{\rm m} + \phi_0$ is the phase of cycloidally modulated pseudospins along the b axis.

While long-range interaction is ignored in the above calculation, it is interesting to notice that $\boldsymbol{k}_2 \to \boldsymbol{k}_1$ when $J_{\rm d} \to 0$, leading to $k_{\rm b} = 0$ for a uniaxial ferroelectric arrangement. In this context, we can consider incommensurate parallel chains of pseudospins along the b direction, which can mutually interact in transversal directions through the coupling $J_{\rm d}$. In fact, for one-dimensional charge-density-waves expressed by $\sigma = \sigma_0 \exp i\phi$ in a uniaxial conductor, Lee et al. [27] discussed dipolar interactions between parallel chains σ_1 and σ_2. Considering that the phase ϕ represents an angle between the classical vector σ and the chain direction, the dipolar interaction potential may be written as

$$-J_{12}\sigma_1\cdot\sigma_2 = -J_{12}\sigma_0^{\,2}\cos(\phi_1 - \phi_2), \tag{3.20}$$

where the parameter J_{12} is assumed to be a scalar. Similarly, if the interaction $J_{\rm d}$ in TSCC is considered as proportional to $\cos(\phi_1 - \phi_2)$, the irrational wavevector $k_{2\rm b}$ can be attributed to the phase difference $\Delta\phi = \phi_1 - \phi_2$ between adjacent chains. As inferred from Fig. 3.5, such a phase difference between pseudospins at $y_{1(\rm m)}$ and $y_{2(\rm m+1/2)}$ on the chains 1 and 2 is $\Delta\phi = \frac{1}{2}\pi$ at the threshold of the critical region, where the interchain correlation $J_{\rm d}$ is maximum. Hence, at a temperature below $T_{\rm c}$ we can write $\Delta\phi = (\frac{1}{2} - \delta_{\rm b})\pi$, where $\delta_{\rm b}$ is the temperature-dependent incommensurability parameter. In terms of $\delta_{\rm b}$, $J_{\rm d} \sim \Delta\phi = 0$, when $\delta_{\rm b} \to \frac{1}{2}$, hence $k_{2\rm b} \to (\pi/b)m = k_{1\rm b}$.

A modulated structure given by solution (ii) is quasi-one-dimensional, where the dominant correlation J_b is counteracted by the interchain interaction J_d. Since $\frac{1}{2}k_b b = \phi = k_b y_m$, (3.19) for $k_a = 0$ and $k_c = 0$ can be written as

$$J(\phi) = 2J_a + 2J_c + 2J_b \cos(2\phi) + 2J_d \cos \phi. \tag{3.21}$$

This is a well-known formula for a quasi-linear chain of classical spins, where J_d and J_b are interaction parameters for nearest-neighbor pairs and next-nearest pairs, respectively, and the phase ϕ represents an angle between adjacent spin vectors [28]. If the displacement $\boldsymbol{e}_m$ is perpendicular to the axis of the chain, the locus of $\boldsymbol{e}_m$ shows a helix around the axis. Such a description of a helical chain of spins is found in standard textbooks on magnetism [29].

Solution (iii) corresponds to another quasi-linear modulation along the *c* axis, however such a modulated structure was not found in TSCC crystals.

3.6. The Curie–Weiss Singularity in Collective Pseudospin Modes

The Weiss molecular field is a useful concept to obtain a singular behavior of the magnetic susceptibility in the mean-field approximation. For ferromagnetic order, the singularity of the Curie–Weiss susceptibility occurs at the origin $\boldsymbol{k} = 0$ in the reciprocal lattice. At a nonzero wavevector $\boldsymbol{k}$ for minimum correlations, the macroscopic susceptibility $\chi(\boldsymbol{k})$ should exhibit a similar singular behavior, if such a collective pseudospin mode initiates a phase transition at $\boldsymbol{k}$. To confirm such a criterion for a structural change, we consider a Fourier transform of the Weiss field to define the susceptibility at $\boldsymbol{k}$.

A collective pseudospin mode $\boldsymbol{\sigma}_m$ during the ordering process is generally nonsinusoidal, as characterized by a finite amplitude due to extended correlations to distant pseudospins. For such a nonlinear problem, we can show that the concept of an effective field is still valid in the mean-field approximation when applied to a nonsinusoidal mode. Considering the correlation energy of the Heisenberg type for pseudospins of finite amplitudes, we write

$$\mathscr{H} = \sum_{mn} J_{mn} \boldsymbol{\sigma}_m \cdot \boldsymbol{\sigma}_n,$$

where the summation may include interactions with distant pseudospins beyond immediate neighbors in the collective mode. Using the variational principle, we assume that the average $\langle \mathscr{H} \rangle$ over the mode can take a minimum value against any variation in pseudospin variables. We impose a condition such that

$$\delta\mathscr{H} = \mathscr{H} - \langle \mathscr{H} \rangle = 0$$

against arbitrary variations $\delta\boldsymbol{\sigma}_m$ under the constraint

$$\boldsymbol{\sigma}_m{}^2 = \boldsymbol{\sigma}_n{}^2 = \sigma_0{}^2 \quad (=\text{const.}).$$

Namely,

$$\delta\mathscr{H} = -\sum_m \delta\boldsymbol{\sigma}_m \cdot \left(\sum_n J_{mn}\boldsymbol{\sigma}_n\right) - \sum_n \left(\sum_m J_{mn}\boldsymbol{\sigma}_m\right) \cdot \delta\boldsymbol{\sigma}_n = 0$$

and

$$\boldsymbol{\sigma}_m \cdot \delta\boldsymbol{\sigma}_m = \boldsymbol{\sigma}_n \cdot \delta\boldsymbol{\sigma}_n = 0.$$

Hence,

$$-\sum_n J_{mn}\boldsymbol{\sigma}_n + \lambda_m \boldsymbol{\sigma}_m = 0 \qquad \text{and} \qquad -\sum_m J_{mn}\boldsymbol{\sigma}_m + \lambda_n \boldsymbol{\sigma}_n = 0,$$

where λ_m and λ_n are the Lagrange multipliers. Here the quantity

$$\boldsymbol{F}_m = \sum_n J_{mn}\boldsymbol{\sigma}_n \tag{3.22}$$

represents the effective field at the site m due to all interacting pseudospins $\boldsymbol{\sigma}_n$. Writing $\boldsymbol{F}_m = N\boldsymbol{f}_m$, it is clear that $\boldsymbol{f}_m = N^{-1}\sum_n J_{mn}\boldsymbol{\sigma}_n$, where $n = 1, 2, \ldots, N$, is the average of field at $\boldsymbol{\sigma}_m$ due to another pseudospin $\boldsymbol{\sigma}_n$ in the mode, and hence $\boldsymbol{F}_m$ represents the total average field at the site m. In this context, the above variational treatment gives results in the mean-field accuracy. Hence

$$\langle \mathscr{H}_m \rangle = -\boldsymbol{\sigma}_m \cdot \boldsymbol{F}_m,$$

where

$$\lambda_m \boldsymbol{\sigma}_m = \boldsymbol{F}_m. \tag{3.23}$$

Equation (3.23) indicates that $|\boldsymbol{F}_m| \propto |\boldsymbol{\sigma}_m|$ and $\boldsymbol{F}_m \parallel \boldsymbol{\sigma}_m$, which are useful results for interpreting the finite amplitude of a pseudospin mode $\boldsymbol{\sigma}_m$.

A collective pseudospin mode at the threshold of a phase transition can be expressed by

$$\boldsymbol{\sigma}_m = N^{-1/2}\{\boldsymbol{\sigma}_{-\boldsymbol{k}} \exp(i\boldsymbol{k}\cdot\boldsymbol{r}_m) + \boldsymbol{\sigma}_{\boldsymbol{k}} \exp(-i\boldsymbol{k}\cdot\boldsymbol{r}_m)\},$$

where N is the number of sites for $\boldsymbol{\sigma}_m$, and the time factor $\exp(-i\varpi t)$ is omitted for brevity. Due to (3.23), the local field $\boldsymbol{F}_m$ can similarly be expressed as

$$\boldsymbol{F}_m = N^{-1/2}\{\boldsymbol{F}_{-\boldsymbol{k}} \exp(i\boldsymbol{k}\cdot\boldsymbol{r}_m) + \boldsymbol{F}_{\boldsymbol{k}} \exp(-i\boldsymbol{k}\cdot\boldsymbol{r}_m)\}.$$

Using (3.22),

$$\boldsymbol{F}_m \exp(\pm i\boldsymbol{k}\cdot\boldsymbol{r}_m) = N^{-1/2}\sum_n J_{mn} \exp\{\pm i\boldsymbol{k}\cdot(\boldsymbol{r}_m - \boldsymbol{r}_n)\}\boldsymbol{\sigma}_{\pm\boldsymbol{k}} = N^{-1/2}J_m(\pm\boldsymbol{k})\boldsymbol{\sigma}_{\pm\boldsymbol{k}}.$$

Therefore,

$$\boldsymbol{F}_{\pm\boldsymbol{k}} = N^{-1/2}\sum_m \boldsymbol{F}_m \exp(\pm i\boldsymbol{k}\cdot\boldsymbol{r}_m) = N^{-1}\sum_m J_m(\boldsymbol{k})\boldsymbol{\sigma}_{\pm\boldsymbol{k}} = J(\boldsymbol{k})\boldsymbol{\sigma}_{\pm\boldsymbol{k}},$$

which is the internal field in the collective mode at $\pm\boldsymbol{k}$. Being equivalent to (3.23) for the Weiss postulate, we can write

$$\lambda(\boldsymbol{k})\boldsymbol{\sigma}_{\pm k} = \boldsymbol{F}_{\pm k} \qquad \text{where} \quad \lambda(\boldsymbol{k}) = J(\boldsymbol{k}). \tag{3.24}$$

The Curie–Weiss law for the mode $\sigma_{\pm k}$ is given by the response to an applied field $\boldsymbol{E}$. Considering the Fourier transform $\boldsymbol{E}_{\pm k}$ in this case,

$$\boldsymbol{\sigma}_{\pm k} = \chi_0(\boldsymbol{E}_{\pm k} + \boldsymbol{F}_{\pm k}) = \left(\frac{C}{T}\right)\{\boldsymbol{E}_{\pm k} + \lambda(\boldsymbol{k})\boldsymbol{\sigma}_{\pm k}\},$$

where $\chi_0 = C/T$ is the Curie law for the static susceptibility above the transition temperature $T_0(k)$. Solving this for $\boldsymbol{\sigma}_{\pm k}$, we obtain

$$\chi(\boldsymbol{k}) = \frac{\boldsymbol{\sigma}_{\pm k}}{\boldsymbol{E}_{\pm k}} = \frac{C}{T - T_0(\boldsymbol{k})},$$

where

$$T_0(\boldsymbol{k}) = \lambda(\boldsymbol{k})C = J(\boldsymbol{k})C.$$

Thus, the collective pseudospin mode at a wavevector $\boldsymbol{k}$ determined by (3.16) is responsible for the singular behavior of $\chi(\boldsymbol{k})$ in the mean-field approximation, giving the transition temperature $T_0(\boldsymbol{k})$. In practice, however $T_0(\boldsymbol{k}) \approx T_0(\boldsymbol{k} = 0)$ for a small $\boldsymbol{k}$ in the critical region, where the mean-field approximation is not valid for the whole crystal.

CHAPTER 4
Soft Lattice Modes and Pseudospin Condensates

4.1. Introduction

The threshold of a continuous-phase transition is dominated by fluctuations of correlated pseudospins, which are responsible for critical anomalies in thermodynamical quantities measured in the transition region. According to neutron inelastic scattering and magnetic resonance results [20], [21], critical fluctuations are not random but sinusoidal in character at very low frequencies and long wavelengths. On the other hand, for instability of displacive crystals, Cochran [30] and Anderson [31] proposed to consider a lattice mode whose characteristic frequency is temperature-dependent, diminishing toward the transition temperature. While entirely different from ordering, such a lattice excitation known as a *soft mode* can interact with pseudospins under specific circumstances of the transition threshold. In an attempt to unify these views, we considered in Chapter 3 a collective mode of pseudospins that may arise from their negative correlation energy, while a soft mode is believed as always accompanying a continuous structural change. It seems logical to consider that these modes are coupled when they are in phase, resulting in transition anomalies. Although conceptually conceivable, such an interaction should concretely be substantiated.

Born and Huang [15] discussed in their general theory of crystal lattices that structural stability is determined not only by minimizing the lattice potential, but also by being free from lattice strains. Applying these principles to a crystal undergoing a spontaneous structural change, we can logically consider that local strains due to collective pseudospins should be offset by displacing active groups from regular lattice points. In this context, pseudospins $\boldsymbol{\sigma}_{\mathrm{m}}$ and the corresponding lattice displacements $\boldsymbol{u}_{\mathrm{m}}$ should be linearly coupled so that the crystal can be free from strains. We may therefore assume that in the critical region the variables $\boldsymbol{\sigma}_{\mathrm{m}}$ and $\boldsymbol{u}_{\mathrm{m}}$ are related as

$$\boldsymbol{\sigma}_{\mathrm{m}} = \mathbf{A}\boldsymbol{u}_{\mathrm{m}}, \tag{4.1}$$

where the factor $\mathbf{A}$ is generally a tensor, independent of the site m. According to the Born–Huang principle, such a displacement mode $\boldsymbol{u}_{\mathrm{m}}$ should occur at a low excitation energy of the lattice as strained by collective pseudospins. Assuming that the tensor $\mathbf{A}$ is a scalar, $\boldsymbol{\sigma}_{\mathrm{m}}$ and $\boldsymbol{u}_{\mathrm{m}}$ can be treated as if representing identical displacements, in which case the phase transition appears to take place at a singular condition of the lattice. However, no such singularity is expected from plain lattice modes, and the phase transition should be attributed to a singular behavior of the pseudospins. At this point we should not be misled by (4.1), which is applicable only to the critical region. In general, the pseudospin $\boldsymbol{\sigma}_{\mathrm{m}}$ represents a finite displacement, whereas the lattice displacement $\boldsymbol{u}_{\mathrm{m}}$ is associated with a harmonic excitation of the lattice, so that these variables do not couple with each other under normal circumstances.

Consider a collective pseudospin mode emerging at the threshold of a phase transition, which is characterized by a wavevector $\boldsymbol{k}$ incommensurate with the lattice periodicity. Writing therefore

$$\boldsymbol{\sigma}_{\mathrm{m}} = \{\boldsymbol{\sigma}_{-\boldsymbol{k}} \exp(i\boldsymbol{k}\cdot\boldsymbol{r}_{\mathrm{m}}) + \boldsymbol{\sigma}_{+\boldsymbol{k}} \exp(-i\boldsymbol{k}\cdot\boldsymbol{r}_{\mathrm{m}})\} \exp(-i\omega t),$$

Equation (4.1) suggests that in the critical region there may be a commensurate lattice mode

$$\boldsymbol{u}_{\mathrm{m}} = \{\boldsymbol{u}_{-\boldsymbol{k}'} \exp(i\boldsymbol{k}'\cdot\boldsymbol{r}_{\mathrm{m}}) + \boldsymbol{u}_{+\boldsymbol{k}'} \exp(-i\boldsymbol{k}'\cdot\boldsymbol{r}_{\mathrm{m}})\} \exp(-i\omega' t),$$

which may be almost in phase with $\boldsymbol{\sigma}_{\mathrm{m}}$. If so, the corresponding Fourier amplitudes are related as

$$\begin{aligned}\boldsymbol{\sigma}_{\pm\boldsymbol{k}} &= \mathbf{A}\boldsymbol{u}_{\pm\boldsymbol{k}'} \exp[i\{\pm(\boldsymbol{k}' - \boldsymbol{k})\cdot\boldsymbol{r}_{\mathrm{m}} \mp ((\omega' - \omega)t)\}]\\ &= \mathbf{A}\boldsymbol{u}_{\pm\boldsymbol{k}'} \exp\{i(\pm\Delta\boldsymbol{k}'\cdot\boldsymbol{r}_{\mathrm{m}} \mp \Delta\omega\cdot t)\},\end{aligned}$$

where $\boldsymbol{k}' - \boldsymbol{k} = \Delta\boldsymbol{k}$ and $\omega' - \omega = \mp\Delta\omega$. These relations imply that both the pseudospin and lattice modes are *amplitude-modulated* in the critical region. It is noted that such a modulation is detectable, if the frequency $\Delta\omega$ is sufficiently low as compared with the timescale of observation.

Figure 4.1 illustrates a one-dimensional chain of active groups that are indicated by the squares, in each of which a hypothetical mass particle is located. Here, the displacement $\boldsymbol{\sigma}_{\mathrm{m}}$ represented by such a particle is by no means the same as $\boldsymbol{u}_{\mathrm{m}}$ that is essentially related to the mass of an active group and the elastic constant of the chain. A transversal vibrational mode $\boldsymbol{u}_{\mathrm{m}}$ of a long wavelength in such a chain near T_{c} is shown in Fig. 4.1(a) assuming that the average particle position is at the center of each square. In Fig. 4.1(b), shown for comparison, is a displacement mode of particles for $\boldsymbol{\sigma}_{\mathrm{m}}$, assuming that the chain remains unchanged. However in reality, the position for the active group minus the particle should be displaced by $\boldsymbol{u}_{\mathrm{m}}$ to compensate the

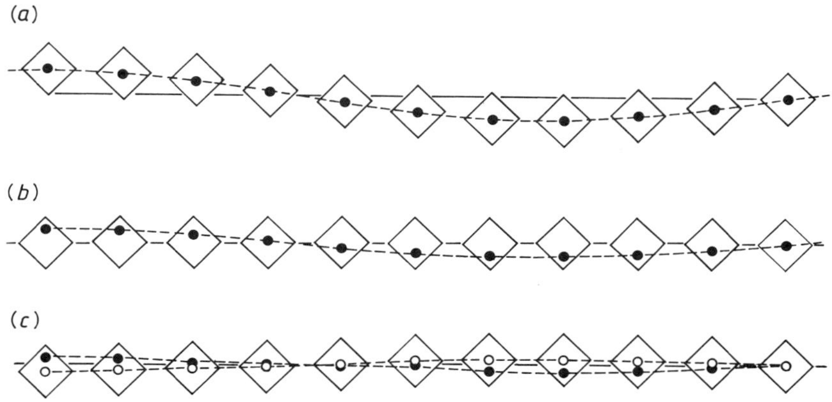

Figure 4.1. A one-dimensional model for a displacive phase transition: (a) the normal phase; (b) the transversal pseudospin mode in a rigid lattice; and (c) the pseudospin mode in a deformable lattice.

lattice strains caused by $\boldsymbol{\sigma}_{\mathrm{m}}$ as shown in Fig. 4.1(c), where a scalar $\mathbf{A}$ and an exact in-phase relation are assumed for simplicity.

Normally, we can write that $\Delta\boldsymbol{k} = \boldsymbol{G} \pm \boldsymbol{q}$, where $\boldsymbol{G}$ is a reciprocal lattice vector, and $|\boldsymbol{q}|$ is small. In this case, a nearly in-phase coupling is specified in the reduced reciprocal space as

$$\Delta\boldsymbol{k} = \pm\boldsymbol{q},$$

implying that the phase transition occurs at the Brillouin zone center $\boldsymbol{G} = 0$. However, there are cases where

$$\Delta\boldsymbol{k} = \boldsymbol{G}_{\mathrm{i}} \pm \boldsymbol{q}.$$

Here the vector $\boldsymbol{G}_{\mathrm{i}}$ represents an arbitrary point in the zone, for which we must look for an additional mechanism. In the former case, the pseudospin arrangement is commensurate with the lattice, whereas in the latter case the crystal phase below T_{c} is signified by an incommensurately modulated structure, being referred to as an *incommensurate phase*. Setting aside the additional mechanism leading to these different cases, it is notable that these continuous-phase transition always accompany incommensurate fluctuations at small wavevectors $\pm\boldsymbol{q}$ and at a low frequency $\mp\Delta\omega$ in the critical region, hence being described in terms of the phases $\phi_{\mathrm{m}} = \pm\boldsymbol{q}\cdot\boldsymbol{r}_{\mathrm{m}} \mp (\Delta\omega)t$. Anisotropic diffuse X-ray patterns [19] near the transition threshold can be interpreted as due to the spatial variation of ϕ_{m} under the condition $\Delta\omega\cdot t_0 < 1$, and soft modes [32] represent temporal fluctuations at a fixed value of $\boldsymbol{q}$.

The coupling expressed in (4.1) is analogous to a phase-matching relation between a charge-density-wave (CDW) and a periodic-lattice-distortion

(PLD), which was first proposed by Peierls [33] for one-dimensional conductors. Such a coupled object as CDW–PLD is called a *condensate*, which is the concept similarly applicable to coupled modes of pseudospins and soft-phonons in the critical region of a phase transition, for which, of course, experimental evidence must be found.

For a phase transition signified by the Curie–Weiss law for the dielectric response, the presence of a soft mode is inferable from the Lyddane–Sachs–Teller (LST) relation, although the concept of soft modes was first introduced by Cochran and Anderson. As will be discussed in Section 4.2, the distinction between polar pseudospins and lattice vibration was explicit in the original derivation for ionic crystals. Being related to the soft-mode frequency $\Delta\omega$ diminishing toward the transition temperature, critical fluctuations in pseudospins are characterized by a very low frequency. Here $\hbar\,\Delta\omega$ represents the amount of energy transfer between soft phonons and pseudospins, which may be expressed as the kinetic energy of fluctuations $\hbar^2 \boldsymbol{q}^2/2m$, where m is the mass of the hypothetical particle. In this context, as proposed in Section 1.7, soft modes can be considered as the "hidden" subsystem when pseudospins are investigated exclusively.

In Section 4.2 the LST relation is derived by assuming $\boldsymbol{q} = 0$. While it is impractical to ignore critical fluctuations in this way, it is instructive to see how differently the variables $\boldsymbol{\sigma}_{\mathrm{m}}$ and $\boldsymbol{u}_{\mathrm{m}}$ are treated in the derivation. In this Chapter 4 the elemental theories of soft modes are reviewed with supporting evidence, and the concept of pseudospin condensates is introduced for subsequent discussions on critical anomalies.

4.2. The Lyddane–Sachs–Teller Relation

We consider here a uniaxially displacive crystal as a simple case, where the order variable is represented by a classical pseudospin characterized by two polar states related by reflection. Assuming further that the tensor $\mathbf{A}$ in (4.1) is a constant scalar, the pseudospin $\boldsymbol{\sigma}_{\mathrm{m}}$ is nothing but an elementary dipole moment $\boldsymbol{p}_{\mathrm{m}}$ at site m, which is regarded as proportional to the displacement $\boldsymbol{u}_{\mathrm{m}}$ of the active group in the critical region. The vector $\boldsymbol{p}_{\mathrm{m}}$ is generally finite in magnitude, while in the critical region it is infinitesimal in the sinusoidal limit.

The behavior of such polar pseudospins can be studied from their response to an externally applied electric field. In practice, a sample crystal is placed in a parallel-plate condenser with its axis normal to the plates to measure the capacitance. While in this case the electric displacement $\boldsymbol{D}$ is uniform over the sample, the corresponding electric field $\boldsymbol{E}$ is distributed according to the polarized charge density $-\mathrm{div}\,\boldsymbol{P}$. It is significant that such a polarized crystal is generally strained, resulting in distributed ionic displacements $\boldsymbol{u}_{\mathrm{m}}$ along the unique axis. According to the Born–Huang theory, such displacements should take place according to a long-wave acoustic excitation of the lattice. When the applied voltage is oscillatory at a frequency ω, $\boldsymbol{D} = \boldsymbol{D}_0 \exp(-i\omega t)$, and the displacement $\boldsymbol{u}_\mathrm{m}$ along the unique direction should be in phase with the electric field

$$\boldsymbol{E} = \{\boldsymbol{E}_{-k} \exp[i(\boldsymbol{k}\cdot\boldsymbol{r})] + \boldsymbol{E}_k \exp(-i\boldsymbol{k}\cdot\boldsymbol{r})\} \exp(-i\omega t),$$

and is written as

$$\boldsymbol{u}(\boldsymbol{r}, t) = \{\boldsymbol{u}_{-k} \exp[i(\boldsymbol{k}\cdot\boldsymbol{r})] + \boldsymbol{u}_k \exp[-i(\boldsymbol{k}\cdot\boldsymbol{r})]\} \exp(-i\omega t). \tag{4.2}$$

The macroscopic polarization density $\boldsymbol{P}$ is due partly to such a displacement, but also induced in ions by the applied electric field $\boldsymbol{E}$, and therefore when $\boldsymbol{E} \parallel \boldsymbol{u}$,

$$\boldsymbol{P} = \varepsilon_0(b'\boldsymbol{u} + \alpha\boldsymbol{E}), \tag{4.3a}$$

where b' is the proportionality factor, and α is the ionic polarizability. If, on the other hand, $\boldsymbol{E} \perp \boldsymbol{u}$,

$$\boldsymbol{P} = \varepsilon_0 \alpha \boldsymbol{E}. \tag{4.3b}$$

It is noted that in this case the polarization $\boldsymbol{P}$ is the order parameter, and (4.3a, b) correspond to the microscopic relation (4.1) for the zero field. Following Elliott and Gibson [34], these two cases are discussed separately for longitudinal and transversal applied fields, $\boldsymbol{E}_l$ and $\boldsymbol{E}_t$.

Dynamically the polar displacement mode $\boldsymbol{u}(\boldsymbol{r}, t)$ is forced to oscillate by the electric field $\boldsymbol{E}(\boldsymbol{r}, t)$, and the equation of motion is generally given by

$$\frac{\partial^2 \boldsymbol{u}}{\partial t^2} - v^2 \nabla^2 \boldsymbol{u} = b\boldsymbol{E}, \tag{4.4}$$

where v is the speed of propagation in the crystal, $b = e/m$, and the damping of the mode is neglected here for simplicity.

For a low frequency ω, the field $\boldsymbol{E}$ behaves as if static, and we have an approximate relation curl $\boldsymbol{E} \approx 0$. Hence, from (4.4)

$$\operatorname{curl}\left(\frac{\partial^2 \boldsymbol{u}}{\partial t^2} - v^2 \nabla^2 \boldsymbol{u}\right) = 0. \tag{4.4a}$$

For the transversal displacement $\boldsymbol{u}_t = \{\boldsymbol{u}_{t-} \exp(i\boldsymbol{k}\cdot\boldsymbol{r} + \boldsymbol{u}_{t+} \exp[i(-\boldsymbol{k}\cdot\boldsymbol{r})]\} \times \exp(-i\omega t)$,

$$\operatorname{curl} \boldsymbol{u}_{t\pm} = \pm i\boldsymbol{k} \times \boldsymbol{u}_{t\pm},$$

and hence (4.4a) is

$$\pm i\boldsymbol{k} \times (-\omega^2 + \omega_0{}^2)\boldsymbol{u}_{t\pm} = 0 \qquad \text{where} \quad \omega_0 = \frac{v}{k},$$

from which, since $\boldsymbol{k} \times \boldsymbol{u}_{t\pm} \neq 0$, for the transversal mode $\omega = \omega_t = \omega_0$.

For the longitudinal mode $\boldsymbol{u}_l$, operating "div" on both sides of (4.4),

$$\operatorname{div}\left(\frac{\partial^2 \boldsymbol{u}_l}{\partial t^2} + \omega_0{}^2 \boldsymbol{u}_l\right) = b \operatorname{div} \boldsymbol{E}_l.$$

On the other hand, using (4.3a),

$$\varepsilon_0 \operatorname{div} \boldsymbol{E}_l = -\operatorname{div} \boldsymbol{P} = -\varepsilon_0 \operatorname{div}(b'\boldsymbol{u}_l + \alpha \boldsymbol{E}_l).$$

Solving for div $\boldsymbol{E}_l$, we obtain

$$\operatorname{div} \boldsymbol{E}_l = -\frac{b' \operatorname{div} \boldsymbol{u}_l}{1+\alpha}.$$

Accordingly, the wave equation for the longitudinal mode is

$$\frac{\partial^2 \boldsymbol{u}_l}{\partial t^2} + \left(\omega_0{}^2 + \frac{bb'}{1+\alpha}\right)\boldsymbol{u}_l = 0,$$

where the longitudinal frequency is given by

$$\omega^2 = \omega_l{}^2 = \omega_0{}^2 + \frac{bb'}{1+\alpha}.$$

Clearly the difference between ω_t and ω_l is attributed to the nonzero constant b' that indicates the significance of the polarizing axis.

The dielectric behavior of polar pseudospins can be described in terms of the dielectric response $\varepsilon(\omega)$. From (4.4)

$$(\omega_0{}^2 - \omega^2)\boldsymbol{u}_l = b\boldsymbol{E}_l.$$

Therefore from (4.3a)

$$\boldsymbol{P} = \varepsilon_0\left(\alpha + \frac{bb'}{\omega_0{}^2 - \omega^2}\right)\boldsymbol{E}_l,$$

and

$$\boldsymbol{D}_l = \varepsilon(\omega)\boldsymbol{E}_l = \varepsilon_0 \boldsymbol{E}_l + \boldsymbol{P} = \varepsilon_0\left(1 + \alpha + \frac{bb'}{\omega_0{}^2 - \omega^2}\right)\boldsymbol{E}_l.$$

Hence the dielectric response function is expressed by

$$\varepsilon(\omega) = \varepsilon_0\left(1 + \alpha + \frac{bb'}{\omega_0{}^2 - \omega^2}\right),$$

which can also be written in the general form

$$\varepsilon(\omega) = \varepsilon(\infty) + \frac{\varepsilon(0) - \varepsilon(\infty)}{1 - (\omega/\omega_0)^2}, \tag{4.5}$$

where

$$\varepsilon(\infty) = \varepsilon_0(1+\alpha) \qquad \text{and} \qquad \varepsilon(0) = \varepsilon(\infty) + \frac{\varepsilon_0 bb'}{\omega_0{}^2}, \tag{4.6}$$

are the values at $\omega = \infty$ and 0, respectively. From (4.5), the following specific values can be obtained, i.e.,

$$\varepsilon(\omega_t) = \pm\infty \qquad \text{and} \qquad \varepsilon(\omega_l) = 0. \tag{4.7}$$

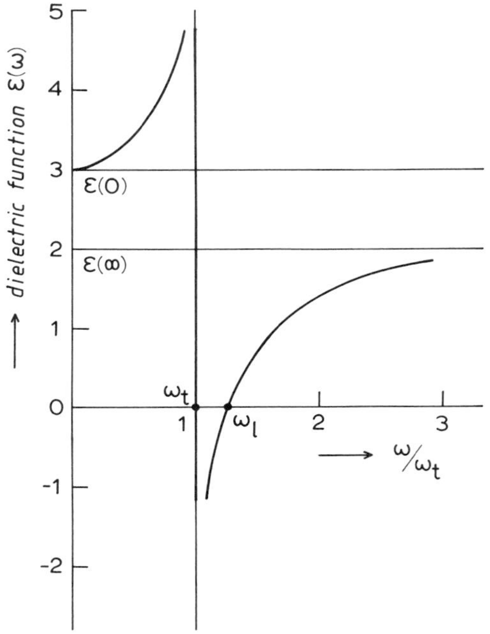

Figure 4.2. The behavior of a dielectric function $\varepsilon(\omega)$ versus frequency ω in a linear polar crystal, where $\varepsilon(\omega_1) = 0$ and $\varepsilon(\omega_t) \pm \infty$.

In Figure 4.2 the response function $\varepsilon(\omega)$ of (4.5) is sketched for varying ω, where there is a forbidden gap $\varepsilon(0) - \varepsilon(\infty)$. The characteristic frequencies ω_t and ω_l signify a polar pseudospin mode coupled with an optical lattice mode, whose response to an applied electric field can be studied at optical frequencies.

The Lyddane–Sachs–Teller (LST) relation can be obtained from (4.5), which is

$$\frac{\omega_l{}^2}{\omega_t{}^2} = \frac{\varepsilon(0)}{\varepsilon(\infty)}. \tag{4.8}$$

From the expression for the response given by (4.6), the static susceptibility is

$$\chi(0) = \alpha + \frac{bb'}{\omega_0{}^2},$$

which goes to $\alpha + \infty$, if $\omega_0 \to 0$. For ferroelectric phase transitions, the static response above the transition temperature T_c is known to exhibit a singular behavior similar to the Curie–Weiss law, i.e.,

$$\varepsilon(0) = \frac{C}{T - T_c} \qquad \text{for} \quad T > T_c, \tag{4.9}$$

which becomes infinite when $T \to T_c$. Therefore, the process for $T \to T_c$ appears to be equivalent to $\omega_0 \to 0$. In fact, using (4.9) in the LST relation (4.8), we obtain

$$\omega_t^2 \propto T - T_c, \tag{4.10}$$

showing the temperature-dependence of the soft-mode frequency.

It is noticed that (4.10) is consistent with the Landau theory, if the presence of harmonic fluctuations is assumed in the vicinity of the minimum of the Gibbs potential. For $T > T_c$, the thermal equilibrium is at $\eta = 0$, where the Gibbs potential is minimum, i.e.,

$$G(\eta) = \tfrac{1}{2}A\eta^2 \qquad \text{where} \quad A = A'(T - T_c).$$

At $T = T_c$, $G(\eta) = 0$ and $\eta = 0$, implying dynamically the presence of a fluctuation $\delta\eta = \eta - 0$. For such a uniform fluctuation arising from a hidden interaction, we can consider a mass particle m with the kinetic energy $\frac{1}{2}m(\mathrm{d}\delta\eta/\mathrm{d}t)^2$ in the parabolic potential $G(\delta\eta) = \frac{1}{2}A(\delta\eta)^2$. In this case, the fluctuation $\delta\eta$ is sinusoidal, and can be attributed to a lattice mode, whose characteristic frequency ω is given by the relation $A = m\omega^2$. Therefore, the frequency of the fluctuation is $\omega \propto (T - T_c)^{1/2}$. Such a lattice mode must be the transversal optical mode, while the fluctuation is spatially uniform as implied by $\boldsymbol{q} = 0$.

4.3. Long-Range Interactions and the Cochran Soft-Mode Theory

Neglecting spatial fluctuations, the LST relation implies that a continuous structural change due to pseudospin ordering always accompanies a soft lattice mode. Hence, the critical temperature T_c and the softening frequency ω_t of the transversal mode both represent the singular behavior at a phase transition. As the result, the soft mode may be observed as if responsible for the structural change, despite the fact that the singularity cannot arise solely from plain phonon spectra.

On the other hand, Cochran [30] showed in his lattice dynamical theory that competing short- and long-range interactions are responsible for softening the frequency ω_t of a transversal lattice mode when T_c is approached. In the Cochran model for an ionic crystal, a pair of positive and negative ions is considered at a site m, although such a site dependence is unnecessary in a uniformly polarized crystal. Displacing a positive ion through a distance $\boldsymbol{u}_m$ corresponds to $+e$ at the displaced position and the addition of an extra charge $-e$ at the original position, while keeping the ion fixed. Similarly, displacing a negative ion is equivalent to creating a pair of charges $(-e, +e)$ where $+e$ is at the original position. Hence an elementary dipole moment $\boldsymbol{p}_m = e(\boldsymbol{u}_{m+} - \boldsymbol{u}_{m-})$ is formed as a result of ionic displacements, acting as the order variable. While $\boldsymbol{p}_{\mathrm{m}}$ can be represented by a pseudospin, in a continuum lattice, $\boldsymbol{u}_{\mathrm{m}+}$ and $\boldsymbol{u}_{\mathrm{m}-}$ are considered as continuous functions of the space–time coordinates $(\boldsymbol{r}, t)$. Further assuming the crystal as uniform, spatial averages $\langle \boldsymbol{u}_{\mathrm{m}+} \rangle = \boldsymbol{u}_+(t)$ and $\langle \boldsymbol{u}_{\mathrm{m}-} \rangle = \boldsymbol{u}_-(t)$ can be considered as functions of t only.

For a uniaxially polar crystal, we consider quasi-one-dimensional ionic chains in parallel with the unique z axis, along which dipolar pseudospins are correlated to form macroscopic polarization in the low-temperature phase, whereas in directions perpendicular to the z axis pseudospins are uncorrelated. Further, we assume that these polar chains can mutually interact via classical dipolar forces. In such a polarized chain, the internal field E_i along the axis can be calculated from the relation $\varepsilon_0\, \mathrm{d}E_i = -(\mathrm{d}P/\mathrm{d}z)\, \mathrm{d}z$, leading to the depolarizing field $E_{\mathrm{dip}} = E_{\mathrm{i}} = -P/\varepsilon_0$ where P is macroscopic polarization. In contrast, rotatable classical dipole moments can be polarized in any direction by an applied external field E, resulting in the internal field $E_{\mathrm{i}} = P/3\varepsilon_0$, known as the Lorentz field.

For transversal displacements of positive and negative ions denoted by u_{t+} and u_{t-}, there is no depolarizing field in any directions perpendicular to z axis, and so the equations of motion can be written as

$$m_+ \frac{\partial^2 u_{t+}}{\partial t^2} + C(u_{t+} - u_{t-}) = e\left(E + \frac{P}{3\varepsilon_0}\right)$$

and

$$m_- \frac{\partial^2 u_{t-}}{\partial t^2} + C(u_{t-} - u_{t+}) = -e\left(E + \frac{P}{3\varepsilon_0}\right),$$

where C is the elastic constant, and m_+ and m_- are effective masses of positive and negative ions. Writing these equations in terms of the transversal component of $\boldsymbol{p}$, i.e., $p_t = e(u_{t+} - u_{t-})$, we obtain

$$m \frac{\partial^2 p_t}{\partial t^2} + C p_t = e^2\left(E + \frac{P_t}{3\varepsilon_0}\right)$$

or

$$m \frac{\partial^2 P_t}{\partial t^2} + C P_t = \left(\frac{e^2}{v}\right)\left(E + \frac{P_t}{3\varepsilon_0}\right), \tag{4.11}$$

where $m = m_+ m_-/(m_+ + m_-)$ is the reduced mass of the ion pair, and $P_t = v p_t$, where v is the unit volume of the crystal. From (4.11), the dielectric susceptibility of P_t associated with the transversal mode is expressed as

$$\chi_t(\omega) = \frac{P_t}{E} = \frac{e^2/v}{C - (e^2/3v\varepsilon_0) - m\omega^2},$$

which is singular when

$$m\omega^2 = m\omega_t^2 = C - \frac{e^2}{3v\varepsilon_0}. \tag{4.12}$$

For the longitudinal mode, the effective field is contributed by the Lorentz field as well as the depolarizing field along the axis. Therefore, the equation of motion for the longitudinal component of the elementary dipole, i.e., $p_l = e(u_{l+} - u_{l-})$, can be written for the polarization P_l as

$$m\frac{\partial^2 P_l}{\partial t^2} + CP_l = \frac{e^2}{v}\left(E - \frac{P_l}{\varepsilon_0} + \frac{P_l}{3\varepsilon_0}\right),$$

and the frequency ω_l for the longitudinal susceptibility $\chi_l(\omega)$ is given by

$$m\omega_l^{\,2} = C + \frac{2e^2}{3v\varepsilon_0}. \tag{4.13}$$

Comparing (4.12) and (4.13), it is noticed that the transversal frequency ω_t can become zero under such specific circumstances that $C = e^2/3v\varepsilon_0$, whereas the longitudinal frequency ω_l remains nonzero. Although such a condition for frequency softening is considered to occur at T_c, in this theory the value of the transition temperature is not predictable. In fact, the singularity in the pseudospin mode originates from their correlations, while the frequency softening takes place as a consequence of its coupling with the lattice mode, and observed from the dielectric response in phases below and above T_c.

For a structural transition, soft modes in these phases are characterized generally by different symmetries, reflecting the change of lattice symmetry at the structural phase transition. Figure 4.3 shows a typical example of soft-mode frequencies changing with temperature, which summarizes infrared and Raman scattering results on the ferroelectric phase transition in TSCC crystals. As indicated in the figure, the soft-mode symmetry is B_{2u} above T_c, whereas it is in A_1 symmetry below T_c, exhibiting different temperature-dependences. Obviously, such differences should be attributed to these transversal modes of different dynamical characters.

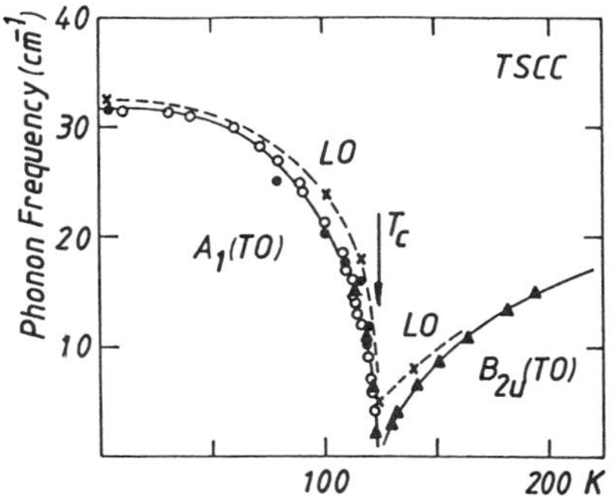

Figure 4.3. Soft-mode frequencies as a function of temperature near $T_c = 130$ K in the ferroelectric phase transition of TSCC crystals. Data were obtained by Raman (o and •)and infrared (▲ and ×) experiments. (From J. F. Scott, Raman Spectroscopy of Structural Phase Transitions: In *Structural Phase Transitions I*, edited by K. A. Müller and H. Thomas (Springer-Verlag, Heidelberg, 1981).)

4.4. Anharmonic Lattice Potentials and the Cowley Theory

The critical region of a structural phase transition is characterized by pseudospin correlations emerging at the threshold, thereby straining the lattice structure. In the critical region, (4.1) for the relation between $\boldsymbol{\sigma}_{\mathrm{m}}$ and $\boldsymbol{u}_{\mathrm{m}}$ suggests that the strained lattice should be described by anharmonic lattice potentials compatible with those for $\boldsymbol{\sigma}_{\mathrm{m}}$ given by (3.1a,b). It is conceivable that such lattice anharmonicity should be consistent with the symmetry change of the soft modes at T_{c}. For example, in TSCC crystals undergoing a ferroelectric phase transition at 120 K, the mirror reflection symmetry on the (010) plane is violated, as signified by observed soft modes showing a symmetry change between B_{2u} and A_1 modes (see Fig. 4.3). Such a symmetry conversion can arise from the perturbation of a quartic anharmonic potential that is given by a combination of $u_{\mathrm{mx}}{}^4$, $u_{\mathrm{mx}}{}^2 u_{\mathrm{my}}{}^2$, and $u_{\mathrm{mx}}{}^2 u_{\mathrm{mz}}{}^2$ in the lowest order, where u_{mx} is the center-of-mass coordinate of the active group at the site m along the *b* axis, and u_{my} and u_{mz} are the coordinates in the *ac* plane. For the commensurate-to-incommensurate phase transition in BCCD crystals (*betaine calcium chloride dihydrate*), the low-temperature phase is signified by a modulated structure along the *a* direction, while the soft mode above the transition temperature was identified as a transversal mode propagating along the *a* axis and vibrating parallel to the *b* axis. In this case, the local anharmonic potential consisting of $u_{\mathrm{mx}}{}^4$ and $u_{\mathrm{mx}}{}^2 u_{\mathrm{my}}{}^2$ can be considered as responsible for the soft-mode conversion in BCCD, where u_{mx} and u_{my} are the coordinates along the *b* and *a* axes, respectively. Such a quartic lattice potential as originating from pseudospin correlations can generally be expressed by

$$V_4(\boldsymbol{u}_{\mathrm{m}}) = \sum_{\mathrm{ij}} u_{\mathrm{mi}}{}^2 V_{\mathrm{ij}} u_{\mathrm{mj}}{}^2 \qquad \text{where} \quad \mathrm{i, j} = \mathrm{x, y, and}\ z,$$

which is inferable from the symmetry change of the soft modes at T_{c}.

Setting microscopic details aside in individual systems, Cowley [35] emphasized the significance of anharmonic perturbations near T_{c}. He showed, among others, that phonon scattering by a quartic potential is responsible for softening of the characteristic frequency, as outlined in the following.

Consider the Fourier transforms of a polar lattice mode $\boldsymbol{u}(\boldsymbol{r}, t)$, that are

$$\boldsymbol{u}_{\pm \boldsymbol{q}}(t) = N^{-1/2} \boldsymbol{u}(\boldsymbol{r}, t) \exp(\pm i\boldsymbol{q} \cdot \boldsymbol{r}),$$

and the applied oscillating field at a frequency ω is also written in the form

$$\boldsymbol{E} \exp(-i\omega t) = \{\boldsymbol{E}_{-\boldsymbol{q}} \exp(i\boldsymbol{q} \cdot \boldsymbol{r}) + \boldsymbol{E}_{+\boldsymbol{q}} \exp(-i\boldsymbol{q} \cdot \boldsymbol{r})\} \exp(-i\omega t).$$

For the lattice modes $\boldsymbol{u}_{\pm \boldsymbol{q}}(t)$, the equations of forced motion by the oscillating field $\boldsymbol{E} \exp(-i\omega t)$ are expressed by

$$\frac{\mathrm{d}^2 \boldsymbol{u}_{\pm \boldsymbol{q}}}{\mathrm{d}t^2} + \gamma \frac{\mathrm{d}\boldsymbol{u}_{\pm \boldsymbol{q}}}{\mathrm{d}t} + \varpi(\boldsymbol{q})^2 \boldsymbol{u}_{\pm \boldsymbol{q}} = \frac{e}{m} \boldsymbol{E}_{\pm \boldsymbol{q}} \exp(-i\omega t),$$

where γ is the damping constant, $\varpi(\boldsymbol{q})$ is the characteristic frequency, and e/m is the effective charge by mass ratio of the lattice modes.

In the presence of anharmonic perturbations Cowley showed that

$$\boldsymbol{u}_{\pm q} = \frac{(e/m)\boldsymbol{E}_{\pm\boldsymbol{q}}}{-\omega^2 + (-\Delta + i\Gamma)\omega + \varpi(\boldsymbol{q})^2} = \frac{(e/m)\boldsymbol{E}_{\pm\boldsymbol{q}}}{-\omega^2 + i\Gamma\omega + \varpi(\boldsymbol{q}, \omega)^2},$$

where the damping is expressed by a complex form $\gamma = \Gamma + i\Delta$. In the perturbed mode, the characteristic frequency $\varpi(\boldsymbol{q}, \omega)$ shifts from the unperturbed frequency $\varpi(\boldsymbol{q})$ as given by

$$\varpi(\boldsymbol{q}, \omega)^2 = \varpi(\boldsymbol{q})^2 + 2\varpi(\boldsymbol{q})\{\Delta(\boldsymbol{q}, \omega) - i\Gamma(\boldsymbol{q}, \omega)\}. \tag{4.14}$$

Here $\Delta(q, \omega)$ and $\Gamma(\boldsymbol{q}, \omega)$ are perturbation corrections causing a frequency shift from $\varpi(\boldsymbol{q})$. Cowley has given the following expressions for these parameters. Namely,

$$\Delta(\boldsymbol{q}, \omega) = \Delta_0(\boldsymbol{q}) + \Delta_1(\boldsymbol{q}) + \Delta_2(\boldsymbol{q}, \omega) \qquad \text{and} \qquad \Gamma(\boldsymbol{q}, \omega) = \Delta_2(\boldsymbol{q}, \omega),$$

where

$$\Delta_0(\boldsymbol{q}) = \left\{\frac{\partial\varpi(\boldsymbol{q})}{\partial V}\right\}\Delta V = -\varpi(\boldsymbol{q})\mathrm{k_B}T\left\{\frac{\phi'''(r_0)^2}{\phi''(r_0)^3}\right\} < 0$$

and

$$\Delta_1(\boldsymbol{q}) = \varpi(\boldsymbol{q})\mathrm{k_B}T\left\{\frac{\phi''''(r_0)}{8\phi''(r_0)}\right\} = \left\{\frac{\hbar}{N\varpi(\boldsymbol{q})}\right\}\sum_{\boldsymbol{q}'}\left(\frac{2n'+1}{2\omega'}\right)V_4(-\boldsymbol{q}, \boldsymbol{q}; \boldsymbol{q}', -\boldsymbol{q}'). \tag{4.15}$$

Here V_4 represents matrix elements of the quartic potential, and Δ_0 and Δ_1 are expressed with derivatives of the interatomic potential $\phi(r_0)$ in the high-temperature approximation. The expressions for $\Delta_2 = \Gamma$ are given by

$$\begin{aligned}\Delta_2(\boldsymbol{q}, \omega) = \Gamma(\boldsymbol{q}, \omega) = &\left\{\frac{\pi\hbar}{16N\varpi(\boldsymbol{q})}\right\}\sum_{\boldsymbol{q}', \boldsymbol{q}''}|V_3(\boldsymbol{q}; \boldsymbol{q}', \boldsymbol{q}'')|^2 \\ &\times [(n' + n'' + 1)\{-\delta(\omega + \omega' + \omega'') + \delta(\omega - \omega' - \omega'')\} \\ &- (n' - n'')\{-\delta(\omega - \omega' + \omega'') + \delta(\omega + \omega' - \omega'')\}],\end{aligned} \tag{4.16}$$

where V_3 represents the cubic anharmonic potential.

The quartic potential V_4 in (4.15) gives a secular perturbation due to symmetric scattering of two phonons $\pm\boldsymbol{q}$, i.e.,

$$\boldsymbol{q} + (-\boldsymbol{q}) \rightarrow \boldsymbol{q}' + (-\boldsymbol{q}'), \tag{4.17}$$

whereas the cubic potential V_3 is responsible for a scattering process

$$\boldsymbol{q} \rightarrow \boldsymbol{q}' + \boldsymbol{q}''. \tag{4.18}$$

The potential V_3 gives a time-dependent perturbation that is responsible for damping, whereas V_4 is secular and hence significant for processes under equilibrium conditions. Corresponding to the wavevectors $\boldsymbol{q}'$ and $\boldsymbol{q}''$, the frequencies and the phonon numbers in (4.16) are designated by ω', n' and ω'', n'', respectively. It is also significant that, owing to the quartic interaction, phonons specified by $\boldsymbol{q}$ and $-\boldsymbol{q}$ are no longer independent, playing a significant role for critical fluctuations as will be discussed in Section 4.6.

It is noted that the frequency shift arising from Δ_0 vanishes under a constant volume condition. In contrast, the shift due to the correction Δ_1 is proportional to the temperature T, if the phonon density at the state n' in (4.15) is given by the high-temperature approximation, i.e., $\hbar\omega' \ll k_B T$, thus prevailing in the frequency shift given by (4.14). In particular, for $\boldsymbol{q} = 0$, we can write $2\varpi(0)\Delta_1 = \pm A'T$, where A' is a positive constant, to be consistent with the frequency shift in $T > T_0$ and $T < T_0$, respectively. Also, in a double-well potential the unperturbed frequency $\varpi(0)$ is imaginary in the harmonic approximation, so that we can write $\varpi(0)^2 = -A'T_0$. Therefore, the characteristic frequencies of the perturbed soft mode at $\boldsymbol{q} = 0$ can be expressed as

$$\varpi(0, \omega)^2 = A'(T - T_0) \qquad \text{for} \quad T > T_0 \tag{4.19}$$

and

$$\varpi(0, \omega)^2 = A'(T_0 - T) \qquad \text{for} \quad T < T_0. \tag{4.19a}$$

To simplify the argument for the frequency shift with temperature, we have assumed in the above that soft modes have the same symmetry above and below T_0, although they are not the same in general as evidenced by the symmetry change at T_0. Nevertheless, for a continuous-phase transition it is logical to consider that the same active groups are responsible for the symmetry change at T_c. In the case of TSCC crystals, pending experimental verification, for the lattice potentials in normal and ferroelectric phases we can assume the following lattice potentials:

$$V_m(B_{2u}) = \tfrac{1}{2}Au_{mz}{}^2 + \tfrac{1}{4}B(u_{mz}{}^4 + u_{mz}{}^2u_{mx}{}^2)$$

and

$$V_m(A_1) = \tfrac{1}{2}Au_{mx}{}^2 + \tfrac{1}{4}B(u_{mx}{}^4 + u_{mx}{}^2u_{mz}{}^2),$$

where the active groups are signified by the identical coefficients A and B for these modes. Here for B_{2u} and A_1 oscillators, the components u_{mz} and u_{mx} are the basic variables along the b and a axes, respectively, and the quartic potential $\frac{1}{4}Bu_{mz}{}^2u_{mx}{}^2$ is considered as responsible for the mode conversion. The A_1 mode does not exist above T_c, whereas due to the coupling between B_{2u} phonons and pseudospins two phonon modes A_1 and B_{2u} should be present below T_c. Hence in the mean-field approximation, we can take $\langle u_{mx}{}^2\rangle = 0$ for the B_{2u} mode in both phases, whereas $\langle u_{mz}{}^2\rangle = -A/B$ for the A_1 mode in the ferroelectric phase. Therefore, the mean-field potentials below T_c are expressed as

$$\langle V(B_{2u})\rangle = \tfrac{1}{2}Au_z{}^2 + \tfrac{1}{4}Bu_z{}^4$$

and

$$\langle V(A_1)\rangle = \tfrac{1}{2}Au_x{}^2 + \tfrac{1}{4}Bu_x{}^4 + \tfrac{1}{4}Bu_x{}^2\left(-\frac{A}{B}\right) = \tfrac{1}{4}Au_x{}^2 + \tfrac{1}{4}Bu_x{}^4.$$

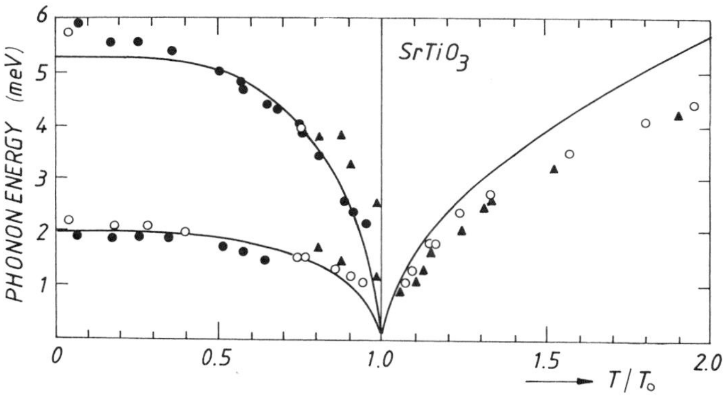

Figure 4.4. Soft-mode frequencies versus temperature near 105 K in $SrTiO_3$. Data: ▲...Cowley et al., ●...Fleury et al., and ○...Shirane et al. From J. Feder and E. Pytte, *Phys. Rev.* **B1**, 4803 (1970).

Accordingly, the characteristic frequency given by (4.19a) of the A_1 soft-mode should be revised as

$$\varpi(0, \omega)^2 = 2A'(T_0 - T) \qquad \text{for} \quad T < T_0. \tag{4.19b}$$

The factor "2" as appears in (4.19b) can be verified to a good approximation, in the curves shown for TSCC in Fig. 4.3 and for $SrTiO_3$ in Fig. 4.4, indicating that the soft-mode conversion by an anharmonic perturbation discussed in the above is the correct mechanism at least qualitatively.

It is realized that the transition temperature T_0 is accurate only in the mean-field approximation, but has never been observed as sufficiently close to T_c. The order parameter cannot be defined with respect to the soft modes, although observed on both sides of T_0, always accompanying the continuous-phase transition.

4.5. Observation of Soft-Mode Spectra

At the threshold of a displacive structural-phase transition, correlated pseudospins in a collective mode in a low-energy excitation are coupled with a lattice mode in a sinusoidal limit, where the coupling is characterized by the wavevector and energy differences, $\Delta \boldsymbol{k}$ and $\Delta\varepsilon$, in the vicinity of a specific point in the first Brillouin zone. As discussed in Chapter 3, such a point can be at the zone center, or at the zone boundary $\frac{1}{2}\boldsymbol{G}$, or at an arbitrary point $\boldsymbol{G}_i$ in the zone. In the first two cases, the lattice periodicity is unchanged or doubled, while in the third case the low-temperature phase is modulated and incommensurate with the lattice. Owing to an excitation in the lattice $\varepsilon' - \varepsilon = \mp\Delta\varepsilon$, phase transitions in all three categories accompany sinusoidal fluctuations signified by $\Delta\boldsymbol{k} = (0, \frac{1}{2}\boldsymbol{G}, \boldsymbol{G}_i) \pm \boldsymbol{q}$ and $\Delta\omega = \mp\Delta\varepsilon/\hbar$, where $\Delta\varepsilon$ is considered as the kinetic energy of the fluctuation. Such a sinusoidal fluctuation is classical with the dispersion relation as $\Delta\omega \propto q$, but the energy $\Delta\varepsilon$ is related to the characteristic frequency ϖ of the classical oscillator as $\Delta\varepsilon \propto \varpi^2$.

In Section 4.1, we consider a uniform dielectric system characterized by $\boldsymbol{q} = 0$, which is represented by a uniform Gibbs potential. Although we only dealt with the temporal profile, the fluctuation characterized by $\boldsymbol{q}$ and ϖ is sinusoidally distributed, where the crystal is by no means uniform. For $\boldsymbol{q} \neq 0$, the Gibbs potential is not sharply defined, fluctuating spatially in the crystal space. When the Gibbs potential is written as

$$G_{\boldsymbol{q},\varpi}(\delta\eta) = \tfrac{1}{2}m\varpi(\boldsymbol{q}, \varpi)^2\delta\eta^2,$$

the characteristic frequency $\varpi(\boldsymbol{q}, \omega)$ can effectively be expressed by

$$\varpi(\boldsymbol{q}, \omega)^2 = A'(T - T_0) + \kappa'\boldsymbol{q}^2 \qquad \text{for} \quad T > T_0 \tag{4.20a}$$

and

$$\varpi(\boldsymbol{q}, \omega)^2 = A''(T_0 - T) + \kappa''\boldsymbol{q}^2 \qquad \text{for} \quad T < T_0, \tag{4.20b}$$

where κ' and κ'' represent kinetic energies in the fluctuation modes above and below T_c, respectively. Furthermore, $A' = A''$ and $\kappa = \kappa'$, if the two phases have the same symmetry, otherwise $A' \neq A''$ and $\kappa \neq \kappa'$ in general. Equations (4.20a, b) are usually employed for analyzing soft-mode frequencies in the mean-field approximation.

For polar crystals, soft-mode spectra are usually obtained by dielectric measurements, while the $\boldsymbol{q}$ dependence at the Brillouin zone center cannot be revealed. In contrast, for phase transitions at zone boundaries or other points, soft modes can be studied as a function of $\boldsymbol{q}$ and $\varpi(0, \omega)$ from neutron inelastic scattering spectra.

In polar crystals, the equation of motion for the elementary dipole moment $\boldsymbol{p}_q = e(\boldsymbol{u}_{+q} - \boldsymbol{u}_{-q})$ in an applied field $\boldsymbol{E}_q \exp(-i\omega t)$ can be written as

$$\frac{d^2\boldsymbol{p}_q}{dt^2} + \gamma\frac{d\boldsymbol{p}_q}{dt} + \varpi^2\boldsymbol{p}_q = \frac{e^2}{m}\boldsymbol{E}_q \exp(-i\omega t).$$

In this case, the complex dielectric susceptibility at a fixed value of $\boldsymbol{q}$ is given by

$$\chi_q(\omega) = \chi_q' - i\chi_q'' = \frac{\boldsymbol{p}_q}{\boldsymbol{E}_q} = \frac{e^2/m}{(\varpi^2 - \omega^2) + i\gamma\omega}, \tag{4.21}$$

where the real and imaginary parts are

$$\chi_q' = \frac{(e^2/m)(\varpi^2 - \omega^2)}{(\varpi^2 - \omega^2) + \gamma^2\omega^2} \tag{4.21a}$$

and

$$\chi_q'' = \frac{(e^2/m)\gamma\omega}{(\varpi^2 - \omega^2)^2 + \gamma^2\omega^2}. \tag{4.21b}$$

These are the basic formula for dielectric analysis in neutron experiments, where the soft mode is identified by a peak at $\omega = \varpi$, if not significantly damped (*underdamped*), i.e., $\gamma < \varpi^{-1}$, otherwise showing a relaxational decay (*overdamped*). It is noticed that $\boldsymbol{q}$ is implicit in (4.21a, b), though derived for its fixed value.

Neutron inelastic scattering is a practical method to observe the soft-mode fluctuations described by $\boldsymbol{q}$ and ϖ. Nevertheless, when applied to pseudospin condensates, the neutron method has a significant difference from dielectric measurements. Namely, neutrons are scattered by heavy active groups, whereas the response from pseudospins can be detected in dielectric measurements. Incident and scattered thermal neutrons are characterized by their wavevectors and energies ($\boldsymbol{k}_1, \varepsilon_1 = \hbar^2 k_1{}^2/2m_n$) and ($\boldsymbol{k}_2, \varepsilon_2 = \hbar^2 k_2{}^2/2m_n$), respectively, where m_n is the neutron mass. For a cell-doubling transition, the monentum and energy relations can be written for the scattering as

$$\boldsymbol{k}_2 - \boldsymbol{k}_1 = \tfrac{1}{2}\boldsymbol{G} \pm \boldsymbol{q} \qquad \text{and} \qquad \varepsilon_2 - \varepsilon_1 = \varepsilon \mp \Delta\varepsilon,$$

where $\mp\Delta\varepsilon$ is the energy imparted to or gained from the lattice mode due to the wavevector change $\pm\boldsymbol{q}$ from the scattering geometry $\boldsymbol{k}_2 - \boldsymbol{k}_1 = \frac{1}{2}\boldsymbol{G}$. In this case, the scattering intensity is generally expressed by the time average

$$I(\tfrac{1}{2}\boldsymbol{G} \pm \boldsymbol{q}, \varepsilon \mp \Delta\varepsilon) = \langle A_{1/2\boldsymbol{G}}{}^* A_{1/2\boldsymbol{G}} \rangle_t = \left\langle \sum_{mn} A_{1/2\boldsymbol{G}m}{}^* A_{1/2\boldsymbol{G}n} \right\rangle_t, \tag{4.22}$$

where $A_{1/2\boldsymbol{G}m}$ is the so-called *scattering amplitude* from the active group at a site m, and the total scattering amplitude $A_{1/2\boldsymbol{G}}$ is given by

$$\begin{aligned} A_{1/2\boldsymbol{G}} &\propto \sum_m \boldsymbol{u}_m \exp\left\{ i(\boldsymbol{k}_2 - \boldsymbol{k}_1 - \tfrac{1}{2}\boldsymbol{G} \pm \boldsymbol{q})\cdot\boldsymbol{r}_m - \frac{i(\varepsilon_2 - \varepsilon_1 - \varepsilon \mp \Delta\varepsilon)t_m}{\hbar} \right\} \\ &= \sum_m \boldsymbol{u}_m \exp\left\{ \mp i\boldsymbol{q}\cdot\boldsymbol{r}_m \pm i\left(\frac{\Delta\varepsilon}{\hbar}\right)t_m \right\} \\ &= \boldsymbol{u}_{\boldsymbol{q}}(t)\exp(-i\omega t) + \boldsymbol{u}_{-\boldsymbol{q}}(t)\exp(i\omega t), \end{aligned} \tag{4.23}$$

where in the last expression $\omega = \Delta\varepsilon/\hbar$, and $\boldsymbol{u}_{\pm\boldsymbol{q}}(t)$ represent Fourier transforms of $\boldsymbol{u}_m$. Using (4.23) in (4.22),

$$I_{1/2\boldsymbol{G}}(\boldsymbol{q}, \Delta\varepsilon) \propto |\boldsymbol{u}_{\boldsymbol{q}}|^2 + |\boldsymbol{u}_{-\boldsymbol{q}}|^2 + 2\langle \boldsymbol{u}_{\boldsymbol{q}}(t)\boldsymbol{u}_{-\boldsymbol{q}}(t')\cos\{i\omega(t - t')\}\rangle_t. \tag{4.24}$$

If we interpret that the two time-dependent modes $\boldsymbol{u}_{\boldsymbol{q}}\exp(-i\omega t)$ and $\boldsymbol{u}_{-\boldsymbol{q}}\exp(i\omega t)$ in (4.23) are forced to be driven by effective impact fields $\boldsymbol{F}_{\boldsymbol{q}}\exp(-i\omega t)$ and $\boldsymbol{F}_{-\boldsymbol{q}}\exp(i\omega t)$, respectively, the equations of motion can be written as

$$\frac{d^2\boldsymbol{u}_{\boldsymbol{q}}}{dt^2} + \gamma\frac{d\boldsymbol{u}_{\boldsymbol{q}}}{dt} + \varpi^2\boldsymbol{u}_{\boldsymbol{q}} = \boldsymbol{F}_{\boldsymbol{q}}\exp(-i\omega t)$$

and

$$\frac{d^2\boldsymbol{u}_{-\boldsymbol{q}}}{dt^2} + \gamma\frac{d\boldsymbol{u}_{-\boldsymbol{q}}}{dt} + \varpi^2\boldsymbol{u}_{-\boldsymbol{q}} = \boldsymbol{F}_{-\boldsymbol{q}}\exp(i\omega t),$$

from which the steady solutions are given by

$$(-\omega^2 + \varpi^2 - i\gamma\omega)(\boldsymbol{u}_q)_0 = \boldsymbol{F}_q \qquad \text{and} \qquad (-\omega^2 + \varpi^2 + i\gamma\omega)(\boldsymbol{u}_{-q})_0 = \boldsymbol{F}_{-q}.$$

Hence

$$\langle \boldsymbol{u}_q(t)\boldsymbol{u}_{-q}(t') \cos\{i\omega(t - t')\}\rangle_t = \frac{\boldsymbol{F}_q\boldsymbol{F}_{-q}\langle\cos\{-i\omega(t - t')\}\rangle_t}{(\varpi^2 - \omega^2)^2 + (\gamma\omega)^2},$$

and the scattering intensity exhibit a variation

$$\Delta I_{1/2\,G}(\boldsymbol{q}, \Delta\varepsilon) \propto |\boldsymbol{F}_q|^2 \Gamma_t \gamma\omega\chi_q''(\omega), \tag{4.25}$$

where the imaginary part of the complex $\chi_q(\omega)$ is given by

$$\chi_q''(\omega) = \frac{\gamma\omega}{(\varpi^2 - \omega^2)^2 + (\gamma\omega)^2},$$

indicating that the maximum energy transfer occurs when $\omega = \varpi$. We have already obtained the expression for the time correlation in (3.14), i.e., $\Gamma_t = \langle\cos\{-i\omega(t - t')\}\rangle_t = \sin \omega t_0/\omega t_0$, which is nearly equal to 1 for a very short neutron impact time and $\omega t_0 < 1$. Thus, for a fixed value of $\boldsymbol{q}$ the scattering intensity $\Delta I_{1/2\,G}$ is proportional to $\chi_q''(0)$, from which the soft mode can be identified as an absorption peak at $\omega = \varpi$. Equation (4.25) is an example of the *fluctuation–dissipation theorem*, since the correlation at a single frequency ω is expressed by $\chi_q''(0)$ representing the energy dissipation.

In the above, we discussed the neutron scattering at zone-boundaries, however, the argument can be repeated for a more general scattering at an arbitrary $\boldsymbol{G}_i$ in the reciprocal lattice. Typical examples of soft-mode spectra are shown in Figs. 4.5 and 4.6(a), which were observed by the Brookhaven group in K_2SeO_4 in the vicinity of $\boldsymbol{G}_i = 0.7\boldsymbol{a}^*$ and in $SrTiO_3$ and $KMnF_3$ at

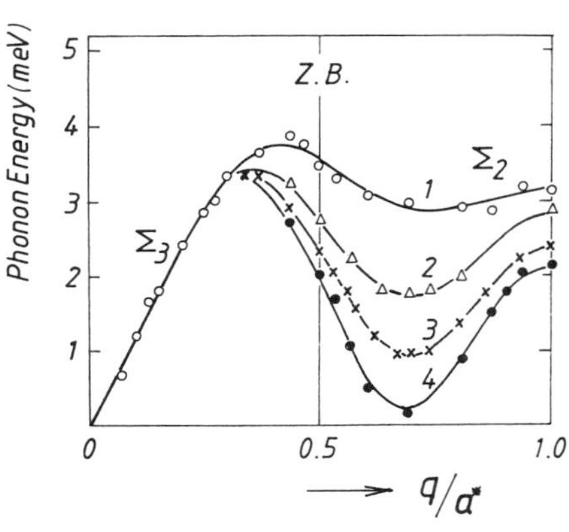

Figure 4.5. Phonon energy in K_2SeO_4 measured by neutron inelastic scattering at $\boldsymbol{G}_i = 0.7\boldsymbol{a}^*$. Curves 1, 2, 3, 4 were obtained at 250, 175, 145, 130 K, respectively. (From M. Iizumi, J. D. Axe, G. Shirane, and K. Shimaoka, *Phys. Rev.* **B15**, 4392 (1977).)

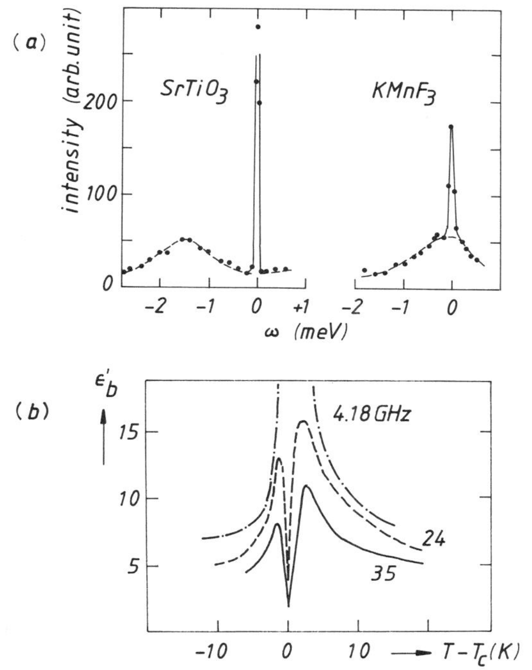

Figure 4.6. (a) Soft-mode spectra from $SrTiO_3$, and $KMnO_3$. (From S. M. Shapiro, J. D. Axe, G. Shirane, and T. Riste, *Phys. Rev.* **B6**, 4332 (1972).) (b) Oscillator–relaxator behavior in the dielectric response from TSCC near $T_c = 130$ K. (From A. Sawada and M. Horioka, *Japan J. Appl. Phys. Suppl.* 24-2, 390 (1985).)

the zone boundaries, respectively. Figure 4.6(b) shows anomalies in dielectric spectra of TSCC crystals observed by Sawada and Horioka [38] for $\varepsilon'(\omega)$ versus $T - T_c$.

4.6. The Central Peak

Pseudospin and phonon susceptibilities provide information about the temporal fluctuations in the critical region. In these results, a phase transition is signified by the soft frequency ϖ that converges to near zero when the transition temperature is approached. On the other hand, Shapiro et al. [36] discovered that the phonon spectra in the cell-doubling transition of $SrTiO_3$ crystals near T_c exhibited a sharp unidentifiable absorption at $\omega = 0$ in addition to the soft-mode peak $\omega = \varpi$ (Fig. 4.6(a)). Being referred to as a *central peak*, the former absorption at zero frequency has since attracted many investigations although its origin was unidentified. Such a sharp peak as representing quasi-elastic scattering can be interpreted as due not only to an intrinsic decay mode, but also to extrinsic interactions with lattice imperfections in practical crystals. Measured decay times are typically of the order of 10^{-9} s, which actually falls in the limit of instrumental resolution. However, recent magnetic resonance studies show that such anomalies can be analyzed at least in part as related to an intrinsic mode at very low frequencies, which appears like a relaxation of the Debye type. In this case, the damping term in the equation of motion for the displacement $\boldsymbol{u}_{\pm q}$ can be attributed to two mechanisms: a coupling with the Debye mode, and an ordinary decay to the lattice. The equations of motion are then modified as

$$\frac{d^2\boldsymbol{u}_{\pm q}}{dt^2} + \gamma\frac{d\boldsymbol{u}_{\pm q}}{dt} + \delta\frac{d\boldsymbol{v}_{\pm q}}{dt} + \varpi^2\boldsymbol{u}_{\pm q} = \boldsymbol{F}_{\pm q}\exp(-i\omega t),$$

where

$$\frac{d\boldsymbol{v}_{\pm q}}{dt} + \frac{\boldsymbol{v}_{\pm q}}{\tau} = \boldsymbol{F}_{\pm q}\exp(-i\omega t)$$

represents the equation for the Debye relaxation mode $\boldsymbol{v}_{\pm q}$, the coefficient δ indicates the coupling with the main mode $\boldsymbol{u}_{\pm q}$, and the coefficient γ describes regular damping of $\boldsymbol{u}_{\pm q}$. The steady solutions of these equations are given by

$$\boldsymbol{u}_{\pm q} = (\boldsymbol{u}_{\pm q})_0\exp(-i\omega t),$$

where

$$(\boldsymbol{u}_{\pm q})_0\{-\omega^2 - i\omega\gamma + \varpi^2\} - i\omega\delta\cdot(\boldsymbol{v}_{\pm q})_0 = \boldsymbol{F}_{\pm q},$$

and

$$\boldsymbol{v}_{\pm q} = (\boldsymbol{v}_{\pm q})_0\exp(-i\omega t) \qquad \text{where} \qquad (\boldsymbol{v}_{\pm q})_0(-i\omega + \tau^{-1}) = \boldsymbol{F}_{\pm q}.$$

Although unidentifiable, such relaxational modes $\boldsymbol{v}_{\pm q}$ are coupled with the phonon mode $\boldsymbol{u}_{\pm q}$ at low frequencies, and so we can assume that $(\boldsymbol{v}_{\pm q})_0 = c(\boldsymbol{u}_{\pm q})_0$ in the critical region, where c is a constant. Under the circumstances, the susceptibility of $\boldsymbol{u}_{\pm q}$ is defined as

$$\chi_q(\omega) \propto \frac{\boldsymbol{u}_{\pm q}}{\boldsymbol{F}_{\pm q}} = \left[\varpi^2 - \omega^2 + i\omega\gamma + \frac{\delta c F_{\pm q}\omega\tau}{1 - i\omega\tau}\right]^{-1},$$

or by rewriting the factor $\delta c F_{\pm q}$ as δ^2

$$\chi_q(\Delta\omega) \propto \left[\varpi^2 - \omega^2 + i\gamma\omega + \frac{i\delta^2\omega\tau}{1 - i\omega\tau}\right]^{-1}. \tag{4.26}$$

Equation (4.26) represents the so-called *oscillator–relaxator* model, which has been used for numerical analysis of observed spectra. If the conditions

$$\gamma \ll \delta^2\tau \qquad \text{and} \qquad \varpi \gg \tau^{-1}$$

are fulfilled, the imaginary part of $\chi_q(\omega)$ can be expressed as

$$\chi_q''(\omega) = \frac{\omega}{\varpi^2 - \omega^2}\left\{\frac{\delta^2}{\varpi^2}\frac{\tau'}{1 + \omega^2\tau'^2} + \left(1 - \frac{\delta^2}{\varpi^2}\right)\frac{\varpi^2\gamma^2}{(\omega^2 - \varpi^2)^2 + \omega^2\gamma^2}\right\}, \tag{4.27}$$

where

$$\tau'^{-1} = \tau^{-1}\left(1 - \frac{\delta^2}{\varpi^2}\right).$$

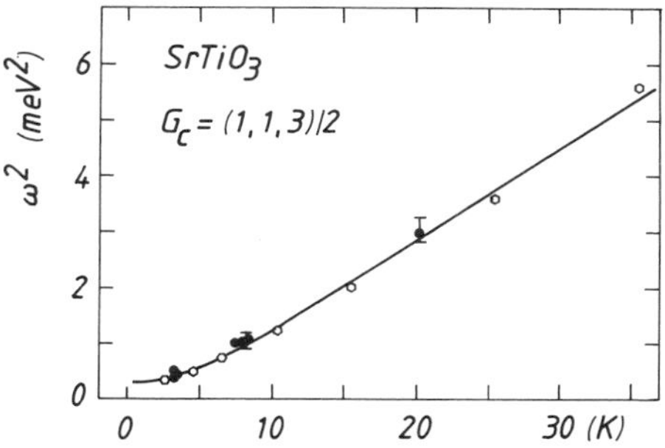

Figure 4.7. A linear plot of the squared soft-mode frequency ϖ^2 versus $T - T_c$ observed by neutron inelastic scattering from $SrTiO_3$ at $G_i = (1, 1, 3)/2$.

The second term on the right-hand side of (4.27) represents an absorption peak of the soft mode at $\omega = \varpi$, whereas the first term is the Debye relaxation that becomes prominent near zero frequency. It is noticed that the dielectric dispersion spectra in TSCC in Fig. 4.6(b) are dominated by such a relaxation term. A notable feature of (4.27) is that the soft mode is terminated at a nonzero frequency $\varpi = \delta$, which is then taken over by the relaxation mode at lower frequencies. While from such spectra it is not immediately clear if ϖ remains finite, as shown in Fig. 4.7, the linear extrapolation of observed ϖ^2 as a function of temperature showed an order-of-magnitude agreement with the value determined from the EPR anomalies. Sawada and Horioka [38] carried out a numerical estimate for the constants δ and τ from the dielectric response observed at $T_c + 6$ K from TSCC crystals, reporting that the values are 0.6 cm^{-1} and 0.9 cm^{-1}, respectively. According to their results, the terminal frequency of the soft mode above T_c can be evaluated at about 20 GHz, which is numerically consistent to the value determined by Fujimoto et al. [21] from the corresponding EPR anomalies in Mn^{2+} spectra. While these results were not sufficiently accurate, the terminal frequency of the soft mode is convincingly nonzero from the estimate of δ. While unidentified from dielectric studies, it is evident that the collective pseudospin mode is responsible for the EPR anomalies in TSCC, as discussed in Chapter 9. On the basis of these consistent results from different measurements, we consider that the condensate model is substantiated at least for the ferroelectric phase transition in TSCC.

4.7. Symmetry-Breaking Fluctuations of Pseudospins

In Section 4.1, we postulated that in the critical region the pseudospin mode can couple with the soft lattice mode as in (4.1), where the energy exchange $\mp\Delta\varepsilon$ is associated with small wavevectors $\pm\boldsymbol{q}$. Accordingly, a low-dimensional collective mode of pseudospins at small incommensurate wavevectors $\pm\boldsymbol{k}$

can interact with commensurate soft-lattice modes at $\pm\boldsymbol{k}'$, resulting in a modulated condensate at $\boldsymbol{k} - \boldsymbol{k}' = \pm\boldsymbol{q}$ and $\Delta\varepsilon$, where the fluctuation $\boldsymbol{q}$ can be in the range $|\boldsymbol{q}| < |\boldsymbol{k}|$, and $\Delta\varepsilon$ can be either positive or negative. For a fixed value of $\boldsymbol{q}$, the pseudospin mode may have wavevectors $\boldsymbol{k} \pm \boldsymbol{q}$ and $-\boldsymbol{k} \pm \boldsymbol{q}$ in the vicinity of $\pm\boldsymbol{k}$, and the corresponding kinetic energies are given by

$$\varepsilon_{+k\pm q} = \left(\frac{\hbar^2}{2m}\right)(\boldsymbol{k} \pm \boldsymbol{q})^2 \quad \text{and} \quad \varepsilon_{-k\pm q} = \left(\frac{\hbar^2}{2m}\right)(-\boldsymbol{k} \pm \boldsymbol{q})^2, \tag{4.28}$$

where m is the effective mass of the pseudospin condensate, and the pseudospin modes can be expressed as

$$\sigma_{+k\pm q} = \sigma_k(0) \exp\left\{i(\boldsymbol{k} \pm \boldsymbol{q})\cdot\boldsymbol{r} - \frac{\varepsilon_{+k\pm q}t}{\hbar}\right\} \quad \text{and}$$

$$\sigma_{-k\pm q} = \sigma_{-k}(0) \exp\left\{i(-\boldsymbol{k} \pm \boldsymbol{q})\cdot\boldsymbol{r} - \frac{\varepsilon_{-k\pm q}t}{\hbar}\right\}.$$

Figure 4.8(a) illustrates such fluctuations in one dimension at the minima of the energy dispersion curve $\varepsilon = \varepsilon_{\pm k}(K)$, where $K = \pm k \pm q$, consisting of two branches $\varepsilon_{\pm k \pm q}$, for which we assume $q \parallel k$ for simplicity. At this point, we realize that these pseudospin modes $\pm k$ represent propagating modes in two opposite directions, which are primarily independent from each other. Therefore, in this approximation there is no mechanism for reversing the wave-

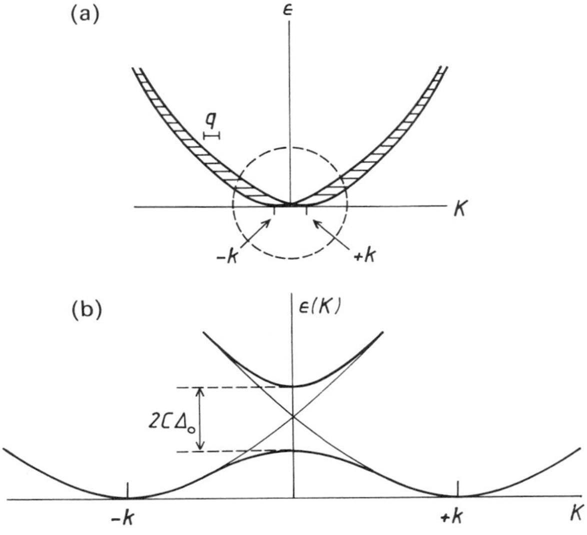

Figure 4.8. (a) Critical fluctuations at the minimum $K = 0$ of an energy dispersion curve $\varepsilon = \varepsilon(q)$. (b) A magnified view of the circled part in (a), showing a gap in the fluctuation energy.

vector direction, i.e., $k \leftrightarrow -k$. However, the two energy curves for $\varepsilon_{+k\pm q}$ and $\varepsilon_{-k\pm q}$ intersect at $K = 0$, where

$$\varepsilon_{+k}(q = -k) = \varepsilon_{-k}(q = +k) = \frac{2\hbar^2 k^2}{m},$$

and this *degeneracy* can be lifted, if perturbed at this point by an interaction with soft phonons. In the following we show that such a perturbation as arising from the quartic lattice potential will lift the degeneracy, resulting in two modes of fluctuations in different symmetries that are separated by an energy gap. One of these fluctuation modes is found as responsible for the propagation reversal.

In the condensate model, the normal modes $u_{\pm Q}$ of lattice vibration are considered to sustain the pseudospin modes $\sigma_{\pm k\pm q}$, when $\pm Q = \pm k \pm q$. The lattice vibrational energy composed of these normal modes is expressed as

$$U = \tfrac{1}{2}A\sum_Q u_Q{}^* u_Q + \tfrac{1}{4}B\sum_Q\sum_{Q'}\sum_{Q''}\sum_{Q'''} u_Q u_{Q'} u_{Q''} u_{Q'''} + \tfrac{1}{2}\kappa\sum_Q\left(\frac{\partial u_Q{}^*}{\partial x}\right)\left(\frac{\partial u_Q}{\partial x}\right) + \cdots, \tag{4.29}$$

where the displacement $u(x, t)$ is given by a combination of normal modes, i.e.,

$$u(x, t) = \sum_Q u_Q(x, t) = \sum_Q u_{\pm Q}(0)\exp\{i(\pm Qx - \omega t)\}.$$

Here each normal mode $u_Q(x, t)$ should be a *real* displacement, which is warranted by the invariance of U under space–time inversion $(x, t) \rightarrow (-x, -t)$. Therefore, the amplitude of the mode $u_Q(0)$ should be equal to its complex conjugate $u_Q(0)^*$, i.e.,

$$u_Q(0)^* = u_{-Q}(0). \tag{4.30}$$

In order for the second quartic terms in (4.29) to give secular perturbations to the unperturbed energy $U_0 = \frac{1}{2}A\sum_Q u_Q{}^* u_Q$, these wavevectors should obey the conservation law

$$Q + Q' + Q'' + Q''' = G$$

at a time t, where G is a reciprocal lattice vector. As in the Cowley theory, we consider quartic perturbations for the phonon scattering $\mathbf{k} + (-\mathbf{k}) = \mathbf{k}' + (-\mathbf{k}')$, where the wavevectors $\mathbf{k}$, $-\mathbf{k}$, $\mathbf{k}'$, and $-\mathbf{k}'$ represent Q, Q', Q'', and Q''' in (4.29), respectively. To evaluate such quartic pertubations, we use the so-called Wick approximation [39] for replacing the product $u_Q u_{Q'} u_{Q''} u_{Q'''}$ by $u_Q u_{Q'}\langle u_{Q''} u_{Q'''}\rangle$, where

$$\langle u_{Q''} u_{Q'''}\rangle = \langle u_{k'}(0)^2\rangle = -\frac{A}{B}$$

is the mean-field average of squared amplitudes of the normal lattice modes $\pm\mathbf{k}'$. In this approximation, only normal modes at Q and Q' are required for fluctuations at $\pm k \pm q$, and the quartic energy for a fixed value of q can be expressed explicitly in terms of many time-independent products, i.e.,

$$\tfrac{1}{4}B\langle u_{k'}(0)^2\rangle[u_{k\pm q}{}^* u_{k\pm q} + u_{-k\pm q}{}^* u_{-k\pm q}$$
$$+ (u_{k+q}{}^* u_{-k+q} + u_{k-q}{}^* u_{-k-q} + u_{k+q}{}^* u_{-k-q} + u_{k-q}{}^* u_{-k+q})$$
$$+ (u_{-k+q}{}^* u_{k+q} + u_{-k-q}{}^* u_{k-q} + u_{-k-q}{}^* u_{k+q} + u_{-k+q}{}^* u_{k-q})],$$

where the normal modes are $u_{k+q} = u_0 \exp\{i(k + q)x - \omega t\}$, $u_{k-q} = u_0 \exp\{i(k - q)x + \omega t\}$, etc. Therefore,

$$U_4 = -\tfrac{1}{2}Au_0{}^2[2 + 2\cos(2kx) + \cos\{2(k+q)x - 2\omega t\}$$
$$+ \cos\{2(k-q)x + 2\omega t\}].$$

Here, the three terms that depend on space–time coordinates (x, t) are considered as responsible for dynamical lattice strains, while the first constant term keeps the lattice uniform. In this context, we may write the strain energy as

$$\Delta(x, t) = \Delta_0[2\cos(2kx) + \cos\{2(k+q)x - 2\omega t\} + \cos\{2(k-q)x + 2\omega t\}], \tag{4.31a}$$

and express its effect to the pseudospin fluctuation by the perturbing potential

$$V(x, t) = C\Delta(x, t). \tag{4.31b}$$

Being typical for *level crossing*, the solution of such a perturbation problem can be found in many standard references, e.g., in Kittel's textbook [8]. For pseudospin fluctuations between the states ε_{+k-q} and ε_{-k+q} in the vicinity of $K = 0$, the pseudospin $\sigma(x, t)$ is expressed by a linear combination of two pseudospin fluctuation modes for the wavevectors $+k - q$ and $-k + q$, i.e.,

$$\sigma(x, t) = c_+ \exp[i\{(+k - q)x - \omega t\}] + c_- \exp[i\{(-k + q)x + \omega t\}]$$

or

$$= c_+ \exp\{i(Kx - \omega t)\} + c_- \exp\{-i(Kx - \omega t)\}, \tag{4.32}$$

where the unperturbed energy difference

$$\varepsilon_{-k+q} - \varepsilon_{k-q} = \varepsilon_{-K} - \varepsilon_K = 2\hbar\omega \tag{4.33}$$

is positive for $K > 0$, and negative for $K < 0$. Here the coefficients c_+ and c_- are to be determined by the variational principle. It is noticed that fluctuation between (K, ω) and $(-K, -\omega)$ can be discussed in terms of the phase $\phi = Kx - \omega t$, and from (4.31a,b) the pertubation potential

$$V(x, t) = C\Delta_0[\cos\{2\phi + 2(qx + \omega t)\} + \cos\{2\phi + 4(qx + \omega t)\} + \cos 2\phi].$$

Clearly, only the last term, $V(\phi) = C\Delta_0 \cos 2\phi$ that synchronizes with $\phi \leftrightarrow$

$-\phi$, is required for the perturbing off-diagonal matrix element, which is calculated as

$$\int_0^{2\pi} \exp(-i\phi) V(\phi) \exp(-i\phi)\} \, d\phi \Big/ \int_0^{2\pi} d\phi = \frac{C\Delta_0}{2\pi} \int_0^{2\pi} \cos(2\phi) \exp(-2i\phi) \, d\phi$$

$$= \frac{C\Delta_0}{2\pi} \int_0^{2\pi} \cos^2(2\phi) \, d\phi = \left(\frac{C}{8\pi}\right) \Delta_0.$$

For brevity, the factor $C/8\pi$ by C' in the following.

The perturbed energies can be calculated from the secular equation

$$\begin{vmatrix} \varepsilon(-\phi) - \varepsilon & C'\Delta_0 \\ C'\Delta_0 & \varepsilon(+\phi) - \varepsilon \end{vmatrix} = 0. \tag{4.34}$$

The roots of (4.34) are given by

$$\varepsilon_\pm = \tfrac{1}{2}\{\varepsilon(-\phi) + \varepsilon(+\phi)\} \pm [\tfrac{1}{4}\{\varepsilon(-\phi) - \varepsilon(+\phi)\}^2 - (C'\Delta_0)^2]^{1/2}. \tag{4.35}$$

Since $\varepsilon(\phi) = \varepsilon(-\phi) = \hbar^2 K^2/2m$, these fluctuation energies are expressed by

$$\varepsilon_\pm = \frac{\hbar^2 K^2}{2m} \pm C'\Delta_0 \qquad \text{and} \qquad \Delta\varepsilon = \varepsilon_+ - \varepsilon_- = 2C'\Delta_0.$$

The ratio c_+/c_- between the coefficients in (4.32) is either $+1$ or -1, expressing symmetric and antisymmetric combinations of the spatial fluctuations $\exp(\pm i\phi)$, corresponding to the $+$ and $-$ signs for the perturbed energies $\varepsilon_\pm$, respectively. At the origin $K = 0$, there is an energy gap between these states, which is given by $\Delta\varepsilon$ as shown in Fig. 4.8(b) illustrating the magnified area around the intersection in Fig. 4.8(a).

In the above, we showed that the spatial fluctuation occurs between two states of propagation $+K$ and $-K$ along the x direction. Considering the corresponding frequencies $-\omega = -\varepsilon_-/\hbar$ and $\omega = \varepsilon_+/\hbar$, it is equivalent to say that the fluctuations are unchanged under space–time inversion $(x, t) \to (-x, -t)$ or phase reversal $\phi \to -\phi$. For a binary transition however, the space–time invariance is not sufficient, and an additional character of the vector pseudospin needs to be taken into consideration. Namely, the vector pseudospin should reverse the direction by reflection on the mirror plane perpendicular to $\boldsymbol{K}$, i.e.,

$$\sigma_K(\phi) \to -\sigma_{-K}(-\phi). \tag{4.36a}$$

Combining this condition with the phase reversal, the phase in a binary fluctuation should change according to

$$\phi \to \pi - \phi. \tag{4.36b}$$

Symmetrical and antisymmetrical fluctuation modes are therefore expressed at the center $\phi = 0$ as

$$\sigma_A = \frac{\sigma(0)\{\exp(i\phi) + \exp[i(\pi - \phi)]\}}{2} = \sigma(0) i \sin\phi \tag{4.37a}$$

and

$$\sigma_{\text{P}} = \frac{\sigma(0)\{\exp(i\phi) - \exp[i(\pi - \phi)]\}}{2} = \sigma(0)\cos\phi, \qquad (4.37\text{b})$$

respectively, where the amplitudes are equal, i.e., $\sigma(0) = \sigma_{+K}(0) = \sigma_{-K}(0)$. Such symmetric and antisymmetric modes of pseudospins σ_{A} and σ_{P} are called the amplitude mode (or *amplitudon*) and the phase mode (or *phason*), respectively.

As is clear from the foregoing argument, critical fluctuations in binary systems are primarily due to pseudospins in the phase mode, where $\sigma_{\text{P}}(\phi) \leftrightarrow -\sigma_{\text{P}}(-\phi)$, whereas in the amplitude mode the fluctuation given by $\sigma_{\text{A}}(\phi) \rightarrow \sigma_{\text{A}}(-\phi)$ is separated in energy by $\Delta\varepsilon$. Both amplitude and phase modes of fluctuations in the vicinity of $K = 0$ are involved in discrete energy variation between ε_+ and ε_- that are proportional to the strain amplitude Δ_0 arising from interaction with the lattice.

In neutron scattering experiments, the intensity is determined by a normal mode of the lattice vibration. It is significant that unlike pseudospins, normal lattice modes are not subjected to reflection symmetry, and therefore symmetric and antisymmetric fluctuations in lattice displacements should be formed as related to invariance under the phase reversal $\phi \rightarrow -\phi$, i.e.,

$$u_{\text{A}} = \frac{u(0)\{\exp(i\phi) + \exp(-i\phi)\}}{2} = u(0)\cos\phi \qquad (4.38\text{a})$$

and

$$u_{\text{P}} = \frac{u(0)\{\exp(i\phi) - \exp(-i\phi)\}}{2} = iu(0)\sin\phi. \qquad (4.38\text{b})$$

Here, the fluctuation frequency given by $\omega = \kappa K^2$ is low but nonzero so that time averages $\langle u_{\text{A}}\rangle_{\text{t}}$ and $\langle u_{\text{P}}\rangle_{\text{t}}$ do not generally vanish in the timescale of neutron impact, giving finite scattering intensities for the two distinct modes of fluctuations in the critical region.

Fluctuating pseudospin modes expressed by (4.37a,b) can be sampled by magnetic resonance probes, as will be discussed in Chapter 9, whereas fluctuating lattice modes (4.38a, b) are responsible for anomalous neutron scattering. In magnetic resonance spectra, the amplitude mode σ_{A} proportional to $\sin\phi$ may be observed as a single line, whereas the phase mode σ_{P} related to $\cos\phi$ exhibits a broadened line when observed with the timescale t_0 shorter than $2\pi/\omega(K)$. On lowering temperatures, we see that K, ω, and ϖ tend to zero with diminishing fluctuations due to increasing correlations with distant pseudospins, resulting in a nonlinear phase mode.

Neutron inelastic scattering results obtained by Bernard et al. [40] for the phase transition in β-ThBr_4 crystals at 81 K (Fig. 4.9(a)) and for biphenyl crystals at 41.5 K [41] were unique among others in that the presence of two modes of fluctuations, u_{A} and u_{P}, were clearly identified in the critical regions. Figure 4.9(b) also shows an example for the two modes near the ferroelectric

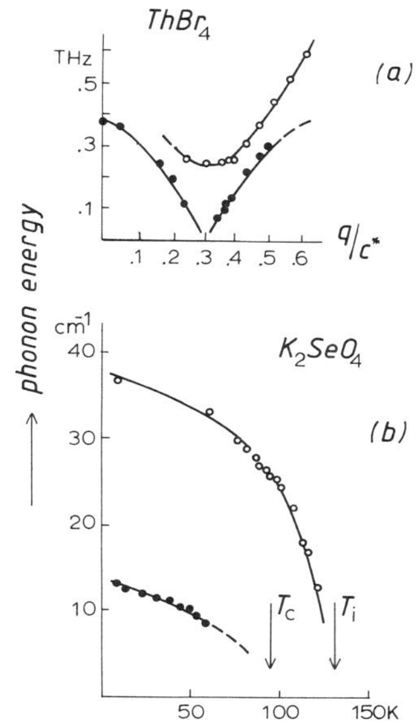

Figure 4.9. (a) The phonon dispersion curve of $ThBr_4$ crystals at 81 K, where the amplitude- and phase-modes of critical fluctuations are clearly resolved. (From L. Bernard, R. Currat, P. Delanoye, C. M. E. Zeyen, S. Hubert, and R. Kouchkovsky, *J. Phys.* **C16**, 433 (1983).) (b) Phonon energies observed by Raman scattering from K_2SeO_4. Between T_i and T_c the phase is incommensurate, while ferroelectric below T_c. (From M. Wada, H. Uwe, A. Sawada, Y. Ishibashi, Y. Takagi, and T. Fakudo, *J. Phys. Soc. Japan*, **43**, 544 (1977).)

phase transition in K_2SeO_4 crystals at 95 K that were identified in Raman scattering experiments by Wada and his group [42]. In their studies the phase mode above 95 K was not detected, presumably due to the fact that the order parameter is not Raman-active in the phase between 95 K–129 K. In spite of these successful experiments, the problem of central peaks is still unsolved, remaining further to be investigated for better understanding of the transition mechanism. While so far limited to only a few cases, the model for pseudospin condensates could be generalized to other structural phase transitions exhibiting critical anomalies.

4.8. Macroscopic Properties of Pseudospin Condensates

A significant feature of pseudospin condensates is their thermal stability in the critical region, thereby being observed as stable objects propagating through the lattice. Such a stability is attributed to low damping of the soft mode that is coupling with the pseudospins in nearly equal phase, where energy and momentum exchanges between them are responsible for critical anomalies. Hence the wavevector of such fluctuations is diverse in the range $0 \leq q \leq k$, and macroscopic properties are determined in relation to integrated fluctuations over q.

Thermodynamically, the equilibrium of pseudospin condensates in the critical region of a phase transition is determined by minimizing the Helmholtz free energy under a constant volume condition. We discussed in Section 4.7 that the soft mode of a long wavelength is responsible for lattice strains described by a potential $\Delta(\phi)$ arising from a quartic anharmonic potential for pseudospin correlations. Owing to low damping of the soft mode, such a condensate is considered as thermally well isolated from the rest of the crystal. Under the circumstances, an approximate equilibrium between pseudospins and soft phonons is established as expressed by

$$\left(\frac{\mathrm{d}}{\mathrm{d}\Delta_0}\right)(U_\sigma + U_{\mathrm{st}}) = 0, \tag{4.39}$$

where U_σ and U_{st} represent the correlation energy of pseudospins and the strain energy in the lattice, respectively. Here, we assume that U_{st} is related to the mean-field average of the periodic potential $\Delta(x)$ of (4.31), namely,

$$U_{\mathrm{st}} = \tfrac{1}{2}\langle \alpha^2 \Delta^2 \rangle = \tfrac{1}{2}\alpha^2 {\Delta_0}^2 \langle \cos^2(2kx) \rangle = \tfrac{1}{4}\alpha^2 {\Delta_0}^2, \tag{4.40}$$

where α is the proportionality constant.

Equation (4.39) deals with the equilibrium between the pseudospin system and the lattice strains, although no detailed mechanism is specified. On the other hand, it is known that the soft-mode energy decaying as the transition is approached from above is eventually converted to another soft mode with an increasing frequency, as T_c is passed through. In the condensate model, the critical region below T_c is dominated by the fluctuations in phase mode, leading eventually to domain structure in the low-temperature phase, while the pseudospin fluctuations in amplitude mode may persist beyond T_c but decay with increasing temperature, as illustrated in Fig. 4.10. In this context, the second-order phase transition can be interpreted as a continuous equilibrium process between pseudospins and soft phonons.

Fluctuation energies of the pseudospin modes between two states $\pm k$ are given by (4.35), which can be reexpressed with abbreviated notations

$$x_q = \frac{\hbar^2 q^2}{m}, \qquad x_k = \frac{\hbar^2 k^2}{m} \qquad \text{and} \qquad x = \frac{\hbar^2 qk}{m},$$

as

$$\varepsilon_\pm(q, \Delta_0) = \tfrac{1}{2}(x_q + x_k) \pm (x_q x_k + C'^2 {\Delta_0}^2)^{1/2}.$$

From this, we obtain

$$\frac{\mathrm{d}\varepsilon_\pm(q, \Delta_0)}{\mathrm{d}\Delta_0} = \pm \frac{C'^2 \Delta_0}{(x_q x_k + C'^2 {\Delta_0}^2)^{1/2}}.$$

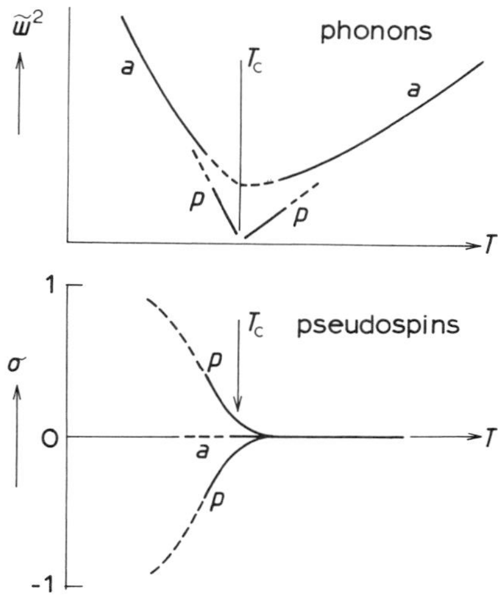

Figure 4.10. Phonon dispersion curves and the corresponding pseudospin mode as a function of temperature. The fluctuations in amplitude and phase modes are indicated by *a* and *p*.

For the calculation of U_σ in (4.38) below T_c we only have to take $\varepsilon_-(q, \Delta_0)$ into account, and hence

$$\frac{dU_\sigma}{d\Delta_0} = 2\int_0^k \frac{d\varepsilon_-(q, \Delta_0)}{d\Delta_0}\frac{dq}{\pi},$$

where dq/π represents the number of states between q and $q + dq$ in a one-dimensional crystal. Therefore,

$$\begin{aligned}\frac{dU_\sigma}{d\Delta_0} &= -\frac{2C'^2\Delta_0}{\pi}\int_0^k (x_q x_k + C'^2\Delta_0{}^2)^{-1/2}\,dq\\ &= -\frac{2C'^2\Delta_0}{\pi}\frac{k}{x_k}\int_0^{x_k}(x^2 + C'^2\Delta_0{}^2)^{-1/2}\,dx\\ &= -\frac{2C'^2\Delta_0}{\pi}\frac{k}{x_k}\sinh^{-1}\left(\frac{x_k}{C'\Delta_0}\right).\end{aligned}$$

Using this result and (4.39) in (4.38) for the equilibrium condition, we obtain the following expression:

$$\tfrac{1}{2}\alpha\Delta_0 - \frac{2C'^2 m\Delta_0}{\pi\hbar^2 k}\sinh^{-1}\left(\frac{\hbar^2 k^2}{mC'\Delta_0}\right) = 0.$$

From here, the value of Δ_0 that corresponds to minimum free energy is given by

$$\frac{\hbar^2 k^2}{mC'\Delta_0} = \sinh\left(\frac{\alpha\pi\hbar^2 k}{4mC'^2}\right) \approx \tfrac{1}{2}\exp\left(\frac{\alpha\pi\hbar^2 k}{4mC'^2}\right),$$

if the argument $\alpha\pi\hbar^2 k/4mC'^2 \gg 1$. In this case,

$$|C'|\Delta_0 \cong \frac{2\hbar^2 k^2}{m} \exp\left(\frac{-\alpha\pi\hbar^2 k}{4mC'^2}\right), \tag{4.41}$$

where $\hbar^2 k^2/2m = \varepsilon_-(k, \Delta_0)$ is the maximum kinetic energy of fluctuations at $K = \pm k$. It is clear from (4.40) that $k = 0$ if $\Delta_0 = 0$, in other words, a binary transition characterized by small but finite $\pm k$ must be associated with a finite gap $2|C'|\Delta_0$ in the pseudospin energy spectrum. Sudden emergence of such an energy gap can explain, at least qualitatively, the specific heat anomaly of λ type observed typically at many continuous-phase transitions, for which however the temperature-dependence of the wavevector k is required to compare the theory with experimental results. It is realized that such an energy gap as described by (4.40) is similar to that predicted in the BCS theory of superconductors.

Figure 4.10 summarizes schematically how pseudospins are coupled with soft phonons in the critical region. While soft modes of two distinct symmetries signify a symmetry change in the phase transition, the low-temperature phase is characterized as continuous ordering from the high-temperature phase. The critical fluctuations in pseudospins are slow as coupled with the soft mode, so that they are responsible for the spatially modulated structure that is nearly static. Thus, phase transitions are caused by coupled soft modes and pseudospins, and neither of these can be solely responsible for the transition mechanism.

CHAPTER 5

Pinning and Dynamics of Pseudospin Condensates in Practical Crystals

5.1. Introduction

In the critical region pseudospin condensates are in such a sinusoidal form as $\sigma_0 \cos \phi$ and $\sigma_0 \sin \phi$, where $\phi = qx - \varpi t + \phi_0$, propagating through the lattice at a speed $v = \varpi/q$. In an idealized "perfect crystal" the phase constant ϕ_0 is undetermined, unless a boundary condition is imposed on the phase ϕ at certain space–time coordinates (x_0, t_0) in the lattice. It is realized that small values of q and ϖ as determined from light and neutron inelastic scattering experiments do not constitute evidence for the condensate as a moving object through the lattice. To substantiate its presence in practical crystals, the condensate must be directly *sampled* by using, for instance, magnetic resonance probes substituted for constituent ions in the active groups. Even so, a propagating condensate could not be properly sampled in the laboratory frame of reference, unless the *observer* is moving at the same speed v as the object.

On the other hand, practical crystals are by no means "perfect" in that there are always boundary surfaces and unavoidable imperfections disrupting the lattice periodicity. In some cases however, practical crystals may be considered as if perfect, if observed with moving probes with a sufficiently high kinetic energy as compared with the depth of imperfection potentials. In contrast, for pseudospin condensates with low energies the imperfections can be significant obstacles to propagation, and condensates are likely immobilized (or *pinned*) in the vicinity of stationary imperfections. Nevertheless, pinned condensates can be sampled in the frame of reference fixed in the crystal. In practice, it is not possible to study pinned condensates in crystals of mediocre quality. The density of defects and impurities has to be reduced to a low level to obtain reliable information about condensates, and sample crystals need to be characterized for the content of imperfections. For example, in ferroelectric crystals, the quality of sample crystals can be specified by the internal bias field to indicate the defect density, which can therefore be utilized for assessing the quality of ferroelectric crystals. Experimental results can normally be justified as related to long-wave condensates pinned by stationary lattice defects, if the sample has a sufficiently low defect density.

We consider that *point imperfections* prevail in structure-sensitive properties of high-quality crystals. By point imperfections we mean that the translational lattice symmetry is disrupted only by well-localized defective sites. Although imperfections of other types cannot be entirely ruled out, their role in a quality sample may be considered as trivial as compared with point defects. The perfect crystal is only a theoretical model, but it is significant that the presence of such imperfections play an essential role in practical crystals for propagation through the lattice structure.

During an ordering process, pseudospin correlations are generally extended beyond nearest neighbors to distant lattice sites, where their collective motion can no longer be sinusoidal. In contrast to the transition threshold, the collective pseudospin mode has a finite amplitude that signifies the degree of ordering, while the propagation is still described by the phase ϕ. We can show with a simplified model that a collective pseudospin mode is generally expressed in a form $\sigma = \sigma_0 f(\phi)$, where both the amplitude σ_0 and the phase ϕ are functions of temperature. Although such a model is primarily designed for mathematical simplification, a collective pseudospin mode in anisotropic crystals can often be found in low dimension, for which even a simplified model is acceptable as a valid approximation. Although difficult to deal with the temperature-dependence of pseudospin correlations, pinning and dynamical aspects of condensates can be reasonably discussed with a model of point defects that disrupt translational symmetry of the lattice.

5.2. The Pinning Potential

In this section we discuss only the critical region, where the collective mode of pseudospins is sinusoidal. In general, a stationary defect at a lattice point $\boldsymbol{r}_i$ can be represented by a local field $\boldsymbol{F}(\boldsymbol{r} - \boldsymbol{r}_i)$ at a point $\boldsymbol{r}$ close to $\boldsymbol{r}_i$, reflecting symmetry of the defect site that constitutes a subgroup of the point group of the crystal. Although only vaguely defined, the field $\boldsymbol{F}(\boldsymbol{r} - \boldsymbol{r}_i)$ represents local distortion in the vicinity of the defect site $\boldsymbol{r}_i$, influencing nearby pseudospins σ_j at sites $\boldsymbol{r} = \boldsymbol{r}_j$ ($j \neq i$). Considering the problem in one dimension, the attractive force between a condensate $\sigma = \sigma(\phi)$ and the defect field $F(x - x_i)$ may be described by the product $\sigma(\phi)F(x - x_i)$, which can be expressed in terms of the potential $V(x, t; x_i)$, i.e.,

$$dV(x, t; x_i) = -\sigma(\phi)F(x - x_i)\,dx.$$

We assume that the field F is symmetrical with respect to the defect coordinate x_i, namely, $F(x - x_i) = F(x_i - x)$, although the defect symmetry may be lower if the defect center x_i is not at its original lattice site. Further, the defect field F is assumed as highly localized in the vicinity of x_i, allowing us to write

$$F(x - x_i) = F\delta(x - x_i), \tag{5.1}$$

where the *delta* function is $\delta(x - x_i) = 1$ for $x = x_i$ or $= 0$ otherwise, and F represents the magnitude of the field at $x = x_i$. We can then define the *pinning potential* at x_i and t due to the distributed defect field by the integral $\int \{\partial V(x, t; x_i)/\partial x\}\, \mathrm{d}x$. For a binary system, we can write such pinning potentials for pseudospins in amplitude mode and in phase mode as

$$V_A(x_i, t) = -\int_{-\infty}^{\infty} \sigma_0 F \sin(qx - \varpi t + \phi_0)\delta(x - x_i)\, \mathrm{d}x$$

and

$$V_P(x_i, t) = -\int_{-\infty}^{\infty} \sigma_0 F \cos(qx - \varpi t + \phi_0)\delta(x - x_i)\, \mathrm{d}x,$$

respectively. Clearly for such a localized defect,

$$V_P(x_i, t) = -\sigma_0 F \cos(qx_i - \varpi t + \phi_0), \tag{5.2a}$$

whereas

$$V_A(x_i, t) = 0. \tag{5.2b}$$

The defect phases $\phi_i = qx_i - \varpi t + \phi_0$ are randomly distributed in the crystal. Owing to a small incommensurate vector q and defect coordinates x_i distributed among regular lattice points, the spatial phase qx_i is virtually continuous over the whole angle 2π repeatedly. Therefore, instead of random phases ϕ_i, we can define a continuous phase $\phi = qx - \varpi t + \phi_0$, where

$$0 \leq \phi \leq 2\pi, \tag{5.3}$$

for the whole system of pinned condensates. The phase ϕ is a convenient variable to express spatial inhomogeneity in a practical crystal, particularly when assumed as a continuum. Using such a phase ϕ "common" to all the defects, we can rewrite (5.2a,b) as

$$V_P(\phi) = -V_0 \cos\phi \qquad \text{and} \qquad V_A(\phi) = 0, \tag{5.4}$$

where $V_0 = \sigma_0 F$, and the negative sign signifies the attractive potential. Equations (5.4) are generally referred to as the *pinning potential.*

The pinning potential $V_P(\phi)$ of (5.4) indicates that condensates in phase mode can be in equilibrium with defects at the minimum given by $\phi = 0$, if the kinetic energies are sufficiently low. On the other hand, the equation $V_A(\phi) = 0$ indicates that there is no equilibrium for condensates in amplitude mode. It is significant that for both modes, pinned condensates are described by the continuous phase ϕ, while behaving as independent of each other.

Dynamically, a pinned phase mode σ_P should be in oscillatory motion fluctuating around the equilibrium, for which the restoring force

$$f_R = -\frac{\partial V_P(\phi)}{\partial x} = -\frac{q\,\partial V_P(\phi)}{\partial \phi} = -q\sigma_0 F \sin\phi$$

is considered as responsible. Therefore, the pinning equilibrium $\phi = 0$ for σ_P can be determined by $f_R = 0$. Consequently, an oscillatory fluctuation of the pinned phase mode in the vicinity of $\phi = 0$ can be described by a small variation $\phi - 0 = \delta\phi$ in the potential

$$V_P(\delta\phi) = -V_0 \cos\delta\phi. \tag{5.5}$$

For an infinitesimal phase fluctuation $\delta\phi$, the variation of the pinning potential is given approximately by

$$\Delta V_P = V_P(\delta\phi) - V_P(0) = \tfrac{1}{2}V_0(\delta\phi)^2. \tag{5.5a}$$

Rice [43] discussed such phase fluctuations of the charge-density-wave condensate pinned by a potential given by (5.5a) in the presence of an applied oscillatory field $E = E_0 \exp(-i\omega t)$, and derived the corresponding susceptibility formula. His result can be transferred directly to pinned pseudospin condensates during a structural-phase transition. Ignoring damping for simplicity, he wrote the equation of oscillatry motion for $\delta\phi$ in the potential of (5.5a) as

$$\frac{m}{q}\frac{d^2(\delta\phi)}{dt^2} = eE_0 \exp(-i\omega t) - \left(\frac{m\omega_0{}^2}{q}\right)\delta\phi,$$

where e is the effective charge of the condensate. Letting $\delta\phi = (\delta\phi)_0 \exp(-i\omega t)$, the steady solution of this equation can be given by the susceptibility formula

$$\chi(\omega) \propto \frac{(\delta\phi)_0}{E_0} = \frac{e/m}{\omega_0{}^2 - \omega^2}. \tag{5.6}$$

In a polar crystal such a susceptibility $\chi(\omega)$ represents the dielectric response of pinned pseudospins, showing a singularity at $\omega = \omega_0$. Pawlaczyk and Unruh [44] discovered a very low-frequency mode at $\omega_0 \approx 0.1$ GHz in dielectric spectra from TSCC at temperatures near T_c, which was attributed to such an oscillatory mode as described by (5.6). Here the frequency ω_0 is not the same as the characteristic frequency ϖ for propagation, but signifies pinned condensates. Due to the small value of $\omega_0 \ll \varpi$, pinned condensates are observed as a standing wave with maximum intensity at $\phi = 0$, however it is realized that such an observation is possible only if the period of oscillation $2\pi/\varpi$ is sufficiently long compared with the sampling timescale t_0. The detail for sampling experiments is to be discussed in Chapters 7 and 9.

In a ferroelectric crystal it is notable that an applied static electric field will influence pinning of polar condensates so that the amplitude mode can be identified by means of its behavior in the applied field. In the presence of a weak uniform electric field E, the pinning potentials in the vicinity of the equilibrium phase should be written as

$$V_P(\delta\phi, E) = -V_0 \cos \delta\phi - \sigma_0 \int_{-\infty}^{\infty} E \cos \delta\phi \, dx$$

and

$$V_A(\delta\phi, E) = -\sigma_0 \int_{-\infty}^{\infty} E \sin \delta\phi \, dx$$

for the pseudospins in phase and amplitude modes, respectively. It is noticed that in a uniform field $E = -dV/dx = -q\, dV/d\phi$, we consider that the integration can be limited to a narrow range $\delta\phi/q > x > -\delta\phi/q$, where the voltage variation is $V_0 + \Delta V > V > V_0 - \Delta V$. Then, the above integrals can be evaluated as

$$\int_{-\infty}^{\infty} E \cos \delta\phi \, dx = -\int_{-\delta\phi}^{\delta\phi} \frac{dV}{d\phi} \cos \delta\phi \, d\phi \approx -\Delta V \cos(\delta\phi) - (-\Delta V) \cos(-\delta\phi)$$
$$= 0$$

and

$$\int_{-\infty}^{\infty} E \sin \delta\phi \, dx = -\int_{-\delta\phi}^{\delta\phi} \frac{dV}{d\phi} \sin \delta\phi \, d\phi \approx -\Delta V \sin(\delta\phi) - (-\Delta V) \sin(-\delta\phi)$$
$$= -2E \sin \delta\phi.$$

Hence, $V_p(\delta\phi, E)$ is virtually unchanged except that $\phi = -\delta\phi$ gives a new equilibrium, whereas the potential $V_A(\delta\phi, E)$ is no longer equal to zero for $\delta\phi \neq 0$ in a field $E \neq 0$. Furthermore, for $E \neq 0$ the equilibrium is specified by $\delta\phi = \pm\frac{1}{2}\pi$, resulting in $\partial V_A(\delta\phi, E)/\partial\phi \propto \cos(\pm\frac{1}{2}\pi) = 0$. Therefore, for σ_A at these new equilibria $\phi = \pm\frac{1}{2}\pi$ in a given field E, the pinning potential can be expressed as

$$V_A(\tfrac{1}{2}\pi \pm \delta\phi, E) = -\sigma_0 E \sin(\tfrac{1}{2}\pi \pm \delta\phi) = -\sigma_0 E \cos(\delta\phi), \qquad (5.7)$$

which is similar in form to $V_P(\delta\phi, E)$, except for the magnitude proportional to E. In fact, the amplitude mode σ_A in ferroelectric TSCC crystals was identified by such pinning by an applied field [21], [45]. The experimental detail will be described in Chapter 9.

5.3. Collective Pseudospin Dynamics in a One-Dimensional Lattice

A quasi-one-dimensional lattice composed of many identical double-well potentials has been discussed as a theoretical model for displacive structural changes. In each double-well potential expressed by (3.1b) for $\sigma_{mx} = \sigma_{my} = 0$, the component σ_{mz} of the pseudospin $\boldsymbol{\sigma}_m$ represents a displacement of the effective mass particle from the center of the site m. In the normal phase, the potential for $\boldsymbol{\sigma}_m$ given by (3.1a) is quadratic, implying that the crystal structure is stable. Above T_c, $\boldsymbol{\sigma}_m$ is therefore harmonic in the restoring potentials $V_>(\boldsymbol{\sigma}_m)$, while a long-wave lattice excitation in $\boldsymbol{u}_m$ is illustrated schematically in Fig. 4.1(a), $\langle \sigma_{mz} \rangle_t = 0$ and $\langle \boldsymbol{\sigma}_m \rangle$ is effectively zero at the center of each active group. In contrast, below T_c a long wave σ_{mz} in the low-temperature phase is illustrated as shown in Fig. 4.1(b), assuming that $\boldsymbol{u}_m = 0$ and the lattice structure remains unchanged from the normal phase. However, in order for such a crystal to be free from strains, each active group should be displaced by $\boldsymbol{u}_m$ to offset the strains due to $\boldsymbol{\sigma}_m$ as shown in Fig. 4.1(c), where out-of-phase displacements $\boldsymbol{u}_m$ and $\boldsymbol{\sigma}_m$ are illustrated for clarity. According to the Born–Huang theory, such an interaction between $\boldsymbol{u}_m$ and $\boldsymbol{\sigma}_m$ is conceivable, suggesting that $\boldsymbol{u}_m$ should represent a low-energy excitation of the lattice. In the following it is shown that the dynamical equation for σ_{mz} in a rigid lattice ($\boldsymbol{u}_m = 0$) has a sinusoidal solution with infinitesimal amplitude. Such a solution was often interpreted as representing a phonon mode, which is however incorrect in the present model, since the motion of $\boldsymbol{\sigma}_m$ is unrelated with the mass of an active group. Nevertheless, dynamical solutions for σ_{mz} should be discussed to see if stable condensates are represented in a strained lattice.

Krumhansl and Schrieffer [46], and Aubry [47], discussed the following Hamiltonian of a dynamical system consisting of an infinite number of particles of mass m in double-well potentials located at every lattice point of a one-dimensional rigid chain. Notice that here m is the effective mass of a particle, and is not the mass of the active group. Namely,

$$\mathscr{H} = \sum_n \left(\frac{p_n{}^2}{2m} + \tfrac{1}{2} a \sigma_n{}^2 + \tfrac{1}{4} b \sigma_n{}^4 \right) + \tfrac{1}{2} C \left\{ \sum_n (\sigma_{n+1} - \sigma_n)^2 + (\sigma_n - \sigma_{n-1})^2 \right\}, \tag{5.8}$$

where the first bracketed term on the right-hand side represents the energy of individual particles, and elastic interactions between σ_n and its nearest neighbors σ_{n+1} and σ_{n-1} are assumed here as expressed by the second term. In (5.8), the coefficients a_n and b_n are just constants for the double-well potentials, but in Section 3.1 we interpreted that these quartic coefficients b_n are related to pseudospin correlations. Furthermore, according to the Landau thermodynamical theory, we can consider that disordered and ordered phases are distinguishable in terms of the signs of these constants, i.e., $a > 0$, $b = 0$ for $T > T_c$ and $a < 0$, $b > 0$ for $T < T_c$. The elastic interactions in this model can be interpreted as equivalent to pseudospin correlations given by (3.3), if the second term in (5.8) is written as

$$\tfrac{1}{2} C \left(\sum_n \sigma_n{}^2 \right) - C \sum_{n,n'} \sigma_n \sigma_{n'},$$

where the first term can be rearranged as included partially in the quadratic potential $\frac{1}{2} a \sigma_n{}^2$. For infinitesimal σ_n's, such interactions cause no anharmonic effects, while finite amplitudes arising from correlations with distant $\sigma_{n'}$'s make the quartic potentials $\frac{1}{4}b\sigma_n{}^4$ significant, as discussed in Section 3.3.

For a continuum chain crystal, the Hamiltonian in (5.8) can be written in the form

$$\mathscr{H} = \int_0^L \frac{\mathrm{d}x}{L} H\left[p(x, t), \sigma(x, t), \frac{\partial \sigma(x, t)}{\partial x}\right],$$

where

$$H = \frac{p(x, t)^2}{2m} + \tfrac{1}{2}a\sigma(x, t)^2 + \tfrac{1}{4}b\sigma(x, t)^4 + \tfrac{1}{2}mc_0{}^2\left\{\frac{\partial \sigma(x, t)}{\partial x}\right\}^2$$

is the Hamiltonian density, $p(x, t)$ is the momentum conjugate to $\sigma(x, t)$, $c_0 = (2LC/m)^{1/2}$, and L represents the length of the chain crystal. By the canonical relations

$$\frac{\mathrm{d}p}{\mathrm{d}t} = -\frac{\partial H}{\partial \sigma} - \frac{\partial}{\partial x}\left\{\frac{\partial H}{\partial(\partial \sigma/\partial x)}\right\} \quad \text{and} \quad \frac{\mathrm{d}\sigma}{\mathrm{d}t} = \frac{\partial H}{\partial p},$$

we can obtain the equation of motion for the order variable $\sigma(x, t)$, i.e.,

$$m\left(\frac{\partial^2}{\partial t^2} - c_0{}^2\frac{\partial^2}{\partial x^2}\right)\sigma(x, t) = -\frac{\partial V_<(\sigma)}{\partial \sigma} = -a\sigma - b\sigma^3.$$

This partial differential equation can be expressed as the ordinary differential equation

$$\frac{\mathrm{d}^2 Y}{\mathrm{d}\phi^2} + Y - Y^3 = 0, \tag{5.9a}$$

if the following rescaled variables are used:

$$Y = \frac{\sigma}{\sigma_0}, \qquad \sigma_0 = \left(\frac{|a|}{b}\right)^{1/2}, \tag{5.9b}$$

$$\phi = \frac{x - vt}{\rho}, \qquad \rho^2 = \frac{m(c_0{}^2 - v^2)}{|a|} = \frac{1 - v^2/c_0{}^2}{k_0{}^2}, \tag{5.9c}$$

and

$$k_0{}^2 = \frac{|a|}{mc_0{}^2}.$$

We solve (5.9a) first for the variable Y with a small amptitude. Ignoring the nonlinear term $Y^3 \ll Y$, the linearized equation

$$\frac{\mathrm{d}^2 Y}{\mathrm{d}\phi^2} + Y = 0$$

represents a harmonic oscillator, for which the solution is written as

$$\mathrm{Y} = \mathrm{Y}_0 \sin(\phi + \phi_0),$$

where the amplitude Y_0 is infinitesimal, and ϕ_0 is a phase constant. This solution represents a sinusoidal wave propagating at a speed v, where the wavevector and the frequency are $k = \rho^{-1}$ and $\omega = vk$, respectively. Using (5.9c), we can derive the relation

$$\omega^2 = c_0{}^2(k^2 - k_0{}^2). \tag{5.10}$$

It is clear that for the real frequency ω, the corresponding wavevector should be restricted to the range $k \geq k_0$. According to the Landau theory, at $T = T_c$, $a = 0$, and hence by definition $k_0 = 0$. However this k_0 may not necessarily be zero if the mean-field approximation is abandoned in the critical region. It is noted that from the relation $\omega = vk$, ω can be equal to zero if $v = 0$, while k can be kept nonzero. Similar to the Lorentz transformation,

$$\phi = \frac{k_0(x - vt)}{(1 - v^2/c_0{}^2)^{1/2}} = kx - \omega t \qquad \text{where} \quad k = \frac{k_0}{(1 - v^2/c_0{}^2)^{1/2}}.$$

In the laboratory frame of reference defined by $v = 0$, the phase $\phi = k_0 x$ is static, representing a spatially modulated structure at T_c.

In fact, the nonlinear equation (5.9a) can be solved analytically, using the Jacobi *elliptic function*. Integrate (5.9a) once, and we can obtain

$$2\left(\frac{dY}{d\phi}\right)^2 = (\lambda^2 - Y^2)(\mu^2 - Y^2), \tag{5.11}$$

where the constants λ and μ are given by

$$\lambda^2 = 1 - (1 - \alpha^2)^{1/2} \qquad \text{and} \qquad \mu^2 = 1 + (1 - \alpha^2)^{1/2}$$

with the constant of integration $\alpha = (dY/d\phi)_{\phi=0}$.

Integrating (5.11) once more, the phase variable ϕ is expressed by the *elliptic integral of the first kind*

$$\frac{\mu\phi}{2^{1/2}} = \int_0^\xi (1 - \xi^2)^{-1/2}(1 - \kappa^2\xi^2)^{-1/2}\, d\xi, \tag{5.12}$$

where $\xi = Y/\lambda$, and the *modulus* $\kappa = \lambda/\mu$. Conversely, the parameters λ and μ can be expressed in terms of κ:

$$\lambda = \frac{2^{1/2}\kappa}{(1 + \kappa^2)^{1/2}} \qquad \text{and} \qquad \mu = \frac{2^{1/2}}{(1 + \kappa^2)^{1/2}}.$$

Equation (5.12) can also be expressed in the inverse form, where the variable Y is written as a function of the phase ϕ, i.e.,

$$\xi = \frac{Y}{\lambda} = \mathrm{sn}\left(\frac{\mu\phi}{2^{1/2}}\right). \tag{5.13}$$

We can also use an integral form using the angular variable Θ defined by $\xi = \sin\Theta$:

$$\frac{\mu\phi}{2^{1/2}} = \int_0^\theta (1 - \kappa^2 \sin^2\Theta)^{-1/2}\, d\Theta, \tag{5.14}$$

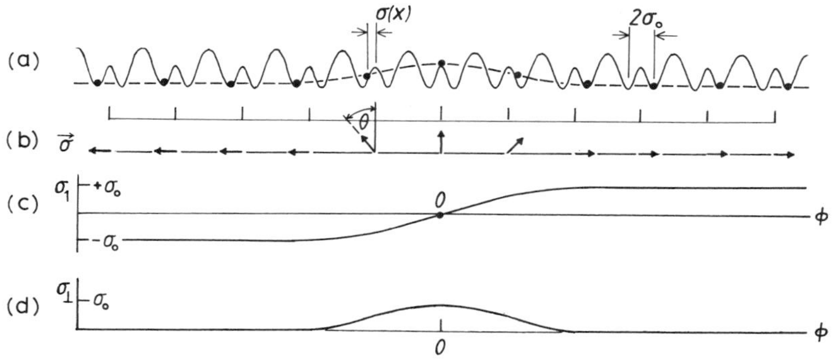

Figure 5.1. (a) A kink solution of the pseudospin mode in a quasi-one-dimensional lattice. (b) The collective kink mode of classical pseudospin vectors. (c) Longitudinal components σ_1, as given by $Y = \sigma_1/\sigma_0 = \tanh(2^{-1/2}\phi)$. (d) The transversal component $\sigma_\perp = \sigma_0 \operatorname{sech}^2(2^{-1/2}\phi)$.

where the angle θ specifies the value of Θ in the upper limit of the integral. Designating the upper limit in (5.12) as the ξ_1, we can write

$$\xi_1 = \sin\theta = \frac{\sigma_1/\sigma_0}{\lambda} = \operatorname{sn}\left(\frac{\mu\phi}{2^{1/2}}\right)$$

and

$$\sigma_1 = \lambda\sigma_0 \sin\theta = \lambda\sigma_0 \operatorname{sn}\left(\frac{\mu\phi}{2^{1/2}}\right),$$

which suggests that the pseudospin $\sigma(x, t)$ can be regarded as a classical vector whose longitudinal component is σ_1, and $\frac{1}{2}\pi - \theta$ represents the angle between σ and the direction of the chain, as illustrated in Fig. 5.1(b).

The Jacobi elliptic sn-function is a periodic function with a period $4K$, where K is defined as the *complete elliptic integral* of a modulus κ:

$$K(\kappa) = \int_0^{\pi/2} (1 - \kappa^2 \sin^2\Theta)^{-1/2}\, d\Theta. \tag{5.15}$$

From the above analysis it is clear that a specific case of $\kappa \to 0$ corresponds to a sinusoidal solution $Y = Y_0 \sin\phi$, where $\lambda \to 0$, $\mu = 2^{1/2}$, and $K = \frac{1}{2}\pi$ (a quarter of the period). On the other hand, $\kappa = 1$ is the maximum of the modulus, corresponding to the maximum $\lambda = 1$ at which also $\mu = 1$. In this case, (5.14) gives a simple expression

$$Y = \tanh\left(\frac{\phi}{2^{1/2}}\right), \tag{5.16}$$

varying from $Y = -1$ to $Y = +1$ at $\phi = 0$, as shown by the curve for $\kappa = 1$ in Fig. 5.1(c). Thus, (5.16) represents a *kink* of the variable $\sigma(\phi)$ between $-\sigma_0$

and $+\sigma_0$, where the phase ϕ reverses its sign, i.e., $\phi \to -\phi$. At such a kink, (4.36a) for mirror reflection is fulfilled, hence the kink is considered to represent a domain wall.

Between these specific solutions, the modulus can take any value in the range $0 \le \kappa < 1$, where the pseudospin mode is periodic with the period $4K(\kappa)$, and described by

$$\sigma_1 = \lambda\sigma_0 \cos(\tfrac{1}{2}\pi - \theta) \qquad \text{and} \qquad 0 \le \lambda < 1.$$

In the present model, the pseudospin mode $\sigma(\phi)$ is represented by a classical vector field, in which $\sigma_1(\phi)$ is the longitudinal component of $\sigma(\phi)$. Therefore, we may consider the transversal component $\sigma_\perp = \lambda\sigma_0 \cos\theta$ for significance, corresponding to σ_1 of the kink solution (5.16). For $\lambda = 1$, the transversal component

$$\sigma_\perp = \sigma_0 \cos\theta = \sigma_0 \operatorname{sech}\left(\frac{\phi}{2^{1/2}}\right)$$

represents a *solitary wave* of a pulse shape as illustrated in Fig. 5.1(d). Physically, $\sigma_\perp$ can be interpreted as being involved in the work required to rotate the direction of a by 180°. Considering the internal field $F_\perp \propto \sigma_\perp$ as given by (3.23), the work expressed by the product $\sigma_\perp F_\perp$ is proportional to $\sigma_\perp{}^2$. Accordingly,

$$V_\sigma(\phi) \propto \sigma_\perp{}^2 = \sigma_0{}^2 \operatorname{sech}^2\left(\frac{\phi}{2^{1/2}}\right) \tag{5.17}$$

is the potential energy for reversing the pseudospin direction in the vicinity of $\phi = 0$. Such a potential $V_\sigma(\phi)$ is known as a *one-soliton* solution of the Korteweg–deVries equation [48], which deals with a potential field where excitation waves of (5.9a) can propagate freely under certain circumstances. Such a potential provides a sharp barrier with a large amplitude for the pseudospin mode to twist by 180°, providing a model for the domain wall. The soliton expressed by (5.17) is mobile freely in an ideal crystal, but may be observable as a domain wall in a practical crystal, if pinned by a defect potential.

The above dynamical theory offers results compatible with fluctuations in pseudospin condensates at long wavelengths and low frequencies, hence supplementing our interpretation of the transition mechanism. Oscillatory solutions (5.13) for $0 \le \kappa < 1$ can represent the pseudospin mode during the process to complete order, when the modulus κ is considered as a function of temperature. However, experimentally, $\kappa = \kappa(T)$ remains to fit in observed results, besides it is not easy to evaluate contributions from long-range interactions. Figure 5.2 shows numerical plots of (5.12) or (5.13) for given values of the modulus κ. Experimental results for supporting such analysis of dynamical condensates are discussed in Chapter 9 for anomalous EPR lineshapes.

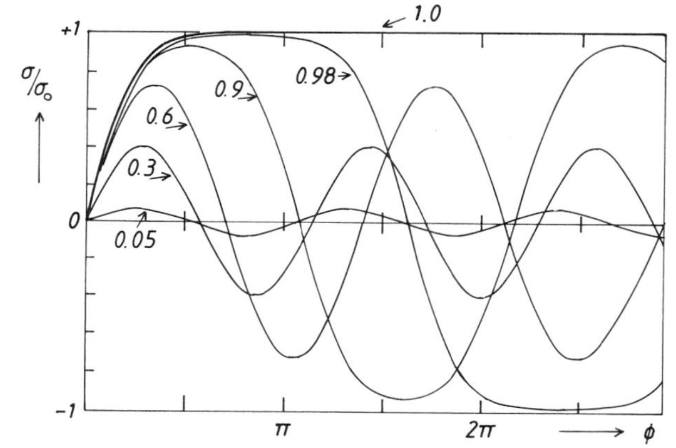

Figure 5.2. Numerical plots of $Y = \lambda\,\mathrm{sn}(2^{-1/2}\mu\phi)$ for various values of the modulus $\kappa = \lambda/\mu$ in $0 < \kappa \leq 1$.

5.4. The Lifshitz Condition for Incommensurability

We discussed in Chapter 3 examples of modulated structures of pseudospins originating from competing short-range correlations, while it was not certain if the presence of such structures signifies a thermodynamical incommemsurate phase. However, now that we know that the pseudospin function is expressed by $\sigma_0 f(\phi)$ where the spatial inhomogeneity represented by a variable ϕ is quasi-static, the system may be regarded as a modulated *thermodynamical phase* as specified by ϕ and $\kappa(T)$. While primarily time-dependent, a pinned pseudospin mode at very low frequencies exhibits a practically "stationary" modulated structure.

While incommensurate fluctuations were first observed at microwave frequencies in the critical region of the ferroelectric phase transition of TSCC [21], incommensurate phases had been known in many systems, for which Lifshitz formulated a macroscopic criterion for the incommensurability. (See [41] for the detail.) In this section, we discuss the Lifschitz condition that emerges naturally from phase fluctuations in the pseudospin mode $\sigma(\phi)$. It is simple to calculate the correlation function, if the fluctuation is sinusoidal, i.e., $\sigma(\phi) = \sigma_0 \exp(i\phi)$. As remarked already, the timescale t_0 of observation should be significantly shorter than the time difference $\delta t = t_2 - t_1$ between two correlated phases ϕ_1 and ϕ_2, to deal with the modulated structure by a phase correlation function

$$\langle\sigma^*(\phi_1)\sigma(\phi_2)\rangle_t = \sigma_0{}^2 \exp\{iq(x_2 - x_1)\}\langle\exp\{-i\varpi(t_2 - t_1)\}\rangle_t,$$

in which their time correlation

$$\langle\exp\{-i\varpi(t_2 - t_1)\}\rangle_t = t_0{}^{-1}\int_0^{t_0}\cos(\varpi\delta t)\,\mathrm{d}(\delta t) = \frac{\sin(\varpi t_0)}{\varpi t_0}$$

is a real function because of time-reversal symmetry (see (3.14)). Such time correlations are very close to 1, when $\varpi t_0 \ll 1$. In the latter case signified as "slow," the binary correlation appears to be quasi-static, and is dominated by the symmetrical spatial factor $\cos q(x_2 - x_1)$. In this case, for two lattice points x_1 and x_2, the spatial factor is zero if $q = \text{integer} \times a$ (lattice const.), and there is no violation of the lattice periodicity in the crystal. If, on the other hand, x_2 and x_1 represent two nonlattice points, spatial correlations are significant.

For nonsinusoidal fluctuations, the nonsinusoidal function $\sigma(x, t)$ is not easily separated into space- and time-dependent factors. However, when slow as characterized by $\varpi t_0 \ll 1$, the correlation $\langle \sigma^*(x_1, t_1)\sigma(x_2, t_2)\rangle_t$ should be dominated by spatial fluctuations between $x_1 = x - \delta x$ and $x_2 = x + \delta x$. For a small $\delta x \ll a$,

$$\langle \sigma^*(x_1, t_1)\sigma(x_2, t_2)\rangle_t = \langle |\sigma(x)|^2\rangle_t + \left\langle \left[\sigma^*(x)\frac{\partial\sigma(x)}{\partial x} - \sigma(x)\frac{\partial\sigma^*(x)}{\partial x} \right] \right\rangle_t \delta x,$$

where $x = \frac{1}{2}(x_1 + x_2)$ and $\delta x = \frac{1}{2}(x_2 - x_1)$. Therefore, the spatial correlation function between x_1 and x_2 can be defined as

$$\begin{aligned} \Gamma(\delta x) &= \langle \{\sigma(x_1, t_1) - \sigma(x, t)\}^* \{\sigma(x_2, t_2) - \sigma(x, t)\rangle_t \\ &= \left\langle \left[\sigma^*(x, t)\frac{\partial\sigma(x, t)}{\partial x} - \sigma(x, t)\frac{\partial\sigma^*(x, t)}{\partial x} \right] \right\rangle_t \delta x. \end{aligned} \tag{5.18}$$

In this expression, $\Gamma(0) = 0$, if $\delta x = 0$. On the other hand, in order for $\Gamma(\delta x)$ to be nonzero, the time average of the quantity in the square brackets [...] should not vanish for a given nonzero variation δx. Omitting all time-dependences from (5.18) for quasi-static cases as justified by the relation $\varpi t_0 \ll 1$, Lifshitz proposed that the Gibbs potential of an incommensurate system should include a term given by

$$G_L = \int \frac{iD}{2} \left\langle \left[\sigma^*(x)\frac{\partial\sigma(x)}{\partial x} - \sigma(x)\frac{\partial\sigma^*(x)}{\partial x} \right] \right\rangle_t \left(\frac{dx}{L}\right), \tag{5.19}$$

where the factor D is proportional to $\delta x \neq 0$, and $\frac{1}{2}i$ is included for convenience. The Lifshitz condition for incommensurability is normally expressed as $G_L \neq 0$ for a given system.

The Gibbs function for an incommensurate crystal can therefore be written as

$$G(\sigma) = G(0) + \int \left[\tfrac{1}{2}a|\sigma(x)|^2 + \tfrac{1}{4}b|\sigma(x)|^4 + \tfrac{1}{2}\kappa\left|\frac{\partial\sigma(x)}{\partial x}\right|^2 \right]\left(\frac{dx}{L}\right) + G_L. \tag{5.20}$$

To solve the equilibrium for this function, we assume a sinusoidal function $\sigma = \sigma_0 \exp(i\phi)$ for the pseudospin, where σ_0 is a variable of x. Equation (5.20) can then be expressed in terms of independent variables σ_0 and ϕ, i.e.,

$$G(\sigma) = G(0) + \int \frac{dx}{L}\left[\tfrac{1}{2}a\sigma_0{}^2 + \tfrac{1}{4}b_0{}^4 + \tfrac{1}{2}\kappa\left(\frac{\partial\sigma_0}{\partial x}\right)^2 + \tfrac{1}{2}\kappa\sigma_0{}^2\left(\frac{\partial\phi}{\partial x}\right)^2 + D\sigma_0{}^2\left(\frac{\partial\phi}{\partial x}\right)\right].$$

For thermal equilibrium determined by the minimum of $G(\sigma)$, equations $\partial G/\partial\sigma_0 = 0$ and $\partial G/\partial\phi = 0$ have to be solved simultaneously, which are, respectively,

$$a\sigma_0 + b_0{}^3 + \kappa\left(\frac{\partial^2\sigma_0}{\partial x^2}\right) + \kappa\sigma_0\left(\frac{\partial\phi}{\partial x}\right)^2 + 2D\sigma_0\left(\frac{\partial\phi}{\partial x}\right) = 0 \qquad \text{(i)}$$

and

$$\left\{\kappa\sigma_0{}^2\left(\frac{\partial\phi}{\partial x}\right) + D\sigma_0{}^2\right\}\left(\frac{\partial}{\partial\phi}\right)\left(\frac{\partial\phi}{\partial x}\right) = 0. \qquad \text{(ii)}$$

From (ii) we obtain immediately that

$$\frac{\partial\phi}{\partial x} = -\frac{D}{\kappa},$$

indicating that the wavevector q of the phase ϕ is given by $q = -D/\kappa$. Since both D and κ are unrelated to the lattice periodicity, the value of calculated q cannot generally be an integral multiple of $a^* = 2\pi/a$. Neglecting the small term $\kappa(\partial^2\sigma_0/\partial x^2)$ in (i), we obtain the relation for determining the amplitude, i.e.,

$$\sigma_0{}^2 = -\frac{a - D^2/\kappa}{b}.$$

For detailed discussions about incommensurate phases, interested readers are referred to the article by Currat and Janssen [41].

5.5. Condensate Locking by a Pseudoperiodic Lattice Potential

Pseudospins represent active groups occupying regular lattice points, and hence in practical crystals their collective motion is significantly influenced by a subtle deviation of any kind from the periodic structure of a crystal. In Section 5.2, we discussed point defects that disrupt translational symmetry, and here we consider as another example the effect of a *superlattice* structure that may exist or emerge in crystals under certain internal or external constraints. In principle, such an pseudostructure should be noticeable crystallographically, but often too small to be recognized in some systems. Nevertheless, pseudospins may be modulated at the period of the superstructure, which is commensurate with the lattice, constituting a distinct thermodynamical phase. A transition to such a commensurate phase takes place normally when the temperature is lowered from an incommensurate phase, which is referred to as an *incommensurate-to-commensurate transition.* In the incommensurate phase, the pseudospin mode is characterized by a phase variable ϕ in the whole angular range $0 \le \phi \le 2\pi$, whereas in such a commensurate phase as transformed from an incommensurate phase the variable ϕ changes to discrete angles, reflecting the presence of a pseudosymmetry in the low-temperature phase.

Typically, such a pseudopotential can be related to internal symmetry of active molecular groups that may not be explicit in observed crystal structure. As an illustrative example, we consider a screw symmetry where active groups are related by successive rotation by a specific angle when advanced by a unit from one cell to the next along a symmetry axis of the crystal. In practice, such a symmetry is often related to two- or three-fold screw axes, along which the molecular orientation rotates by π or $2\pi/3$ in sequence over two or three unit cells, respectively. In these cases, the period of the pseudopotential is two or three times longer than the original lattice spacing. Such a pseudoperiodicity exists generally in relation to m-fold screw symmetry in crystals under specific circumstances, while absent in the incommensurate phase above T_i.

On the axis of such a pseudopotential, classical displacement vectors σ are related by rotation and translation along the m-fold screw axis. Components of these vectors may be written as $\sigma_\perp = \sigma_{\perp 0}\exp(i\theta)$ and $\sigma_\parallel = \sigma_{\parallel 0}\exp(i\phi)$, in which the angle θ takes values expressed by $\theta = 2\pi \mathrm{p}/\mathrm{m}$, where $\mathrm{p} = 0, 1, 2, \ldots, \mathrm{m} - 1$, as required by the m-fold rotation, while the pseudopotential for $\sigma_\parallel$ should be a real function, and expressed as a function of ϕ, i.e.,

$$V_{\mathrm{m}}(\phi) = \frac{\rho\{\sigma_\parallel{}^{\mathrm{m}} + (\sigma_\parallel{}^{\mathrm{m}})^*\}}{\mathrm{m}} = \left(\frac{2\rho}{\mathrm{m}}\right)\sigma_{\parallel 0}{}^{\mathrm{m}}\cos(\mathrm{m}\phi), \tag{5.21}$$

where ρ is the proportionality constant.

Considering that the fluctuations are extremely slow near the transition temperature, the equilibrium process between the pseudospins and the pseudopotential can be treated as quasi-static. Disregarding the time-dependence, the Gibbs potential in an incommensurate phase can be written as

$$G(\sigma) = G(0) + \int \frac{\mathrm{d}x}{L}\left[\tfrac{1}{2}a|\sigma|^2 + \tfrac{1}{4}b|\sigma|^4 + \tfrac{1}{2}\kappa\left|\frac{\partial\sigma}{\partial x}\right|^2 + V_{\mathrm{m}}\right],$$

which takes a minimimum value in thermal equilibrium. Writing $\sigma = \sigma_0\exp(i\phi)$ for a situation where a sinusoidal approximation is acceptable, the function $G(\sigma)$ can be expressed as

$$G(\sigma_0, \phi) = G(0) + \int_0^L \frac{dx}{L}\left[\tfrac{1}{2}a\sigma_0^2 + \tfrac{1}{4}b\sigma_0^4 + \tfrac{1}{2}\kappa\left(\frac{\partial\sigma_0}{\partial x}\right)^2 + \tfrac{1}{2}\kappa\sigma_0^2\left(\frac{\partial\phi}{\partial x}\right)^2 + \left(\frac{2\rho}{m}\right)\sigma_0^{m}\cos(m\phi)\right],$$

where σ_0 and ϕ are both functions of x, and is considered as adequate approximation if there are no significant long-range correlations. For equilibrium specified by $dG(\sigma_0, \phi) = 0$, we solve equations $\partial G/\partial\sigma_0 = 0$ and $\partial G/\partial\phi = 0$ simultaneously, i.e.,

$$a\sigma_0 + b\sigma_0^3 + 2\rho\sigma_0^{m-1}\cos(m\phi) + \kappa\sigma_0\left(\frac{\partial\phi}{\partial x}\right)^2 + \kappa\left(\frac{\partial^2\sigma_0}{\partial x^2}\right) = 0$$

and

$$\kappa\sigma_0^2\left(\frac{\partial^2\phi}{\partial x^2}\right) + 2\rho\sigma_0^{m}\sin(m\phi) = 0. \qquad (5.22)$$

Assuming, now, $\sigma_0 = \text{const.}$ in these equations, (5.22) is the so-called *sine–Gordon equation*, which is simplified as

$$\frac{d^2\psi}{dx^2} + \zeta\sin\psi = 0, \qquad (5.22a)$$

when abbreviations $\psi = m\phi$ and $\zeta = (2m\rho/\kappa)\sigma_0^{m-2}$ are used.

Writing (5.22a) in the form

$$\frac{\partial}{\partial\phi}\left[\tfrac{1}{2}\kappa\sigma_0^2\left(\frac{\partial\phi}{\partial x}\right)^2 + V_m\right] = 0, \qquad (5.22b)$$

equation (5.22b) can be interpreted so that the fluctuating pseudospin mode and the pseudopotential are in dynamical equilibrium. The sine–Gordon equation signifies that the fluctuating motion is nonlinear, depending on the magnitude ρ of the pseudopotential V_m. Such a dynamical problem was originally discussed by Frank and van der Merwe [49], and the results can be directly transferred to the present model for nonlinear fluctuations, as described in Böttger's textbook [39].

Integrating (5.22a) once, we obtain

$$\tfrac{1}{2}\left(\frac{d\psi}{dx}\right)^2 - \zeta\cos\psi = E, \qquad (5.23)$$

where E is the constant of integration, and is determined if the values of ψ and $d\psi/dx$ are specified at a certain point $x = x_0$, being similar to the "initial condition" for a pendulum. Such a classical problem has an oscillatory solution with a finite amplitude if $E < \zeta$, whereas the motion is not oscillatory, but hopping multiple potential valleys, if $E \geq \zeta$. When E is very large, the constant ζ is negligible, the pseudopotential V_{m} is no obstacle to the propagating pseudospins.

Equation (5.23) can be integrated and expressed by an elliptic integral

$$x - x_0 = \int_0^{\psi} [2(E + \zeta \cos \psi)]^{-1/2} \, d\psi,$$

which can be written in the standard form by using the modulus defined by

$$\kappa^2 = \frac{2\zeta}{E + \zeta}. \tag{5.24a}$$

Namely,

$$x - x_0 = \frac{\zeta^{1/2}}{\kappa} \int_0^{\varphi} (1 - \kappa^2 \sin^2 \Theta)^{-1/2} \, d\Theta \qquad \text{for} \quad 0 < \kappa \leq 1, \tag{5.24b}$$

where $\Theta = \frac{1}{2}\psi$, and the angle φ for the upper integration limit corresponds to the variable x. This is exactly the same integral as (5.14), and written in the inverse form

$$\sin \varphi = \operatorname{sn}\left[\frac{\kappa(x - x_0)}{\zeta^{1/2}}\right],$$

representing a periodic function with a quarter period $K(\kappa)$ given by (5.15) for $0 < \kappa < 1$. It is clear from the argument that in the presence of the potential V_{m} the phase ϕ of the pseudospin σ is determined by the relation $\varphi = \mathrm{m}\phi$, indicating that $0 \leq \phi \leq 2\pi/\mathrm{m}$. A useful parameter is the spatial distance $\Lambda(\kappa)$ between two adjacent nodal points that are given by $\operatorname{sn}[\kappa(x - x_0)/\zeta^{1/2}] = 0$, or by $\varphi = 0$ and $\frac{1}{2}\pi$, and is written as

$$\Lambda(\kappa) = \frac{2\kappa K(\kappa)}{\zeta^{1/2}}.$$

As seen from Fig. 5.2, $\Lambda(\kappa) \to \infty$ when $\kappa \to 1$. For $\kappa = 1$. we have a specific expression

$$\psi = 2\theta = 2 \tan^{-1} \sinh\left[\frac{x - x_0}{\zeta^{1/2}}\right], \tag{5.25}$$

representing the "kink" solution of the sine–Gordon equation (5.22a). It is noted that (5.25) is identical to $\sin \phi = \tanh[(x - x_0)/\zeta^{1/2}]$ as expressed previously by (5.16). In terms of the phase ϕ, such a kink appears when

$$\phi = \left(\frac{2\pi}{\mathrm{m}}\right)\mathrm{p}, \qquad \text{where} \quad \mathrm{p} = 0, 1, \ldots, \mathrm{m} - 1, \tag{5.26}$$

as illustrated in Fig. 5.3.

An incommensurate-to-commensurate phase transition is a phase locking of the pseudospin mode to a periodic pseudopotential V_{m}. In this context, we

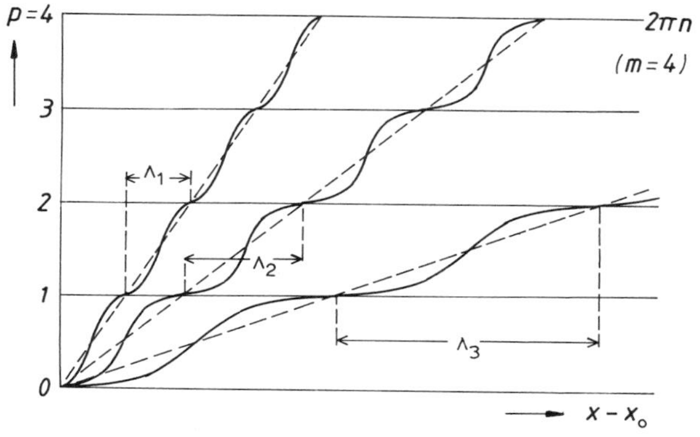

Figure 5.3. Periodic discommensuration lines for p = 4.

can describe the transition as matching the wavelength λ with the repeated translation of V_m, i.e., $\lambda = mb$, where b is the lattice constant. Using wavevectors, this matching condition is expressed as

$$\boldsymbol{G}_b = m\boldsymbol{b}^* \qquad \text{where} \quad G_b = \frac{2\pi}{\lambda}$$

and

$$V_m = V_0 \sum_p \cos(G_b x_p).$$

Here $V_0 = (2\rho/m)\sigma_0{}^m$, and $x_p = pb$, p given by (5.26). The pseudopotential is perturbed by the propagating pseudospin mode, and consequently expressed in terms of effective phase shifts $\Delta\phi_p$, i.e.,

$$V_m{}' = V_0 \sum_p \cos(G_b x_p - \Delta\phi_p) = V_0 \sum_p \cos\phi_p(\delta) \tag{5.27a}$$

where the effective phase $\phi_p(\delta)$ is defined as

$$\phi_p(\delta) = \frac{2\pi}{m} p(1 - \delta). \tag{5.27b}$$

Here the parameter δ indicates a deviation from the unperturbed phase $(2\pi/m)p$, which is discrete in contrast to the continuous shift in an incommensurate phase. Such a pseudopotential in low dimension may exhibit a ripple-like pattern of kinks called *discommensuration lines* near a lock-in phase transition, where $E = \zeta$.

In the above we have discussed an anomaly expressed by discommensuration phases near incommensurates–commensurate phase transitions. In practical crystal, such anomalous patterns may not be so simple as described by a soliton ripple in a one-dimensional model, and should be considerably modified by complex boundaries in practical crystals. However, laminar patterns as discussed in the one-dimensional model were reported by Pan

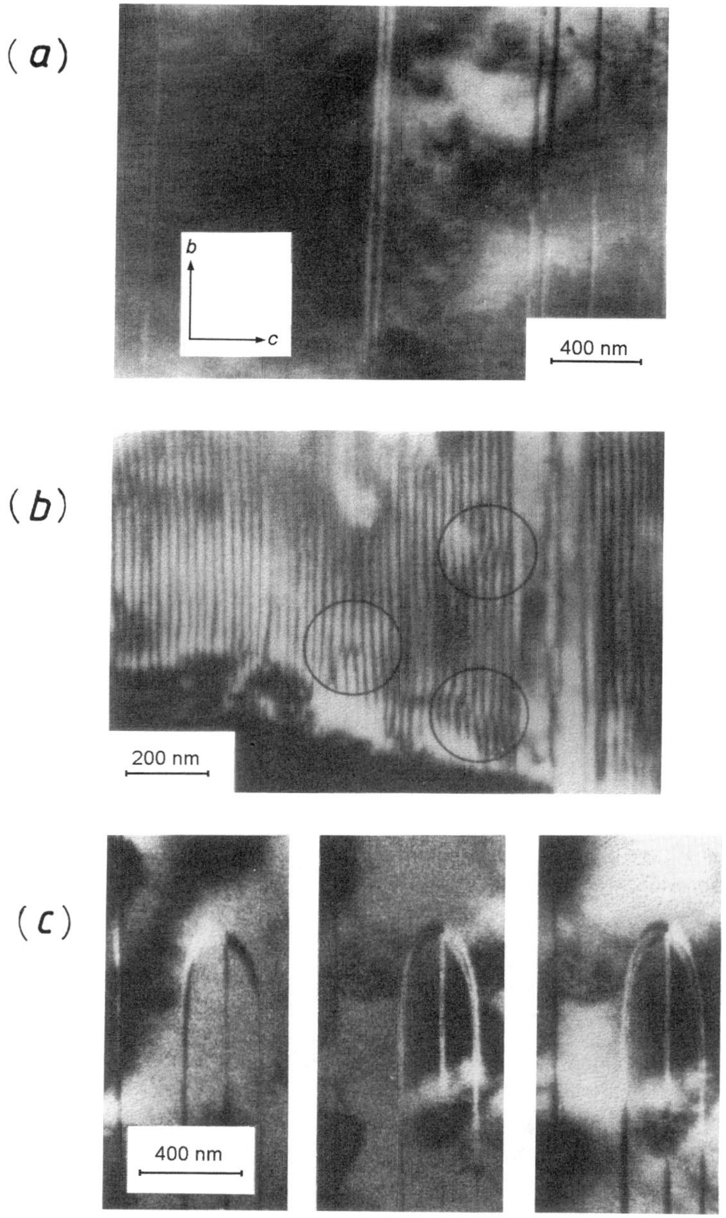

Figure 5.4. (a) A dark field image from satellite reflections from the (100) plane of a K_2ZnCl_4 crystal at 205 K, showing discommensuration stripes parallel to the *b* direction. (b) A dark field micrograph at 208 K. In the circled areas, discommensuration lines showed evidence for a pair structure. (c) The "vortex" of three pairs of discommensuration lines, where a splitting of outer pairs is visible from the left to the right with increasing electron irradiation. Courtesy of Professor H.-G. Unruh. (From Xiaoquing Pan and H.-G. Unruh, *J. Phys. Cond. Matter* **2**, 323 (1990).)

and Unruh [50] and other workers for lock-in phase transitions in K_2ZnCl_4 and Rb_2ZnCl_4 crystals, respectively. The photographs in Figs. 5.4((a), (b), and (c)) show such patterns consisting of soliton stripes observed by Pan and Unruh in *a* plates of K_2ZnCl_4 using transmission electron microscopy (TEM). In principle, such discommensuration lines along the *b* axis (Fig. 5.4(a)) can be understood with the theory, while "splitting" and "vortex"-like behaviors of these stripes were also observed in dark fields of electron diffractions (Fig. 5.4(c)). Among photo recordings presented in [50], notable are those that at low density of discommensuration stripes isolated groups of three stripes were observed in Fig. 5.4(b), signifying the presence of pseudopotentials $V_3(\phi_i)$, where ϕ_i are phases along the three-fold screw axis, and that these lines were terminated by what the authors called "vortices." Further, in their detailed studies, each stripe showed a double-line structure, which appeared as related to mirror-symmetry breaking at the transition to the ferroelectric phase, while not so suggested by the authors. For such deviations from laminar structure, the authors considered possible thermal and dynamical inhomogeneities that are inevitable during observations under electron irradiation. Such an interpretation may remain only qualitative, but providing useful information for soliton dynamics in the crystal. Quite independent of such visible evidence, Abe et al. and Fujimoto et al. have reported on a change in EPR lineshapes, from broadened lines to a discrete set of lines representing discommensurations, in transitions of Ke_2SeO_4 and BCCD crystals, respectively. The experimental details are discussed in Chapter 9.

PART TWO

EXPERIMENTAL STUDIES

Structural phase transitions in crystals are complex phenomena, originating from an interplay between order variables and their hosting lattice. In the harmonic approximation, basic excitations in these subsystems are independent from each other, and therefore it is not surprising that results of experiments performed exclusively for one part of the composite system may lead us to an interpretation that appears to be incompatible with the other. For example, some ferroelectric phase transitions can be judged to be displacive on the basis of soft-mode results, while dipolar ordering characterizes polarized low-temperature phases. However, it is noted that neither of these views deals with critical anomalies that are essentially due to interactions between order variables and soft phonons. In this context, such a classification as displacive versus order–disorder is not logical, unless critical anomalies are interpreted properly.

In Part One we discussed the order-variable condensate that prevails in the critical region of typically displacive systems. The condensate is considered as a mobile object of long life in crystals, owing to low damping of soft modes on both sides of T_c, whereas the pseudospins below T_c are in phase mode fluctuating between binary states. The condensate motion is so slow that in the timescale of microwave measurements the lattice appears to be quasi-statically modulated, while only the temporal fluctuations are revealed in the phonon spectra.

We expressed the pseudospin condensate by $\sigma = \sigma_0 f(\phi)$ where the phase ϕ is determined by soft phonons in the critical region, while the amplitude σ_0 and the phase ϕ are both functions of temperature in the low-temperature phase. The nature of the wavevector $\boldsymbol{q}$ and the frequency ϖ in the phase ϕ can be studied by neutron and light-scattering experiments, whereas σ_0 and its spatial phase distribution can be determined from magnetic resonance sampling. In addition, nuclear spin relaxation studies provide evidence for the coupling between order variables and soft phonons. Dielectric relaxation and Brillouin scattering experiments provide useful information as to the dynamic behavior of condensates in the lattice. Results from these different experiments are complementary, substantiating condensates in the critical region.

In Part Two, principles for these basic measurements are outlined, and experimental results on representative systems are discussed in the light of the condensate model. Since many articles reviewing existing data are available in the literature, we discuss mainly phase transitions in selected systems for which both soft-mode and magnetic-resonance studies were already performed, in order to see if the results in the critical region can be consistently interpreted. At the end of Part Two, we also discuss some structural phase transitions of other types for comparison, which are believed to be helpful in enlightening our knowledge of the collective mechanism of order variables.

CHAPTER 6

Diffuse X-Ray Diffraction and Neutron Inelastic Scattering from Modulated Crystals

6.1. Modulated Crystals

Idealized crystals consisting of a large number of identical atoms or molecules can be considered as macroscopically homogeneous, if in equilibrium with isotropic surroundings specified by external pressure p and temperature T. In this case, their thermodynamical properties are described by the Gibbs potential $G(p, T)$. While the presence of surfaces cannot be ignored entirely, bulk properties prevail in a large uniform crystal that is signified by translational symmetry and periodic boundary conditions. For a crystal undergoing a structural change, the crystalline phases above and below the transition temperature T_c are primarily so idealized in the first approximation. In the transition region. in contrast, the crystal becomes spatially inhomogeneous, arising from locally violated translational symmetry.

In an idealized crystal phase, where the basic translational vectors are $\boldsymbol{a}_1$, $\boldsymbol{a}_2$, and $\boldsymbol{a}_3$, a periodic function $f(\boldsymbol{r})$ in the lattice is invariant under a translation

$$\boldsymbol{R} = n_1\boldsymbol{a}_1 + n_2\boldsymbol{a}_2 + n_3\boldsymbol{a}_3 \quad (n_1, n_2, n_3 \text{ are integers}),$$

i.e.,

$$f(\boldsymbol{r}) = f(\boldsymbol{r} + \boldsymbol{R}). \tag{6.1a}$$

A *perfect* crystal can also be characterized by the Fourier transform $g(\boldsymbol{k})$ defined by

$$g(\boldsymbol{k}) = \int f(\boldsymbol{r}) \exp(-i\boldsymbol{k}\cdot\boldsymbol{r})\, \mathrm{d}^3\boldsymbol{r} \qquad \text{and} \qquad f(\boldsymbol{r}) = \int g(\boldsymbol{k}) \exp(i\boldsymbol{k}\cdot\boldsymbol{r})\, \mathrm{d}^3\boldsymbol{k}.$$

Using (6.1a) in the last transform, the equation $\exp(i\boldsymbol{k}\cdot\boldsymbol{R}) = 1$ can be derived, from which the wavevector $\boldsymbol{k}$ can take only a specific $\boldsymbol{G}$ value defined by the relation $\boldsymbol{G}\cdot\boldsymbol{R} = 2\pi \times$ integer. We can express such a vector $\boldsymbol{G}$ in the reciprocal lattice as

$$\boldsymbol{G} = h\boldsymbol{a}_1{}^* + k\boldsymbol{a}_2{}^* + l\boldsymbol{a}_3{}^*,$$

where

$$\boldsymbol{a}_1{}^* = \frac{2\pi}{\Omega}\boldsymbol{a}_2 \times \boldsymbol{a}_3, \qquad \boldsymbol{a}_2{}^* = \frac{2\pi}{\Omega}\boldsymbol{a}_3 \times \boldsymbol{a}_1, \qquad \boldsymbol{a}_3{}^* = \frac{2\pi}{\Omega}\boldsymbol{a}_1 \times \boldsymbol{a}_2,$$

and

$$\Omega = (\boldsymbol{a}_1, \boldsymbol{a}_2, \boldsymbol{a}_3)$$

is the volume of a unit cell, and the indices h, k, l are integers to specify the vector $\boldsymbol{G}$. Here $\boldsymbol{a}_1{}^*$, $\boldsymbol{a}_2{}^*$, and $\boldsymbol{a}_3{}^*$ are the basic translational vectors in the reciprocal lattice, in which a lattice point is specified by a vector $\boldsymbol{G}(h, k, l)$. In a perfect crystal all unit cells are identical, and the macroscopic uniformity is warranted by invariance of the functions $f(\boldsymbol{r})$ and $g(\boldsymbol{k})$ under all translations $\boldsymbol{R}$ and $\boldsymbol{G}$ in the crystal space and the corresponding reciprocal lattice, respectively. Similar to (6.1b), the periodicity in the reciprocal lattice is expressed by

$$g(\boldsymbol{k}) = g(\boldsymbol{k} + \boldsymbol{G}). \tag{6.1b}$$

On the other hand, in a *modulated* crystal the order variable $\boldsymbol{\sigma}_{\mathrm{m}}$ is no longer periodic in the lattice, but periodic in relation to a wavevector $\boldsymbol{G}_{\mathrm{i}}$ which cannot be expressed by a set of three integral indices (h, k, l) in the reciprocal units $\boldsymbol{a}_1{}^*$, $\boldsymbol{a}_2{}^*$, and $\boldsymbol{a}_3{}^*$. For such a vector $\boldsymbol{G}_{\mathrm{i}}$, at least one of these indexes must be *irrational* in the unit of a basic vector, say $\boldsymbol{a}_1{}^*$, giving an incommensurate modulation along the $\boldsymbol{a}_1$ axis of the crystal. Such a lattice modulation can alternatively be specified by an additional vector

$$\boldsymbol{G}_{\mathrm{i}} - \boldsymbol{G} = \boldsymbol{Q} = m\boldsymbol{a}_4{}^*,$$

where $\boldsymbol{a}_4{}^*$ is defined as the unit vector in the direction from the nearest lattice point $\boldsymbol{G}$ to the point $\boldsymbol{G}_{\mathrm{i}}$, and m is an additional index that is generally an irrational number. de Wolff and his group [51] have developed a group-theoretical method for such a four-dimensional space spanned by four indexes (h, k, l, m), which is called the *superspace*. The supergroup theory provides a convenient method to deal with *aperiodic* crystals that have been reported in recent literature. However, in this monograph, where one of the main objectives is to study disrupted translational symmetry, we shall stay on with the traditional method, using traditional reciprocal space $(\boldsymbol{a}_1{}^*, \boldsymbol{a}_2{}^*, \boldsymbol{a}_3{}^*)$ plus a specific modulation vector $\boldsymbol{a}_4{}^*$.

In a perfect crystal, a lattice modulation at an incommensurate wavevector $\boldsymbol{Q} = m\boldsymbol{a}_4{}^*$ implies the presence of an excitation of the crystal at an energy $\varepsilon(\boldsymbol{Q})$ above the ground state. Considering that such an incommensurate excitation is caused by the order variable mode, the energy $\varepsilon(\boldsymbol{Q})$ must be offset at least in part by a commensurate lattice mode, in order for the strains in the crystal to be minimum in thermal equilibrium. Such interactions do not occur exactly in phase, hence in a binary system, for example, the equilibrium between pseudospins and the lattice mode accompanies fluctuations that are characterized by small $\pm \boldsymbol{q}$ and $\mp \Delta\varepsilon$.

Such an excitation energy $\varepsilon(\boldsymbol{Q})$ is typically smaller than the order of thermal energy, and is difficult to be identified in the thermal spectra of a modulated crystal, whereas the vector $\boldsymbol{Q}$ can be determined by inelastic scattering of thermal neutrons. In contrast, X-ray diffraction at $\boldsymbol{Q} = 0$ is primarily elastic, and hence incapable for observing modulated structures, while diffused spots observed in the Bragg diffraction pattern can be attributed to fluctuations $\pm \boldsymbol{q}$. The corresponding energy $\varepsilon(0)$ associated with a collective mode of order variables cannot be determined from such diffused spots, however it may be determined from a Raman shift of a quantum transition $E_1 \to E_0$ of a constituent atom in the crystal, i.e., from the relation

$$h\nu = E_1 - E_0 \pm \varepsilon(\boldsymbol{q}).$$

On the other hand, for a finite $\boldsymbol{Q}$, the frequency $\Delta\omega = \varepsilon(\boldsymbol{q})/\hbar$ can be estimated from the characteristic frequency ϖ of a soft mode observed in neutron inelastic scattering. Normally, the soft mode is considered as a classical oscillator, so that $\Delta\omega \propto \varpi^2$.

6.2. The Bragg Law for X-Ray Diffraction

In this section, the principle of the Bragg diffraction from a "perfect" crystal is reviewed, prior to discussing the effect of modulated structures. In the diffraction of a collimated X-ray beam by a crystal, the three-dimensional arrangement of identical atoms can be identified as reflection from a large number of parallel planes of atoms. The diffraction pattern can be described in terms of the distance between these planes and their orientation with respect to the incident beam. The Bragg law for X-ray diffraction is such a simple rule, based on the concept of atomic planes that behave like optical mirror reflectors. For diffraction, the interaction between incident electromagnetic X-ray and orbiting electrons can be described as elastic scattering, and the Bragg law allows us to consider the crystal as if consisting of a large number of such crystal planes.

It is significant in the Bragg law for X-ray diffraction that the normal direction of such a group of parallel crystal planes can be specified by a translational vector $\boldsymbol{G}$ in the reciprocal lattice. Such a crystal plane can be considered as an optical planes for reflection, whose distance from the origin is given by $\boldsymbol{d} = d\boldsymbol{n}$, where $\boldsymbol{n}$ is the unit vector along the "normal." Indicating lattice points on the plane by vectors $\boldsymbol{R}$ from the origin, the equation of the plane is expressed as

$$\boldsymbol{n} \cdot (\boldsymbol{R} - \boldsymbol{d}) = 0 \qquad \text{or} \qquad \boldsymbol{n} \cdot \boldsymbol{R} = d = \text{const.}$$

Therefore the normal $\boldsymbol{n}$ should be parallel to the reciprocal lattice vector $\boldsymbol{G}$, as compared with the relation $\boldsymbol{G}\cdot\boldsymbol{R} = 2\pi$ (= const.), and we can write

$$\boldsymbol{G} = G\boldsymbol{n} \qquad \text{where} \quad \boldsymbol{n} = \frac{2\pi}{\Omega}(h\boldsymbol{a}_2 \times \boldsymbol{a}_3 + k\boldsymbol{a}_3 \times \boldsymbol{a}_1 + l\boldsymbol{a}_1 \times \boldsymbol{a}_2). \tag{6.2}$$

For a simple cubic lattice, as an example, $\boldsymbol{a}_1 = \boldsymbol{a}_2 \times \boldsymbol{a}_3$, etc. and $\Omega = 1$, and therefore $\boldsymbol{n} = 2\pi(h\boldsymbol{a}_1 + k\boldsymbol{a}_2 + l\boldsymbol{a}_3)$, and the crystal planes are represented by indices (h, k, l) with respect to the basic vectors $\boldsymbol{a}_1$, $\boldsymbol{a}_2$, and $\boldsymbol{a}_3$.

Specifying incident and reflected X-ray beams by the wavevectors $\boldsymbol{K}_0$ and $\boldsymbol{K}$, respectively, the Bragg law can be expressed by

$$\boldsymbol{K}_0 - \boldsymbol{K} = \boldsymbol{G} \qquad \text{where} \quad |\boldsymbol{K}_0| = |\boldsymbol{K}| \tag{6.3}$$

for elastic scattering. The lattice constant is typically of the order of 1 Å ~ 10 Å, and hence the X-ray energies required for diffraction studies of crystals should be in the range of 10 keV to 50 keV. Orbiting electrons are excited to a higher level, but subsequently fall into the ground state by emitting a photon with the same energy as the incident one, whereas heavy ions remain at lattice sites, being unchanged by X-ray impact. In this context, X-ray scattering is regarded as "elastic," so that the lattice is virtually independent of the impact.

While the classical diffraction theory is generally complex, the mathematical consequences can be simplified for a distant point of observation, where the radiation process can be interpreted in a simple manner. The incident X-ray represented by a plane wave $\boldsymbol{E}_0 \exp\{i(\boldsymbol{K}_0 \cdot \boldsymbol{r} - \omega t)\}$ interacts with a target atom located at a point $\boldsymbol{r}_0$ and at a time t_0, inducing an oscillating electric dipole moment $\boldsymbol{p}$ in the excited state as expressed by

$$\boldsymbol{p} \propto \int \rho(\boldsymbol{r}_0)\, \mathrm{d}^3\boldsymbol{r}_0 \boldsymbol{E}_0 \exp\{i(\boldsymbol{K}_0 \cdot \boldsymbol{r}_0 - \omega t_0)\},$$

where $\rho(\boldsymbol{r}_0)$ is the density of electrons at $\boldsymbol{r}_0$, representing the result integrated over the electron cloud. Such an excited dipole moment can radiate a spherical wave whose amplitude at a distant point $\boldsymbol{r}$ is proportional to

$$\frac{\exp[i\{\boldsymbol{K}\cdot(\boldsymbol{r} - \boldsymbol{r}_0) - \omega(t - t_0)\}]}{|\boldsymbol{r} - \boldsymbol{r}_0|}.$$

Therefore, the scattering amplitude at $\boldsymbol{r}$ is expressed by

$$A_0 \propto \left[\int \mathrm{d}^3\boldsymbol{r}_0\, \rho(\boldsymbol{r}_0) \frac{[\exp(i\boldsymbol{K}_0 \cdot \boldsymbol{r}_0)]\{\exp[i\boldsymbol{K}\cdot(\boldsymbol{r} - \boldsymbol{r}_0)]\}}{|\boldsymbol{r} - \boldsymbol{r}_0|}\right] \exp(-i\omega t).$$

At a large distance $|\boldsymbol{r}| \gg |\boldsymbol{r}_0|$, $|\boldsymbol{r} - \boldsymbol{r}_0| \sim r$ is a reasonable approximation for the calculation of A_0, and so

$$A_0 \propto \int \mathrm{d}^3\boldsymbol{r}_0\, \rho(\boldsymbol{r}_0)[\exp(i\boldsymbol{K}\cdot\boldsymbol{r})/r][\exp\{i(\boldsymbol{K}_0 - \boldsymbol{K})\cdot\boldsymbol{r}_0\}] \exp(-i\omega t). \tag{6.4}$$

Practically, a collimated X-ray beam strikes a finite area of the target crystal, where there are a number of identical scatterers at lattice points $\boldsymbol{r}_0 + \boldsymbol{R}_{\mathrm{m}}$, where m = 1, 2, ..., n, and n is the total number of target atoms. Hence, the total scattering amplitude due to these scatterers is given by

$$A = \sum_{\mathrm{m}} A_0 \sum_{\mathrm{m}} \exp\{i(\boldsymbol{K}_0 - \boldsymbol{K}) \cdot \boldsymbol{R}_{\mathrm{m}}\}. \tag{6.5}$$

When the Bragg condition, (6.3) is fulfilled, (6.5) indicates that the total amplitude is maximum, i.e., $A_0\mathrm{n}$, since $\exp(i\boldsymbol{G} \cdot \boldsymbol{R}_{\mathrm{m}}) = 1$ for all m. In this case, the interference of scattered photons is said to be *constructive*, and (6.5) manifests the Bragg diffraction from the group of parallel crystal planes represented by the vector $\boldsymbol{G}$. As shown in Fig. 6.1, the Bragg law suggests a simple geometrical interpretation for elastic scattering.

When the Bragg condition is met, we write the total scattering amplitude from the planes that are specified by $\boldsymbol{G}$ as $A_{\boldsymbol{G}} = A_0\mathrm{n}$. It is noted that for the scattering intensity given by $A_{\boldsymbol{G}}{}^*A_{\boldsymbol{G}}$, some complex exponential factors in (6.3) and (6.4) are insignificant. Therefore, we rewrite (6.4)

$$A_{\boldsymbol{G}0} \propto r^{-1}\{\exp[i(\boldsymbol{K} \cdot \boldsymbol{r} - \omega t)]\} f(\boldsymbol{G}), \quad \text{where} \\ f(\boldsymbol{G}) = \int \mathrm{d}^3\boldsymbol{r}_0\, \rho(\boldsymbol{r}_0) \exp(i\boldsymbol{G} \cdot \boldsymbol{r}_0) \tag{6.6}$$

is called the *atomic form factor*, which gives the scattering amplitude from a crystal whose unit cell contains one scatterer. Equation (6.4), can be expressed in a generalized form for scattering from a unit cell consisting of more than one scatterers, j = 1, 2, ..., p. Namely,

$$A_{\boldsymbol{G}} = \mathrm{n}\{r^{-1} \exp[i(\boldsymbol{K} \cdot \boldsymbol{r} - \omega t)]\} S(\boldsymbol{G}), \quad \text{where} \\ S(\boldsymbol{G}) = \sum_{\mathrm{j}} f_{\mathrm{j}}(\boldsymbol{G}) \exp(i\boldsymbol{G} \cdot \boldsymbol{r}_{0\mathrm{j}}) \tag{6.7}$$

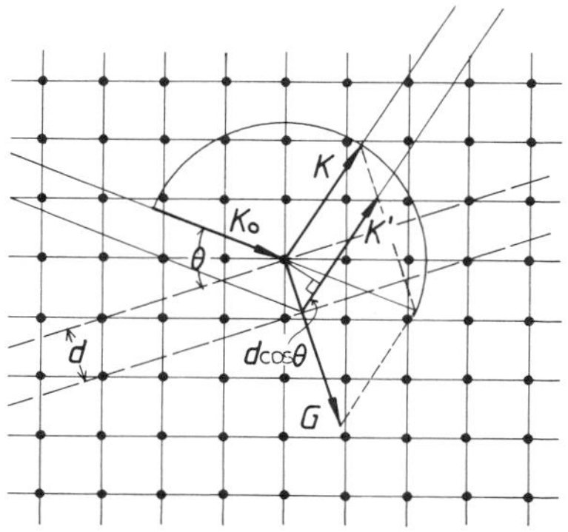

Figure 6.1. Bragg's diffraction from a periodic lattice. Interference of reflected beams from crystal planes $\boldsymbol{G}$ that are shown here by parallel broken lines. The path difference between two reflected beams, $\boldsymbol{K}$ and $\boldsymbol{K}'$, is given by $d \cos \theta$, which determines the condition for constructive interference.

is called the *structural form factor*. Writing $\boldsymbol{r}_{0j} = x_j\boldsymbol{a}_1 + y_j\boldsymbol{a}_2 + z_j\boldsymbol{a}_3$ for atoms in the unit cell, the structural factor can be expressed as a useful formula

$$S(\boldsymbol{G}) = \sum_j f_j(\boldsymbol{G}) \exp\left[\frac{2\pi i}{\Omega}(x_j h + y_j k + z_j l)\right]. \tag{6.8}$$

The scattering intensity is then expressed by

$$I(\boldsymbol{G}) = A_{\boldsymbol{G}}{}^* A_{\boldsymbol{G}} \propto \mathrm{n}^2 r^{-2} S^*(\boldsymbol{G}) S(\boldsymbol{G}). \tag{6.9}$$

In the above argument, the lattice was considered as rigid, however, each atom should be in thermal motion. The thermal effect of X-ray scattering is expressed in terms of scattering intensities that are modified by the so-called Debye–Waller factor, which is derived as follows.

In the presence of thermal motion, the position of a scatterer can be written as $\boldsymbol{r}_0(t) = \boldsymbol{r}_0 + \boldsymbol{u}(t)$. If thermal displacements $\boldsymbol{u}(t)$ are assumed to be uncorrelated as in the Einstein model, such random fluctuations can be characterized by a Gaussian distribution around the average $\langle \boldsymbol{u}(t) \rangle_t = 0$. In this case, the exponential factor in the atomic form factor can be expanded as

$$\langle \exp\{i\boldsymbol{G}\cdot\boldsymbol{r}_0(t)\}\rangle = \exp(i\boldsymbol{G}\cdot\boldsymbol{r}_0)[1 - \tfrac{1}{2}\langle\{\boldsymbol{G}\cdot\boldsymbol{u}(t)\}^2\rangle + \cdots],$$

where

$$\langle\{\boldsymbol{G}\cdot\boldsymbol{u}(t)^2\}\rangle = \frac{G^2\langle u(t)^2\rangle}{3}$$

for isotropic fluctuations. Therefore, the scattering intensity is given by

$$I(\boldsymbol{G}) = I_0(\boldsymbol{G}) \exp\left[-\frac{\langle u(t)^2\rangle G^2}{3}\right],$$

where $I_0(\boldsymbol{G})$ is for the rigid lattice. The squared average $\langle u(t)^2\rangle$ can be evaluated by the equipartion theorem for a harmonic oscillator, i.e., $\frac{1}{2}M\omega^2\langle u(t)^2\rangle = \frac{3}{2}\mathrm{k_B}T$, and the Debye–Waller factor is defined by

$$W = \frac{I(\boldsymbol{G})}{I_0(\boldsymbol{G})} = \exp\left(-\frac{\mathrm{k_B}TG^2}{M\omega^2}\right). \tag{6.10}$$

From (6.10), it is noticed that scattering intensities decrease with increasing temperature, and that diffraction from low $\boldsymbol{G}$ planes are less affected by thermal motion than high $\boldsymbol{G}$ planes.

6.3. Diffraction from Weakly Modulated Structures

In a crystal modulated at a wavevector $\boldsymbol{Q}$ unit cells are not all identical, and hence the wavevector $\boldsymbol{G}_i = \boldsymbol{G} + \boldsymbol{Q}$ does not represent crystal planes. When the magnitude of $\boldsymbol{Q}$ is very small, anisotropically broadened diffraction spots may be observed, or related *satellite spots* may be resolved if the modulation is sufficiently uniform over the crystal for a clear pattern. However, for phase transitions at Brillouin zone boundaries, or at other nonlattice points, where $\boldsymbol{Q}$ is not a small vector, the X-ray diffraction is not a logical method to obtain the modulation vector $\boldsymbol{Q}$. Instead, neutron inelastic scattering experiments provide direct information about the vector $\boldsymbol{Q}$ and its related energy $\varepsilon(\boldsymbol{Q})$.

Near the transition temperature of a crystal undergoing a binary structural change at the Brillouin zone center, the Bragg diffraction spots are diffuse because of critical fluctuations $\Delta\boldsymbol{Q} = \pm\boldsymbol{q}$, as shown in Fig. 6.2, which accompany the corresponding energies $\mp\Delta\varepsilon$. Such small fluctuations, $\boldsymbol{q}$ and $\Delta\varepsilon$, arise from interactions between pseudospins and phonons in a condensate, for which the Bragg diffraction condition is approximately fulfilled, while the scattering is inelastic. The conservation laws of energy and momentum for such a binary system can be written as

$$E(\boldsymbol{K}_0) - E(\boldsymbol{K}) = \mp\Delta\varepsilon(\boldsymbol{q}) \tag{6.11a}$$

and

$$\boldsymbol{K}_0 - \boldsymbol{K} = \boldsymbol{G} \pm \boldsymbol{q}, \tag{6.11b}$$

where $\varepsilon(\boldsymbol{G}) = 0$, and $E(\boldsymbol{K}_0)$ and $E(\boldsymbol{K})$ represent X-ray energies before and

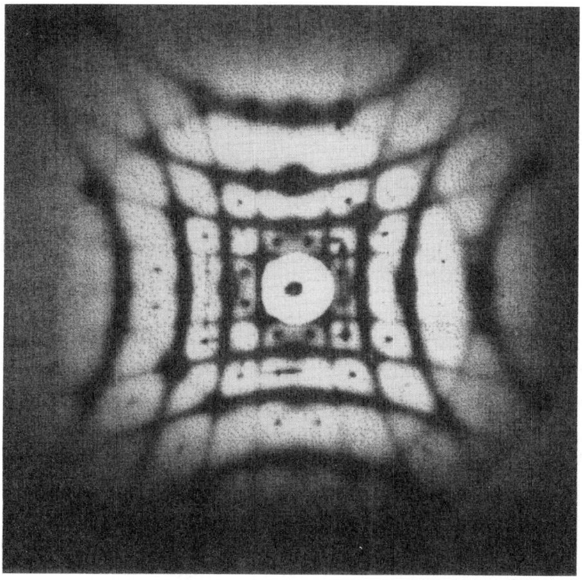

Figure 6.2. A diffuse X-ray scattering pattern from a $NaNbO_3$ crystal at 700° C. The incident beam is parallel to the c axis, being scattered from chain pseudospin modes coexisting in the a and b directions. (From R. Comès, R. Currat, F. Denoyer, M. Lambert, and A. M. Quittet, *Ferroelectrics* **12**, 3 (1976).)

after the impact. In this case, the argument in Section 6.2 can be modified by writing $\omega = E(\boldsymbol{K}_0)/\hbar$, $\omega' = E(\boldsymbol{K})/\hbar$, and $\omega - \omega' = \Delta\omega$, for which the scattering amplitude is expressed as

$$A_0 \propto \int d^3\boldsymbol{r}_0\, \rho(\boldsymbol{r}_0) \exp\{i(\boldsymbol{K}_0\cdot\boldsymbol{r}_0 - \omega t_0)\}\left[\frac{\exp\{i\{\boldsymbol{K}\cdot(\boldsymbol{r}-\boldsymbol{r}_0)\} - \omega'(t-t_0)\}}{|\boldsymbol{r}-\boldsymbol{r}_0|}\right]$$

$$= \int d^3\boldsymbol{r}_0\, \rho(\boldsymbol{r}_0) \exp\{i(\boldsymbol{K}_0\cdot\boldsymbol{r}_0 - \Delta\omega\cdot t_0)\}\left[\frac{\exp\{i\boldsymbol{K}\cdot(\boldsymbol{r}-\boldsymbol{r}_0)\}}{|\boldsymbol{r}-\boldsymbol{r}_0|}\right]\exp(-i\omega' t)$$

$$\approx \int d^3\boldsymbol{r}_0\, \rho(\boldsymbol{r}_0)[r^{-1}\exp\{i(\boldsymbol{K}\cdot\boldsymbol{r} - \omega' t)\}]\exp\{i[(\boldsymbol{K}_0 - \boldsymbol{K})\cdot\boldsymbol{r}_0 - \Delta\omega\cdot t_0]\}.$$

Using (6.11b) and the translational invariance (6.1a), the total scattering amplitude from atoms in the target area is expressed as

$$A_{\boldsymbol{G}} = r^{-1}\exp\{i(\boldsymbol{K}\cdot\boldsymbol{r} - \omega' t)\}\sum_{\mathrm{m}}[f(\boldsymbol{G}+\boldsymbol{q})\exp\{i(\boldsymbol{q}\cdot\boldsymbol{R}_{\mathrm{m}} - \Delta\omega\cdot t_0)\}$$
$$+ f(\boldsymbol{G}-\boldsymbol{q})\exp\{i(-\boldsymbol{q}\cdot\boldsymbol{R}_m + \Delta\omega\cdot t_0)\}], \tag{6.12}$$

where

$$f(\boldsymbol{G}\pm\boldsymbol{q}) = \int d^3\boldsymbol{r}_0\, \rho(\boldsymbol{r}_0)\exp\{i(\boldsymbol{G}\pm\boldsymbol{q})\}\cdot\boldsymbol{r}_0.$$

If $\boldsymbol{q}\perp\boldsymbol{G}$, $f(\boldsymbol{G}+\boldsymbol{q}) = f(\boldsymbol{G}-\boldsymbol{q}) = f(G')$ where $G' = |\boldsymbol{G}\pm\boldsymbol{q}|$, otherwise these atomic form factors are unequal in principle. However, when $\boldsymbol{q}\cdot\boldsymbol{r}_0 \ll 2\pi$, these two factors are practically identical for all the scatterers, and accordingly the scattering intensity is written as

$$I(\boldsymbol{G}\pm\boldsymbol{q}) = r^{-2}|f(G')|^2\sum_{\mathrm{mn}}[\exp[i\{\boldsymbol{q}\cdot(\boldsymbol{R}_{\mathrm{m}} - \boldsymbol{R}_{\mathrm{n}}) - \Delta\omega(t_{0\mathrm{m}} - t_{0\mathrm{n}})\}]$$
$$+ \exp[i\{-\boldsymbol{q}\cdot(\boldsymbol{R}_{\mathrm{m}} - \boldsymbol{R}_{\mathrm{n}}) + \Delta\omega(t_{0\mathrm{m}} - t_{0\mathrm{n}})\}]]$$
$$= I(\boldsymbol{G}) + 2r^{-2}|f(G')|^2\sum_{\mathrm{m}\neq\mathrm{n}}\cos\{\boldsymbol{q}\cdot(\boldsymbol{R}_{\mathrm{m}} - \boldsymbol{R}_{\mathrm{n}}) - \Delta\omega(t_{0\mathrm{m}} - t_{0\mathrm{n}})\}, \tag{6.13}$$

where the second correlation terms on the right in (6.13) are responsible for intensity anomalies due to a modulated structure. In this expression, it is clear that the intensity $I(\boldsymbol{G}\pm\boldsymbol{q})$ is maximum when $\boldsymbol{q}\|(\boldsymbol{R}_{\mathrm{m}} - \boldsymbol{R}_{\mathrm{n}})$, i.e., $\boldsymbol{q}\perp\boldsymbol{G}$, and in this case the modulation effect should be most appreciable. For binary fluctuations, we can consider a one-dimensional mode $\sigma(\phi)$ of scatterers, where the phase $\phi = \boldsymbol{q}\cdot(\boldsymbol{R}_{\mathrm{m}} - \boldsymbol{R}_{\mathrm{n}}) - \Delta\omega(t_{0\mathrm{m}} - t_{0\mathrm{n}}) = qx - \Delta\omega\cdot t$ is virtually a continuous function of space–time coordinates $x = x_{\mathrm{m}} - x_{\mathrm{n}}$ and $t = t_{0\mathrm{m}} - t_{0\mathrm{n}}$. Therefore, such distributed scatterers are represented by a condensate, $\sigma_{\mathrm{P}} \propto \cos\phi$, referring $\phi = 0$ to the pinning equilibrium. It is significant that the intensity is independent of the impact coordinates $(\boldsymbol{R}_{\mathrm{m}}, t_{0\mathrm{m}})$, but related only to relative coordinates (x, t) between correlated pseudospins, and that the diffused intensity $\Delta I(\boldsymbol{G}\pm\boldsymbol{q}) = I(\boldsymbol{G}\pm\boldsymbol{q}) - I(\boldsymbol{G})$ is determined by the distributed phase $0 \le \phi \le 2\pi$. Such modulation broadening of a X-ray diffraction spot is expressed by

$$I(\boldsymbol{G} \pm \boldsymbol{q}) = I(\boldsymbol{G}) + 2r^{-2}|f(G)|^2 \sum_{\mathrm{m} \neq \mathrm{n}} \cos(qx - \Delta\omega \cdot t) \tag{6.14a}$$

and

$$\Delta I(\boldsymbol{G}) = 2r^{-2}|f(G)|^2 \left\langle \int \cos\phi \, \mathrm{dA} \right\rangle_{\mathrm{t}},$$

where A is the impact area of the X-ray beam including all scatterers designated by indices m and n. Since $I(\boldsymbol{G}) = Nr^{-2}|f(G)|^2$ represents the scattering intensity from the unmodulated crystal, where N is the number of scatterers in the area, the broadened intensity is written as

$$\begin{aligned} \Delta I &= I(\boldsymbol{G} \pm \boldsymbol{q}) - I(\boldsymbol{G}) \\ &= 2N^{-1} I(\boldsymbol{G}) {t_0}^{-1} \int_0^{t_0} \mathrm{d}t (L_x L_y)^{-1} \int_0^{L_x} (\cos\phi) \, \mathrm{d}x \int_0^{L_y} \mathrm{d}y, \end{aligned} \tag{6.14b}$$

where we assumed that the target is a rectangular area $L_x L_y$, and t_0 is the timescale of measurement. We have assumed in (6.14b) that condensates $\cos\phi$ are all parallel with the x direction, and uncorrelated in the y direction. Also, such broadening is observable only if $\Delta\omega < {t_0}^{-1}$, otherwise $\langle \cos\phi \rangle_{\mathrm{t}}$ is averaged out.

Here the integration is elementary. That is,

$$\begin{aligned} \int_0^{L_x} \cos\phi \, \mathrm{d}x &= q^{-1} \int_{\phi_1}^{\phi_2} \cos\phi \, \mathrm{d}\phi = q^{-1}(\sin\phi_2 - \sin\phi_1) \\ &= q^{-1} \sin(\tfrac{1}{2}\Delta\phi) \cos \boldsymbol{\phi}, \end{aligned}$$

where $\Delta\phi = \frac{1}{2}(\phi_2 - \phi_1)$ and $\boldsymbol{\phi} = \frac{1}{2}(\phi_2 + \phi_1) = qx - \Delta\omega \cdot t$, the average phase of the scattered X-ray. Since $L_x = q^{-1} 2\Delta\phi$, this result can be expressed as

$${L_x}^{-1} \int_0^{L_x} \cos\phi \, \mathrm{d}x = \frac{\sin(\frac{1}{2}\Delta\phi)}{(\frac{1}{2}\Delta\phi)} \cos \boldsymbol{\phi},$$

which is a familiar expression for Fraunhofer's light diffraction through a slot of width $2\Delta x$. The time integration in (6.14b) can also be performed similarly, i.e.,

$${t_0}^{-1} \int_0^{t_0} \cos \boldsymbol{\phi} \, \mathrm{d}t = {t_0}^{-1} \int_0^{t_0} \cos(\Delta\omega \cdot t) \, \mathrm{d}t = \frac{\sin(\Delta\omega \cdot t_0)}{(\Delta\omega \cdot t_0)} \cos(\boldsymbol{\phi} - \tfrac{1}{2}\Delta\omega \cdot t_0).$$

Combining the above results, the intensity anomaly ΔI of (6.14b) is given by

$$\Delta I \propto \frac{\sin(q\,\Delta x)}{(q\,\Delta x)} \frac{\sin(t_0\,\Delta\omega)}{(t_0\,\Delta\omega)} \cos(\boldsymbol{\phi} - \tfrac{1}{2}\Delta\omega \cdot t). \tag{6.15}$$

Here, using the well-known formula $\lim_{\theta \to 0} \sin\theta/\theta \to 1$, the two front factors are practically equal to 1, if $q\,\Delta x \ll 1$ and $t_0\,\Delta\omega \ll 1$. The phase of cos function in (6.15) can be redefined as $\phi = \boldsymbol{\phi} - \frac{1}{2}\Delta\omega \cdot t_0$, indicating that $\Delta I(\phi)$

is related to the distributed phase $0 \le \phi \le 2\pi$. Although the diffused diffraction is clearly due to the distributed phase, no accurate information can be obtained from such a small diffraction spot. A similar *lineshape* analysis is significant for neutron inelastic scattering spectra, as will be discussed in Section 6.5.

6.4. Diffuse Laue Diffraction from Perovskite Crystals in the Critical Region

Comès et al. [19] carried out X-ray studies on structural phase transitions in various perovskite crystals, and observed diffuse diffraction patterns at critical temperatures. To explain the anomalous diffraction patterns, these authors have suggested that low-dimensional modulations of order variables are responsible for critical anomalies. We discussed in Chapter 3 collective modes of pseudospins in perovskite crystals, and obtained possible two-dimensional incommensurate modes in the plane perpendicular to the tetragonal axis, which are consistent with the X-ray and magnetic resonance results [20]. In this section, we show that the diffraction pattern in Fig. 6.2, which Comès et al. obtained from $NaNbO_3$ crystals at 700° C, can be explained with the model of fluctuating condensates at least qualitatively.

It is realized that the Bragg law is applicable to a rigid crystal, showing the exact direction of the diffracted X-ray beam. Diffraction spots can also be figured out by the Laue conditions, which are derivable from the Bragg law. Multiplying both sides of the Bragg equation

$$\Delta \boldsymbol{K} = \boldsymbol{K}_0 - \boldsymbol{K} = G$$

by

$$\boldsymbol{R} = n_1 \boldsymbol{a}_1 + n_2 \boldsymbol{a}_2 + n_3 \boldsymbol{a}_3,$$

we obtain

$$\boldsymbol{R} \cdot \Delta \boldsymbol{K} = \boldsymbol{G} \cdot \boldsymbol{R} = 2\pi,$$

or equivalently

$$\Delta \boldsymbol{K} \cdot \boldsymbol{a}_1 = \frac{2\pi}{n_1}, \qquad \Delta \boldsymbol{K} \cdot \boldsymbol{a}_2 = \frac{2\pi}{n_2}, \qquad \text{and} \qquad \Delta \boldsymbol{K} \cdot \boldsymbol{a}_3 = \frac{2\pi}{n_3}. \tag{6.16}$$

Equations (6.16) are known as the Laue conditions, allowing us to determine diffraction spots geometrically. (For details, see Chapter 2 in Kittel's textbook.) In each of (6.16), the locus of the end point of the vector $\boldsymbol{K}$ shows a cone with the axis along the basic translational vector $\boldsymbol{a}_i$, and crossing points between these hyperbolic curves on a plane for photographic observation represent the so-called Laue spots. That is only a graphical procedure, however, it is significant to realize that scattered radiations are on these conical surfaces, and each of the intersections is where the radiations interfere constructively. The Laue conditions are more practical than the Bragg law, when dealing with diffraction from a modulated structure in low dimension, where the concept of crystal planes cannot be used.

Consider that a collimated X-ray beam is incident perpendicularly onto a linear chain of the lattice constant a, i.e., $\boldsymbol{K}_0 \perp \boldsymbol{a}$. At a diffraction angle θ, the condition for constructive interference is specified by the relation

$$a \cos \theta = h\lambda,$$

where $\lambda = 2\pi/K$ is the wavelength of the X-ray, and h is an integer. This condition can be expressed in terms of the corresponding reciprocal lattice unit $a^* = 2\pi/a$, i.e.,

$$|\boldsymbol{K}| \cos \theta = ha^* = G,$$

which is identical to a Laue condition applied to the linear chain. As shown in Fig. 6.3(a), in this case, the diffracted beam is on a conical surface of the apex angle 2θ, so that the wavevector $\boldsymbol{K}$ of the scattered X-ray has compo-

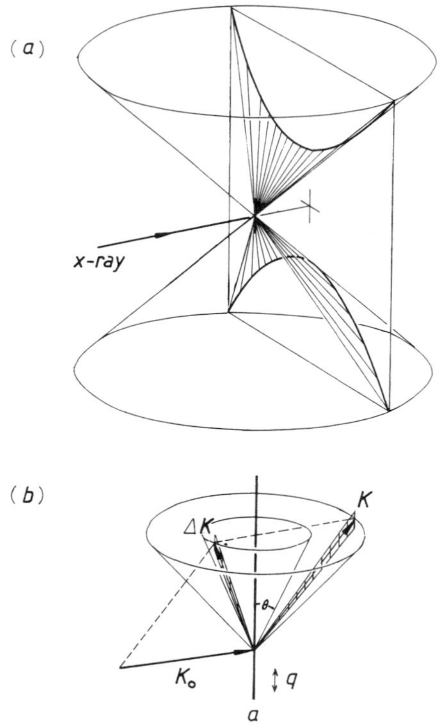

Figure 6.3. (a) X-ray diffraction observed in the plane of pseudospin chains. The incident beam is perpendicular to the plane. (b) Diffuse diffraction geometry for chains with longitudinal fluctuations.

nents

$$K_{\parallel} = G \quad \text{and} \quad K_{\perp} \text{ in any direction perpendicular to the axis.}$$

The atomic form factors in such a chain, when modulated at $\pm q$, are given by

$$f(G \pm q) = \int d^3 r_0 \, \rho(r_0) \exp\{i(G \pm q) r_0\}.$$

Following (6.12), the total amplitude from all scatterers in the impact area is

$$A_G \propto \sum_{\mathrm{m}} \{f(G + q) \exp\{i(qx_{\mathrm{m}} - \Delta\omega \cdot t_{0\mathrm{m}})\} + f(G - q) \exp\{i(-qx_{\mathrm{m}} + \Delta\omega \cdot t_{0\mathrm{n}})\}\},$$

where $x_{\mathrm{m}} = \mathrm{m}a$, and $\mathrm{m} = 1, 2, \ldots, \mathrm{p}$. Since $f(G + q) = f(G - q) \approx f(G)$, we have

$$I(G \pm q) = A_G{}^* A_G \propto I(G) + |f(G)|^2 \sum_{\mathrm{mn}} \cos\{q(x_{\mathrm{m}} - x_{\mathrm{n}}) - \Delta\omega(t_{0\mathrm{m}} - t_{0\mathrm{n}})\}$$

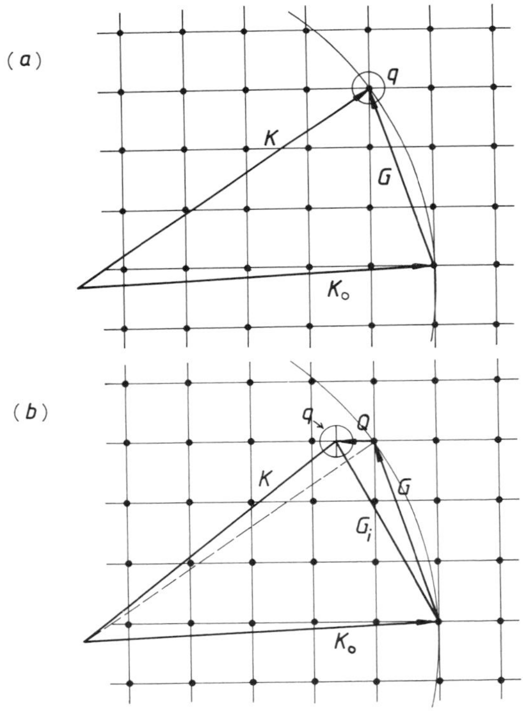

Figure 6.4. Scattering geometries in the reciprocal lattice, $\boldsymbol{K}_0 = \boldsymbol{K} + \boldsymbol{Q} \pm \boldsymbol{q}$. (a) Elastic scattering, $\boldsymbol{Q} = 0$. (b) Inelastic scattering where $\boldsymbol{Q} \neq 0$ and incommensurate. In both eases, the fluctuations $\pm \boldsymbol{q}$ are small and incommensurate.

for the scattering intensity. Therefore, considering that $x = x_m - x_n$ and $t = t_{0m} - t_{0n}$ are continuous space–time variables, we can use the phase variable $\phi = qx - \Delta\omega \cdot t$ for the intensity anomaly, i.e.,

$$\Delta I \propto |f(G)|^2 t_0^{-1} \int_0^{t_0} dt \cdot L^{-1} \int_0^L \cos\phi \, dx \sim \cos \boldsymbol{\phi} \qquad \text{where} \quad 0 \le \phi \le 2\pi.$$

Figure 6.3(b) shows how scattered X-ray is spread by distributed phases along the a direction.

In the photograph shown in Fig. 6.2, the diffraction pattern consists of two families of hyperbolic lines due to modulations along the b and c axes, which are appreciably broadened as attributed to by irrational wavevector components q_b and q_c, respectively. In the above argument, we assumed that phases of those parallel condensates are uncorrelated, while interchain correlations are responsible for incommensurability. In the tetragonal phase, the b and c directions are equivalent, so that such parallel condensates occur in the b and c directions with an equal probability, leading to two perpendicular families of hyperbolic diffraction lines.

6.5. Neutron Inelastic Scattering

A structural phase transition arises generally from the singular behavior of the collective order-variable mode, which occurs at a specific transition temperature. Originating from short-range correlations in the crystal, such a collective mode is primarily independent of the lattice vibration, but coupled with a soft lattice mode in the critical region. Although neutrons are scattered by heavy ions, scattered neutrons reflect the nature of condensates. Generally, inelastic scattering occurs at a specific wavevector $\boldsymbol{Q}$ and an energy $\varepsilon(\boldsymbol{Q})$, when pseudospin and soft modes are in near phase in the sinusoidal limit, where their dispersion relation $\varepsilon(\boldsymbol{Q})$ versus $\boldsymbol{Q}$ in the critical region can be determined by neutron inelastic scattering experiments. It is noted that neutrons are generally probing the lattice mode of heavy atoms, which constitutes a part of the transition mechanism responsible for a structural change. While coupled in the critical region, neutrons cannot fully sample the pseudospin mode showing a nonlinear character with decreasing temperature.

On the dispersion curve $\varepsilon = \varepsilon(\boldsymbol{Q})$, determined by neutron scattering experiments, the phase transition can be identified by a temperature-dependent dip at a particular $\boldsymbol{Q}$, corresponding to the soft-mode energy, as seen typically from an example of the phase transition in K_2SeO_4 at $\boldsymbol{Q} \sim 2\boldsymbol{a}^*/3$ in Fig. 4.5. For the inelastic scattering of neutrons, the conservation laws are written as

$$\boldsymbol{K}_0 - \boldsymbol{K} = \boldsymbol{G}_i = \boldsymbol{G} + \boldsymbol{Q}$$

and

$$E(\boldsymbol{K}_0) - E(\boldsymbol{K}) = \varepsilon(\boldsymbol{G} + \boldsymbol{Q}) = \varepsilon(\boldsymbol{Q}).$$

It is noted that the energy $\varepsilon(\boldsymbol{Q})$ at the transition temperature should correspond thermodynamically to the minimum of the Gibbs thermodynamical potential at $\boldsymbol{Q}$. Such a smooth minimum in the dispersion curve signifies that there are fluctuations in the wavevector as expressed by $\boldsymbol{Q} \pm \boldsymbol{q}$. Considering binary fluctuations, the conservation laws should be revised as

$$\boldsymbol{K}_0 - \boldsymbol{K} = \boldsymbol{G} + \boldsymbol{Q} \pm \boldsymbol{q} \tag{6.16a}$$

and

$$E(\boldsymbol{K}_0) - E(\boldsymbol{K}) = \varepsilon(\boldsymbol{Q} + \boldsymbol{q}) = \varepsilon(\boldsymbol{Q}) \mp \Delta\varepsilon(\boldsymbol{q}). \tag{6.16b}$$

In Fig. 6.4, the wavevector geometry to neutron inelastic scattering (b) is composed with elastic scattering (a). If $\boldsymbol{q} \ll \boldsymbol{Q}$, the vicinity of the minimum is approximately parabolic, so that the curve is represented by $\Delta\varepsilon \approx \kappa \boldsymbol{q}^2$, where κ is a constant.

The basic setup for neutron scattering is sketched in the following. Figure 6.5 illustrates schematically a *triple axis* spectrometer that is commonly used for inelastic neutron scattering experiments. A thermally moderated neutron flux from a nuclear reactor is admitted to a monochromator crystal (A) of a known lattice constant d, from which a neutron beam of a wavelength λ_0 can be selected by adjusting the reflection angle θ to fit the Bragg formula $n\lambda_0 = d \sin\theta$. A sample crystal is mounted on a goniometer with a known axis of rotation in the cryostat (B). The wavelength λ, of scattered neutrons by the sample, is then analyzed by the analyzer (C) consisting of a crystal with the lattice constant d'. Using the formula $n'\lambda = d' \sin\theta'$ again, λ can be calculated with the known d' and the measured angle θ'. By plotting scattered neutron intensity at each setting of $\boldsymbol{Q}$ in the scattering geometry, the dispersion curve can be drawn. When an anomalous dip is found in the curve, the energy loss of neutrons $\varepsilon(\boldsymbol{Q})$ can be determined.

Similar to the atomic form factor for X-ray diffraction, we can define the nuclear scattering amplitude by active groups as

$$n_{\mathrm{m}}(\boldsymbol{G}_{\mathrm{i}} \pm \boldsymbol{q}) = n_0 \exp\left[i \left\{ (\boldsymbol{G}_{\mathrm{i}} \pm \boldsymbol{q}) \cdot \boldsymbol{r}_m - \frac{\Delta\varepsilon(\boldsymbol{q}) t_{0\mathrm{m}}}{\hbar} \right\} \right]$$

$$= n_0 \exp(i\boldsymbol{G}_{\mathrm{i}} \cdot \boldsymbol{r}_{\mathrm{m}}) \exp\{ i(\pm \boldsymbol{q} \cdot \boldsymbol{r}_{\mathrm{m}} - \Delta\omega \cdot t_{0\mathrm{m}}) \},$$

where $\omega = \varepsilon(\boldsymbol{q})/\hbar$. The scattering intensity is then given by

$$I(\boldsymbol{G}_{\mathrm{i}} \pm \boldsymbol{q}) = \left\langle \sum_{\mathrm{mn}} [n_{\mathrm{m}}^*(\boldsymbol{G}_{\mathrm{i}} + \boldsymbol{q}) n_{\mathrm{n}}(\boldsymbol{G}_{\mathrm{i}} + \boldsymbol{q}) + n_{\mathrm{m}}^*(\boldsymbol{G}_{\mathrm{i}} - \boldsymbol{q}) n_{\mathrm{n}}(\boldsymbol{G}_{\mathrm{i}} - \boldsymbol{q})] \right\rangle_{\mathrm{t}}$$

$$= \left\langle \sum_{\mathrm{m}} |n_{\mathrm{m}}(\boldsymbol{G}_{\mathrm{i}})|^2 \right\rangle_{\mathrm{t}}$$

$$+ \left\langle \sum_{\mathrm{m} \neq \mathrm{n}} [n_{\mathrm{m}}^*(\boldsymbol{G}_{\mathrm{i}} + \boldsymbol{q}) n_{\mathrm{n}}(\boldsymbol{G}_{\mathrm{i}} + \boldsymbol{q}) + n_{\mathrm{m}}^*(\boldsymbol{G}_{\mathrm{i}} - \boldsymbol{q}) n_{\mathrm{n}}(\boldsymbol{G}_{\mathrm{i}} - \boldsymbol{q})] \right\rangle_{\mathrm{t}}.$$

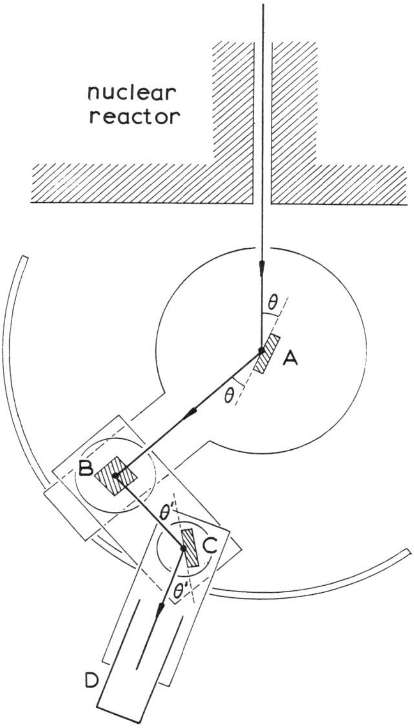

Figure 6.5. A typical experimental set-up for neutron inelastic scattering experiments with a triple-axis spectrometer. A: a shielded monochromator for high-flux thermal neutron beam from a nuclear reactor. B: a sample crystal in a cryostat. C: an analyzer crystal. These three crystals in A, B, and C are rotatable around parallel axes that are all perpendicular to the page.

Therefore,

$$\begin{aligned} & I(\boldsymbol{G}_{\mathrm{i}} \pm \boldsymbol{q}) - I(\boldsymbol{G}_{\mathrm{i}}) \\ &= \Delta I = n_0{}^2 \Big\langle \sum_{\mathrm{m}\neq\mathrm{n}} \exp\{i\boldsymbol{G}_{\mathrm{i}}\cdot(\boldsymbol{r}_{0\mathrm{m}} - \boldsymbol{r}_{0\mathrm{n}})\} \\ &\quad \times [\exp[i\{\boldsymbol{q}\cdot(\boldsymbol{r}_{0\mathrm{m}} - \boldsymbol{r}_{0\mathrm{n}}) - \Delta\omega(t_{0\mathrm{m}} - t_{0\mathrm{n}})\}] \\ &\quad + \exp[i\{-\boldsymbol{q}\cdot(\boldsymbol{r}_{0\mathrm{m}} - \boldsymbol{r}_{0\mathrm{n}}) + \Delta\omega(t_{0\mathrm{m}} - t_{0\mathrm{n}})\}]\Big\rangle_{\mathrm{t}} \\ &= 2n_0{}^2 \Big\langle \sum_{m\neq\mathrm{n}} \cos \boldsymbol{G}_{\mathrm{i}}\cdot(\boldsymbol{r}_{0\mathrm{m}} - \boldsymbol{r}_{0\mathrm{n}}) \cos\{\boldsymbol{q}\cdot(\boldsymbol{r}_{0\mathrm{m}} - \boldsymbol{r}_{0\mathrm{n}}) - \Delta\omega(t_{0\mathrm{m}} - t_{0\mathrm{n}})\}\Big\rangle_{\mathrm{t}}. \end{aligned} \tag{6.17}$$

Here, we have considered that the summation in (6.17) is symmetrical with respect to the indices m and n.

The above argument is made for binary fluctuations of $\boldsymbol{Q}$, where $n_0(\boldsymbol{G}_i \pm \boldsymbol{q}) \approx n_0(\boldsymbol{Q})$ is assumed, or ΔI is expressed for symmetrical fluctuations $|\boldsymbol{G}_i + \boldsymbol{q}| = |\boldsymbol{G}_i - \boldsymbol{q}|$. However, the fluctuations may not necessarily be so in general, particularly when $\boldsymbol{Q}$ is not in the direction of the reciprocal lattice vector $\boldsymbol{G}$. In this case, (6.17) should be modified as

$$\Delta I \propto \left\langle \sum_{m \neq n} [|n_0(\boldsymbol{G}_i + \boldsymbol{q})|^2 \cos\{\boldsymbol{G}_i \cdot (\boldsymbol{r}_{0m} - \boldsymbol{r}_{0n})\} \right.$$
$$\times \exp[i\{\boldsymbol{q} \cdot (\boldsymbol{r}_m - \boldsymbol{r}_n) - \Delta\omega(t_{0m} - t_{0n})\}]$$
$$+ |n_0(\boldsymbol{G}_i - \boldsymbol{q})|^2 \cos\{\boldsymbol{G}_i \cdot (\boldsymbol{r}_{0m} - \boldsymbol{r}_{0n})\}$$
$$\left. \times \exp[i\{-\boldsymbol{q} \cdot (\boldsymbol{r}_{0m} - \boldsymbol{r}_{0n}) + \Delta\omega(t_{0m} - t_{0n})\}] \right\rangle_t,$$

where

$$|n_0(\boldsymbol{G}_i \pm \boldsymbol{q})|^2 \approx |n_0(\boldsymbol{G}_i)|^2 \pm 2i \operatorname{grad} |n_0(\boldsymbol{G}_i)|^2 \cdot \boldsymbol{q} \qquad \text{for} \quad |\boldsymbol{q}| \ll |\boldsymbol{G}_i|.$$

Therefore

$$\Delta I \propto |n_0(\boldsymbol{G}_i)|^2 \sum_{m \neq n} \cos\{\boldsymbol{G}_i \cdot (\boldsymbol{r}_{0m} - \boldsymbol{r}_{0n})\}$$
$$\times \langle \cos\{\boldsymbol{q} \cdot (\boldsymbol{r}_{0m} - \boldsymbol{r}_{0n}) - \Delta\omega(t_{0m} - t_{0n})\}\rangle_t$$
$$+ 2(\operatorname{grad} |n_0(\boldsymbol{G}_i)|^2 \cdot \boldsymbol{q}) \sum_{m \neq n} \cos\{\boldsymbol{G}_i \cdot (\boldsymbol{r}_{0m} - \boldsymbol{r}_{0n})\}$$
$$\times \langle \sin\{\boldsymbol{q} \cdot (\boldsymbol{r}_{0m} - \boldsymbol{r}_{0n}) - \Delta\omega(t_{0m} - t_{0n})\}\rangle_t. \tag{6.18}$$

Here, the space–time coordinates, $\boldsymbol{r}_{0m} - \boldsymbol{r}_{0n} = \boldsymbol{R}$ and $t_{0m} - t_{0n} = t$, are virtually continuous for small $\boldsymbol{q}$, and hence the phase $\phi = \boldsymbol{q} \cdot (\boldsymbol{r}_m - \boldsymbol{r}_n) - \Delta\omega(t_{0m} - t_{0n})$ may be considered as representing the continuous phase of condensates distributed in the range $0 \leq \phi \leq 2\pi$. With regard to the time average in (6.18), we have already discussed in Section 6.3 the same in relation to the timescale of observation t_0. It is noted that the time-independent factor

$$f(\boldsymbol{G}_i) = \cos(\boldsymbol{G}_i \cdot \boldsymbol{R})$$

is related to the intensity, and equal to 1 when $\boldsymbol{G}_i$ is a reciprocal lattice vector, otherwise expressing a broadening in the unmodulated lattice. The time integrals in (6.18) can be expressed as

$$t_0^{-1} \int_t^{t+t_0} \cos(\boldsymbol{q} \cdot \boldsymbol{R} - \Delta\omega \cdot t)\,\mathrm{d}t = \frac{\sin(t_0\,\Delta\omega)}{(t_0\,\Delta\omega)} \cos\{\boldsymbol{q} \cdot \boldsymbol{R} - \Delta\omega(t + \tfrac{1}{2}t_0)\},$$

when calculated for a short interval t_0. Redefining the phase by $\phi = \boldsymbol{q} \cdot \boldsymbol{R} - \Delta\omega(t + \frac{1}{2}t_0)$, the result is still expressed as proportional to $\cos\phi$ with somewhat reduced amplitude when $t_0\,\Delta\omega \ll 1$, and a similar result can be obtained for $\langle \sin\phi \rangle_t$. Thus, the intensity given by (6.18) consists of two contributions in the critical region, namely, referring to the soft phonon modes

$$\Delta I = \Delta I_A + \Delta I_P,$$

where

$$\Delta I_A = A \cos \phi \qquad \text{and} \qquad \Delta I_P = P \sin \phi. \tag{6.19a}$$

These represent scattering intensity anomalies due to fluctuations of the lattice in amplitude and phase modes, respectively. Notice that here sin and cos assigned to these modes of fluctuations A and P are opposite to the corresponding pseudospin modes, as discussed in Chapter 4. The factors A and P in (6.19a) are given by

$$A \propto |n_0(\boldsymbol{G}_i)|^2 f(\boldsymbol{G}_i)^2 \alpha_t \qquad \text{and} \qquad P \propto \boldsymbol{q} \cdot (\text{grad}\, |n_0(\boldsymbol{G}_i)|^2) f(\boldsymbol{G}_i)^2 \alpha_t, \tag{6.19b}$$

where

$$\alpha_t = \frac{\sin(t_0\, \Delta\omega)}{(t_0\, \Delta\omega)}$$

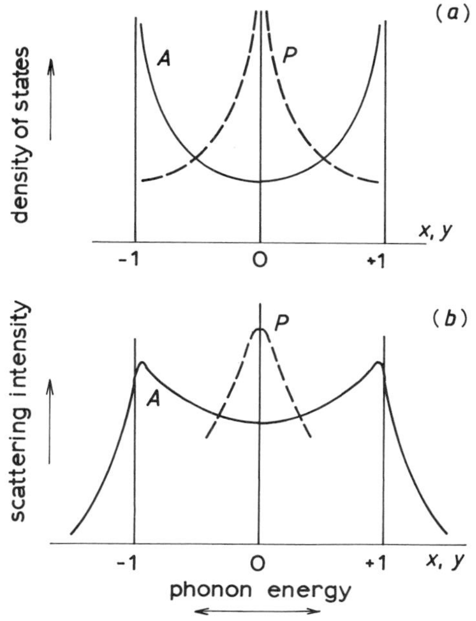

Figure 6.6. Intensity distribution of scattered neutrons from sinusoidally modulated condensates in the critical region. A and P indicate neutron intensity distributions scattered from fluctuating pseudospins in amplitude and phase modes, respectively, as a function of phonon energies. (a) Densities of phonon states. (b) Simulated distributions.

is the intensity reduction factor. Applying (6.19b) to low-dimensional ordering in crystals, both A and P are anisotropic, and P can vary between a maximum and zero for $\boldsymbol{q}$ in parallel and perpendicular to $\text{grad}\,|n_0(\boldsymbol{G}_i)|^2$, respectively.

Scattering intensities due to these fluctuation modes are distributed as a function of the condensate phase ϕ, and described as $f(\Delta I_A)\,d\phi$ and $g(\Delta I_P)\,d\phi$ between ϕ and $\phi + d\phi$. Hence, such distributed intensities are expressed

$$f(\Delta I_A)\,d\phi = \frac{f(\Delta I_A)\,d(\Delta I_A)}{|\sin\phi|} = \frac{f(Ax)\,dx}{(1-x^2)^{1/2}} \tag{6.20a}$$

and

$$g(\Delta I_P)\,d\phi = \frac{g(\Delta I_P)\,d(\Delta I_P)}{|\cos\phi|} = \frac{g(Py)\,dy}{(1-y)^{1/2}}, \tag{6.20b}$$

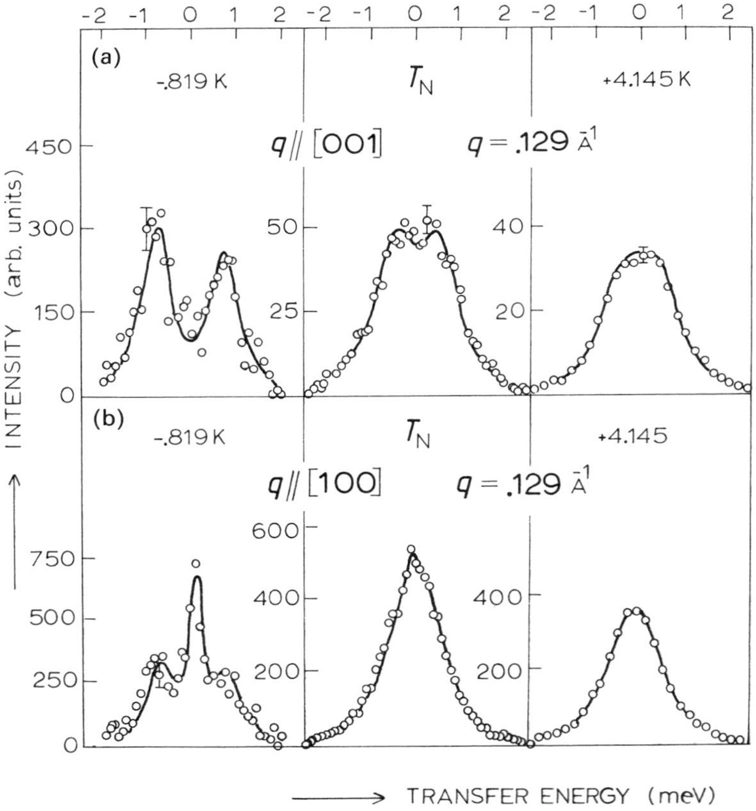

Figure 6.7. Anomalously scattered neutron intensities from antiferromagnetic MnF_2 crystals near the Neél temperature $T_N = 67.459$ K. (a) $\boldsymbol{q}\|[001]$, showing only the transversal mode of fluctuations (b) $\boldsymbol{q} \perp [100]$, where two distinct modes of fluctuations are clearly seen. (From M. P. Schlhof, R. Nathans, and A. Linz, *Phys. Rev. Lett.* **24**, 1184 (1970).)

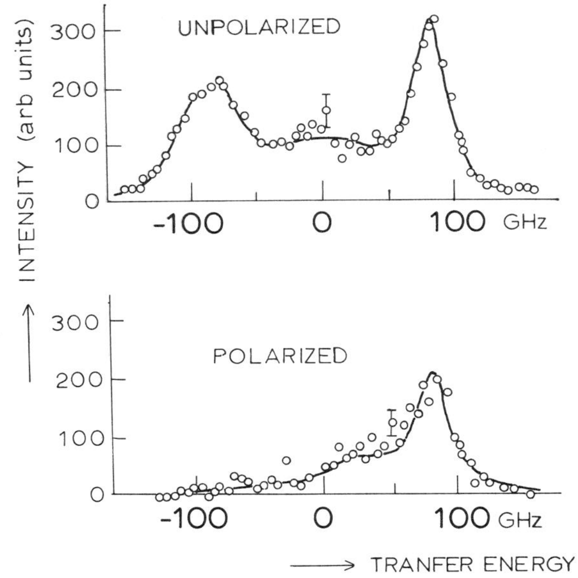

Figure 6.8. Anomalous intensity distributions of unpolarized and polarized scattered neutrons from magnetic spin solitons in $CsNiF_3$ crystals. Here, the scattering was dominated by the amplitude mode. (From R. Pinn and B. E. F. Fender, *Phys. Today*, **38**, 47 (1985).)

where $x = \Delta I_A/A$ and $y = \Delta I_P/P$. Equations (6.20a, b) are useful formula for "shape analysis" of scattering lines when scanned as a function of the neutron transfer energy during the inelastic scattering process, which is proportional to $\Delta I_{A,P}$. Figure 6.6 illustrates such distribution densities as in (6.20a, b), which are characterized as being spread between two edges for the amplitude mode and centralized for the phase mode, respectively.

In the above we discussed neutron inelastic scattering for studies of structural phase transitions. The results can also be applied to magnetic systems, where scattering is due to interactions with magnetic spins. In this case, similar to binary dielectric systems, the fluctuations in a cos mode must be assigned to the phase mode, and those in a sin mode represent the amplitude mode. In Figs. 6.7 and 6.8 are shown typical examples of scattering intensity anomalies observed from magnetic crystals. Schulhof et al. [52] reported such transition anomalies near the Neél temperature T_N of antiferromagnetic MnF_2 crystals, and a similar anomaly from a one-dimensional magnet $CsNiF_3$ was also discussed by Pinn and Fender [53]. In MnF_2 crystals, where fluctuations are anisotropic, the spectra for $\boldsymbol{Q} \| \boldsymbol{a}$ below T_N consist of a broadened doublet typical of (6.20a), but representing spin fluctuations in the phase mode, and a single line arising from the amplitude mode given by (6.20b), whereas there is no single line for $\boldsymbol{Q} \perp \boldsymbol{a}$. In a uniaxial magnet $CsNiF_3$, the reported spectrum consists virtually of a doublet of spin fluctuations, though visibly asymmetrical due perhaps to the cycloidal distribution of classical spin vectors with a finite magnitude.

A remark for binary fluctuations. In scattering experiments, fluctuations $(\boldsymbol{q}, -\Delta\omega) \leftrightarrow (-\boldsymbol{q}, \Delta\omega)$ are detectable as a consequence of the conservation laws of momentum and energy. It is noted that such fluctuations are consistent with space–time fluctuations $(x, t) \leftrightarrow (-x, -t)$, and hence are characterized by $\phi \leftrightarrow -\phi$ discussed in Chapter 4.

CHAPTER 7

Light Scattering and Dielectric Studies on Modulated Crystals

7.1. Raman Scattering Studies on Structural Phase Transitions

Neutron inelastic scattering is a straightforward method for investigating the energy and wavevector exchange between pseudospins and their hosting lattice in the critical region of structural phase transitions. However at the Brillouin zone center, $\boldsymbol{G} = 0$, the scattering geometry $\boldsymbol{K}_0 = \boldsymbol{K}$ demands a measurement at a very small angle, making the experiment extremely difficult. Instead, it is more practical to perform light-scattering experiments with a right-angle geometry, i.e., $\boldsymbol{K}_0 \perp \boldsymbol{K}$, by using intense coherent light from a laser oscillator.

In crystals, Raman scattering takes place if incident light is scattered inelastically from scatterers, where a part of the photon energy is imparted to or gained from the lattice vibration with certain probabilities. In such a process, scattered photons carry an energy equal to the incident photon energy minus or plus a vibrational quantum energy. Therefore, Raman spectra can be useful to study pseudospin condensates composed partly of a lattice mode, provided that a constituent ion can be active in the inelastic scattering process. In general, however, pseudospins may not be Raman-active targets, and Raman-active ions are not necessarily associated with the pseudospin mode either. Nevertheless, being modulated by the soft mode in the critical region, its optical transitions in the active ions can be utilized to study the condensate mechanism. Although generally weak in intensity as compared with elastic Rayleigh scattering, inelastic Raman scattering can be observed with enhanced intensities, when stimulated by a strong coherent radiation.

Consider optical transitions between the ground state "0" and an excited state "1" of an ion in a pseudospin condensate. Regardless of the relation to the pseudospins, the ionic states in the critical region should be modified by the soft lattice mode. Placzek discussed the general problem of perturbed ionic states with adiabatic approximation (see Chapter 4 in [15]), and his results can be used for the present discussion of condensates. Following his theory, we write the wavefunctions of these electronic states as

$$\psi_0 = \varphi_0 \chi_n \qquad \text{and} \qquad \psi_1 = \varphi_1 \chi_n,$$

where φ_0 and φ_1 are eigenfunctions of unmodified ionic states 0 and 1, i.e., for the ionic Hamiltonian $\mathscr{H}$, $\mathscr{H}\varphi_0 = \varepsilon_0 \varphi_0$ and $\mathscr{H}\varphi_1 = \varepsilon_1 \varphi_1$; and the function χ_n represents a vibrational state n of the lattice mode. Here, assuming $\hbar\varpi \ll \varepsilon_1 - \varepsilon_0$, where ϖ is the frequency of the lattice mode, in each of these perturbed states the ion can be considered as "polarized" by means of these vibrational substates. The pertubed ionic energies are therefore given by $\varepsilon_0 + n\hbar\varpi$ and $\varepsilon_1 + n\hbar\varpi$, where the induced dipole moment is denoted by $\boldsymbol{p}$.

Coherent light from a laser oscillator is normally linearly polarized for spectroscopic studies, which is expressed as equivalent to oppositely rotating circular waves, namely,

$$\boldsymbol{E} \cos \omega t = \tfrac{1}{2}\boldsymbol{E}_+ \exp(i\omega t) + \tfrac{1}{2}\boldsymbol{E}_- \exp(-i\omega t),$$

where the target ion is considered to be at the fixed position $\boldsymbol{r} = 0$. Here $\boldsymbol{E}_\pm$ are amplitudes of these circularly polarized components of the incident electric waves that propagate in the direction of the wavevector $\boldsymbol{k}$. The ionic dipole moment $\boldsymbol{p}$ can interact with the electric field as $-\boldsymbol{p}\cdot\boldsymbol{E} \cos \omega t$, and hence we have to solve the time-dependent Schrödinger equation

$$[\mathscr{H} - \tfrac{1}{2}\boldsymbol{p}\cdot\boldsymbol{E}_+ \exp(i\omega t) - \tfrac{1}{2}\boldsymbol{p}\cdot\boldsymbol{E}_- \exp(-i\omega t)]\Psi = \frac{i\hbar\partial\Psi}{\partial t}. \tag{7.1}$$

Applying (7.1) to the ground state 0, the perturbed ionic wavefunction can be expressed as

$$\Psi_0 = \psi_0 \exp\left\{-\frac{i(\varepsilon_0 + n\hbar\varpi)t}{\hbar}\right\}$$
$$+ [\psi_0{}^+ \exp(i\omega t) + \psi_0{}^- \exp(-i\omega t)] \exp\left\{-\frac{i(\varepsilon_0 + n\hbar\varpi)t}{\hbar}\right\}.$$

Here the first term represents the unperturbed ground state, whereas the second term is the correction due to the perturbation. In the first-order approximation, (7.1) can be converted to a pair of equations that are independent of time t, i.e.,

$$\{\mathscr{H} - (\varepsilon_0 + n\hbar\varpi) \pm \hbar\omega\}\psi_0{}^\pm = \tfrac{1}{2}\boldsymbol{p}\cdot\boldsymbol{E}_\pm\psi_0. \tag{7.2}$$

The functions $\psi_0{}^\pm$ represent polarized ionic states $\varepsilon_0{}^\pm = \varepsilon_0 + (n \pm 1)\hbar\varpi$ by the applied circular fields $\boldsymbol{E}_\pm$, respectively, which are connected with the ground state ε_0 by nonvanishing matrix elements $\langle\psi_0{}^{\pm}|\boldsymbol{p}|\psi_0\rangle$, which are abbreviated as $\langle\pm|\boldsymbol{p}|0\rangle$ in the following calculation.

As assumed, the energy difference between such polarized and unpolarized ionic states is much smaller than the ionic excitation energy, i.e.,

$$\varepsilon_0{}^{\pm} - (\varepsilon_0 + \mathrm{n}\hbar\varpi) = \pm\hbar\varpi \ll \varepsilon_1 - \varepsilon_0,$$

so that these vibrational substates can be considered as degenerate, forming a complete set. We can therefore write

$$\psi_0{}^{\pm} = \varphi_0(c_-\chi_{n-1} + c_0\chi_\mathrm{n} + c_+\chi_{\mathrm{n}+1}) \qquad \text{and} \qquad \psi_0 = \varphi_0\chi_\mathrm{n}.$$

From (7.2), the coefficients $c_{\pm}$ are

$$c_{\pm} = \frac{(2\hbar)^{-1}\langle\pm|\boldsymbol{p}|0\rangle\cdot\boldsymbol{E}_{\pm}}{\pm\varpi\pm\omega}. \tag{7.3}$$

In such a modulated ground state, the induced dipole moment by the field $\boldsymbol{E}\cos\omega t$ is time-dependent, as expressed by

$$\begin{aligned}\int \Psi_0{}^*\boldsymbol{p}(t)\Psi_0\,\mathrm{d}v = \int \psi_0{}^*\boldsymbol{p}\psi_0\,\mathrm{d}v \\ &+ \exp(i\omega t)\int[\psi_0\langle+|\boldsymbol{p}|0\rangle\psi_0{}^{-*} + \psi_0{}^*\langle 0|\boldsymbol{p}|+\rangle\psi_0{}^+]\,\mathrm{d}v \\ &+ \exp(-i\omega t)\int[\psi_0{}^*\langle 0|\boldsymbol{p}|-\rangle\psi_0{}^- + \psi_0\langle 0|\boldsymbol{p}|-\rangle\psi_0{}^{+*}]\,\mathrm{d}v \\ = \langle 0|\boldsymbol{p}|0\rangle c_0 &+ \exp(i\omega t)[\langle+|\boldsymbol{p}|0\rangle c_-{}^* + \langle 0|\boldsymbol{p}|+\rangle c_+] \\ &+ \exp(-i\omega t)[\langle 0|\boldsymbol{p}|-\rangle c_- + \langle 0|\boldsymbol{p}|-\rangle c_+{}^*],\end{aligned}$$

where dv is the volume element for integration. The first term in the last expression represents a permanent dipole moment if nonzero, whereas the second and third terms express dipole moments induced by the circular fields $\boldsymbol{E}_{\pm}$. Using (6.3), components of the vectors $\boldsymbol{p}$ can be written as

$$p_i(t) = \langle 0|p_0|0\rangle + \tfrac{1}{2}\sum_\mathrm{j}[\alpha_{\mathrm{ij}}(\omega)E_{+\mathrm{j}}\exp(i\omega t) + \alpha_{\mathrm{ij}}(-\omega)E_{-\mathrm{j}}\exp(-i\omega t)], \tag{7.4a}$$

where

$$\begin{aligned}\alpha_{\mathrm{ij}}(\omega) = \hbar^{-1}&\left[\frac{\langle 0|p_i|+\rangle\langle+|p_\mathrm{j}|0\rangle}{\varpi+\omega} + \frac{\langle 0|p_\mathrm{j}|+\rangle\langle+|p_\mathrm{i}|0\rangle}{\varpi-\omega}\right] \\ + \hbar^{-1}&\left[\frac{\langle 0|p_\mathrm{i}|-\rangle\langle-|p_\mathrm{j}|0\rangle}{-\varpi+\omega} + \frac{\langle 0|p_\mathrm{j}|-\rangle\langle-|p_\mathrm{i}|0\rangle}{-\varpi-\omega}\right].\end{aligned} \tag{7.4b}$$

Here $\alpha_{\mathrm{ij}}(\omega)$ constitute the *polarizability* tensor with regard to the indices i, j = x, y in the plane of $\boldsymbol{E}$. It is noted that these tensor elements obey the following relations:

$$\alpha_{ij}(\omega) = \alpha_{ij}^*(-\omega) \qquad \text{and} \qquad \alpha_{ij}(\omega) = \alpha_{ji}^*(\omega). \tag{7.5}$$

The first relation states that the electric moment $\boldsymbol{p}(t)$ is real, whereas the second one indicates that the tensor $\alpha_{ij}(\omega)$ is *Hermitian*.

We can obtain similar expressions for the upper ionic state, where the wavefunction is

$$\Psi_1 = \exp\left\{-\frac{i(\varepsilon_1 + n\hbar\varpi)t}{\hbar}\right\}[\psi_1 + \psi_1^+ \exp(i\omega t) + \psi_1^- \exp(-i\omega t)],$$

where

$$\psi_1^\pm = \frac{\psi_1(\langle \pm|\boldsymbol{p}|0\rangle \cdot \boldsymbol{E}_\pm\rangle}{2\hbar(\pm\varpi \pm \omega)}. \tag{7.6}$$

However, as illustrated in Fig. 7.1(a) for Raman transitions we only need matrix elements for the emission $1 \to 0$, $\Delta n = \pm 1$, after the induced photon absorption $0 \to 1$, $\Delta n = 0$. The transition probability for the electric dipole transition $1 \to 0$ is

$$\begin{aligned}\int \Psi_0^* \boldsymbol{p}(t)\psi_1 \, dv &= \int \exp\left\{\frac{i(\varepsilon_0 + n\hbar\varpi)t}{\hbar}\right\}[\psi_0^* + \psi_0^{+*}\exp(i\omega t) \\ &\quad + \psi_0^{-*}\exp(-i\omega t)]\boldsymbol{p}\psi_1 \exp\left\{-\frac{i(\varepsilon_1 + n\hbar\varpi)t}{\hbar}\right\} dv \\ &= \int [\psi_0^*\boldsymbol{p}\psi_1 + \psi_0^{+*}\boldsymbol{p}\psi_1 \exp(-i\omega t) + \psi_0^{-*}\boldsymbol{p}\psi_1 \exp(i\omega t)] \, dv. \\ &= \langle 0|\boldsymbol{p}|1\rangle \exp(\pm i\omega_{10}t) \\ &\quad + \hbar^{-1}\left[\frac{\langle 0|\boldsymbol{p}^*|\pm\rangle\langle\pm|\boldsymbol{p}|1\rangle}{\pm\varpi + \omega}\right] E_+ \exp\{i(\omega \pm \omega_{10})t\} \\ &\quad + \hbar^{-1}\left[\frac{\langle 0|\boldsymbol{p}^*|\pm\rangle\langle\pm|\boldsymbol{p}|1\rangle}{\pm\varpi - \omega}\right] E_- \exp\{i(-\omega \pm \omega_{10})t\},\end{aligned}$$

where $\omega_{10} = (\varepsilon_1 - \varepsilon_0)/\hbar$. The first term in the last expression gives the transition probabilities for $\omega = \omega_{10}$ and $\Delta n = 0$, whereas the second and third ones present Raman transitions for $\omega = \omega_{10} \pm \varpi$ and $\Delta n = \pm 1$. The coefficients in the square brackets can be expressed in terms of the polarizability tensor components α_{ij} of (7.4).

If α_{ij} is not zero, Raman transitions occur, constituting necessary conditions for cases called "Raman-active," otherwise no Raman effect is expected. The Raman emission lines for $\omega = \omega_{10} - \varpi$ and $\omega = \omega_{10} + \varpi$ are traditionally referred to as Stokes and anti-Stokes lines, respectively, as illustrated schematically in Fig. 7.1(a). The principle for a typical Raman scanning spectrometer is shown in Fig. 7.1(b), where a double (or triple) grating

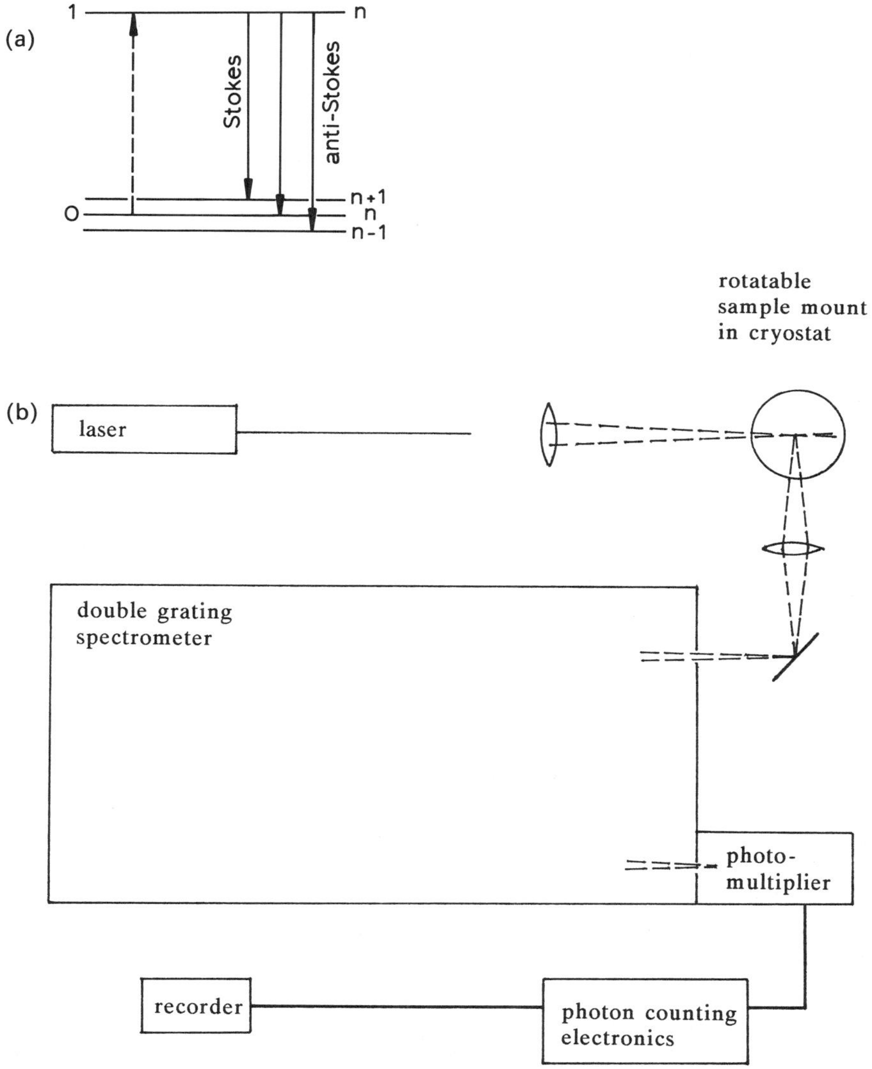

Figure 7.1. (a) An energy diagram for stimulated Raman emissions. The ground and excited ionic states are designated by 0 and 1, respectively, and n represents a quantum number for the phonon mode. (b) A basic Raman spectrometer.

spectrometer constitutes an essential part to obtain an automatic frequency scanning with high resolution.

According to the above theory, frequency shifts of Stokes or anti-Stokes lines from the central emission line called the Rayleigh line should correspond to the characteristic frequency of the soft mode, which shows a

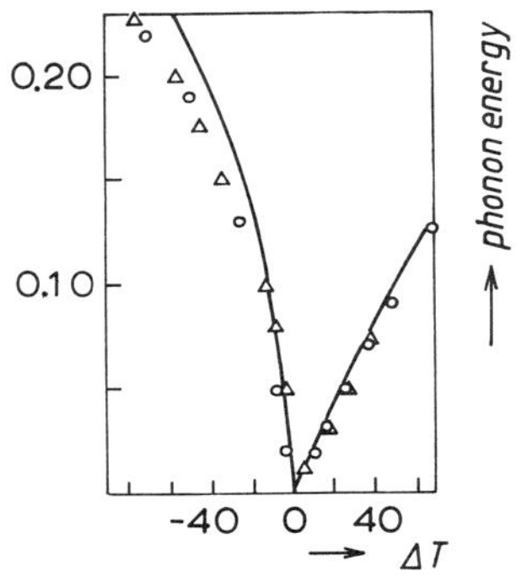

Figure 7.2. Temperature-dependence of the soft-phonon energy in the ferroelastic phase transition of LaP_5O_{14} and $La_{0.5}Nd_{0.5}P_5O_{14}$ measured by Raman scattering. (From P. S. Peercy, *5th Int. Conf. Raman Spectroscopy*, p. 571 (1976); and J. C. Toledana, E. Errandonea, and I. P. Jaguin, *Solid State Commun.* **20**, 905 (1976).)

temperature-dependence that is expressed empirically as $(T_c - T)^\beta$. Typical examples of Raman results of soft modes were shown already in Figs. 4.3 and 4.9(b) for phase transitions in TSCC and K_2SeO_4, respectively. Figure 7.2 shows another example of soft-mode results obtained by Toledano et al. [54] from ferroelastic phase transitions in crystals of the lanthanum phosphate family.

As indicated in (7.4a, b), the Raman activity depends on the polarizability tensor (α_{ij}) arising from interactions between ions and optical phonons that are induced by the electric field of light. We have considered only the ion located at $\boldsymbol{r} = 0$ for the scattering of light, but all ions modulated by the same soft mode scatter the incident radiation simultaneously. In the condensate model, the lattice wavefunction χ in the Placzek wavefunction $\psi = \varphi\chi$ at the site m should be related to the pseudospin σ_m in the critical region, while the ionic function φ is independent of lattice translation. Optical transitions determined by $\Psi_0{}^*\boldsymbol{p}\psi_1$ should therefore be modulated by σ_m if the excited state ψ_1 can remain unmodulated, otherwise contributed by an additional term proportional to $\sigma_m{}^2$. Therefore, for the system of Raman-active ions the redefined dynamic polarizability (per unit volume to be consistent with macroscopic arguments) can generally be expressed as

$$\alpha_{ij} = C_{ij}\sigma(\phi) + D_{ij}\sigma(\phi)^2,$$

although the second-order Raman shifts due to the term D_{ij} were not discussed in the above argument. While the Raman activity is determined by "selection rules" of active ions, the Raman intensities from a modulated crystal are modified by the order variable $\sigma(\phi)$, as given by the above expression for α_{ij}. Hence, the scattering spectrum is not informative about the spatial distribution of σ, and only the temporal fluctuations reflect on the frequency shifts $\pm\varpi$. In Raman spectroscopy, the optical absorption–emission process can be regarded as quasi-elastic scattering of light observed at the right-angle geometry. In this context, Raman scattering shares common grounds with Brillouin scattering from dielectric fluctuations, as will be discussed in the following section.

For a group theoretical discussion of the tensor α_{ij}, interested readers are referred to the book by Wilson et al. [55]. In this monograph, only the basic principle of Raman scattering is discussed, leaving many experimental results to review articles such as those written by Scott [56].

7.2. Rayleigh and Brillouin Scattering from Polar Crystals

Rayleigh and Brillouin scattering of light from liquids has been well known as evidence for macroscopic fluctuations. Optically transparent crystals may also exhibit similar light scattering, if slow dielectric fluctuations are present in the lattice undergoing a structural change. We discussed in Chapter 5 that sinusoidal critical fluctuations are pinned by lattice defects, resulting in spatial inhomogeneity in practical crystals. Light scattering from dielectrically modulated crystals is due to fluctuating polarization.

Since wavelengths of both light and fluctuating scatterers are much longer than lattice spacings, the crystal can be regarded as a continuum macroscopically. The dielectric inhomogeneity in such a crystal can be expressed by a dielectric function $\kappa(\boldsymbol{r}, t)$. For spontaneous inhomogeneity in the critical region can be described by a change $\delta\kappa(\boldsymbol{r}, t)$ which is directly associated with the pseudospin mode $\sigma(\boldsymbol{r}, t)$.

While induced by incident radiation in any polarizable material, dielectric fluctuations may also occur spontaneously in crystals undergoing structural changes. In the former case, $\delta\kappa(\boldsymbol{r}_0, t_0)$ depends on the molecular polarizability, hence resulting in an isotropic scattering independent of the lattice, known as Rayleigh scattering. In contrast, scattering from spontaneous fluctuations involves an excitation of the lattice, since they are coupled as condensates in the critical region. Furthermore, it is notable that light scattering from a crystal may also be induced generally in association with acoustic phonons, as in polar liquids, and therefore observed from any crystal. To observe the Brillouin scattering exclusively from the critical fluctuations, the scattering geometry must be specifically arranged in relation to their symmetry.

For the right angle geometry, $\boldsymbol{K} \perp \boldsymbol{K}_0$, for an arbitrary $\boldsymbol{Q}$, the scattering $\boldsymbol{K}_0 = \boldsymbol{K} \pm \boldsymbol{Q}$ is inelastic, even if $|\boldsymbol{K}| = |\boldsymbol{K}_0|$, and the frequency ω of the scattered light is characterized by a shift from the incident frequency ω_0, where the differences $\omega - \omega_0$ and $\omega_0 - \omega$ correspond to $\boldsymbol{Q}$ and $-\boldsymbol{Q}$, respectively, of the phonon mode in the direction specified by the geometry. In general, the incident field $\boldsymbol{E}_0 \exp\{i(\boldsymbol{K}_0, \boldsymbol{r} - \omega t)\}$ induces a polarization fluctuation $\delta\boldsymbol{P}(\boldsymbol{r}_0, t_0)$ at $\boldsymbol{r}_0$ and t_0, i.e.,

$$\delta\boldsymbol{P}(\boldsymbol{r}_0, t_0) = \delta\kappa(\boldsymbol{r}_0, t_0)\varepsilon_0\boldsymbol{E}_0 \exp\{i(\boldsymbol{K}_0\cdot\boldsymbol{r}_0 - \omega t_0)\}, \qquad (7.7)$$

and the scattered wave at a large distance $|\boldsymbol{r} - \boldsymbol{r}_0| \gg |\boldsymbol{r}_0|$ is expressed as

$$\begin{aligned}\boldsymbol{E}(\boldsymbol{r}, t) &\simeq r^{-1}\exp\{i(\boldsymbol{K}\cdot\boldsymbol{r} - \omega t)\}\iint \varepsilon_0\delta\kappa(\boldsymbol{r}_0, t_0)\boldsymbol{E}_0 \exp\Big\{i[(\boldsymbol{K}_0 - \boldsymbol{K})\cdot\boldsymbol{r}_0 \\ &\quad - (\omega_0 - \omega)t_0]\Big\}\, d^3\boldsymbol{r}_0\, dt_0 \\ &\propto \frac{\boldsymbol{E}_0}{r}\iint \varepsilon_0\delta\kappa(\boldsymbol{r}_0, t_0)\{\exp\{i(\boldsymbol{Q}\cdot\boldsymbol{r}_0 - \Delta\omega\cdot t_0]\} \\ &\quad + \exp\{i(-\boldsymbol{Q}\cdot\boldsymbol{r}_0 + \Delta\omega\cdot t_0)\}\}\, d^3\boldsymbol{r}_0\, dt,\end{aligned}$$

where $\Delta\omega = \omega - \omega_0$. Therefore the Brillouin scattering intensities at $\omega_\pm = \omega_0 \pm \Delta\omega$ as calculated from $\boldsymbol{E}^*(\boldsymbol{r}_0, t_0)\boldsymbol{E}(\boldsymbol{r}_0, t_0)$ are given by

$$\begin{aligned}\Delta I_B(\boldsymbol{K}, \pm\Delta\omega) \propto r^{-2}E_0{}^2 &\iint \delta\kappa^*(\boldsymbol{r}_0', t_0')\delta\kappa(\boldsymbol{r}_0, t_0)\exp[i\{\pm\boldsymbol{Q}\cdot(\boldsymbol{r}_0' - \boldsymbol{r}_0) \\ &\mp \Delta\omega(t_0' - t_0)\}]\, d^3(\boldsymbol{r}_0' - \boldsymbol{r}_0)\, d(t_0' - t_0). \qquad (7.8)\end{aligned}$$

Using the Fourier transforms defined by

$$\delta\kappa(\pm\boldsymbol{Q}; t_0) = \int \delta\kappa(\boldsymbol{r}_0, t_0)\exp(\pm i\boldsymbol{Q}\cdot\boldsymbol{r}_0)\, d^3\boldsymbol{r}_0,$$

(7.8) is written as

$$\Delta I_B(\boldsymbol{K}, \pm\Delta\omega) \propto r^{-2}E^2\frac{1}{2\tau}\int_{-\tau}^{\tau} \delta\kappa^*(\mp\boldsymbol{Q}; t_0')\delta\kappa(\mp\boldsymbol{Q}; t_0)\exp(\pm it\Delta\omega)\, dt, \quad (7.9)$$

where τ is the characteristic time of fluctuations, and the integration should be performed in the range $-\tau \le t \le \tau$ for $t = t_0' - t_0$. Equation (7.9) is further modified by the fluctuation–dissipation theorem (3.14)

$$\delta\kappa^*(\pm\boldsymbol{Q}; t_0')\,\delta\kappa(\pm\boldsymbol{Q}; t_0) \propto \frac{\gamma\Omega}{(\Delta\omega^2 - \Omega^2)^2 + \Omega^2\gamma^2}\exp i\Omega t, \qquad (7.10)$$

where $-\omega_0 \le \Omega \le \omega_0$. As the result, the Brillouin intensities are expressed as

$$\Delta I_B(\pm\boldsymbol{Q}, \pm\Delta\omega) \propto \frac{1}{2\tau}\int_{-\tau}^{\tau}\left[\frac{\gamma\Omega}{(\Delta\omega^2 - \Omega^2)^2 + \Omega^2\gamma^2}\right]\exp i(\Omega \mp \Delta\omega)t\, dt, \quad (7.11)$$

which are maxima at $\Omega = \pm\Delta\omega$.

Figure 7.3(a) shows such a scattering spectrum schematically, when the frequency is scanned.

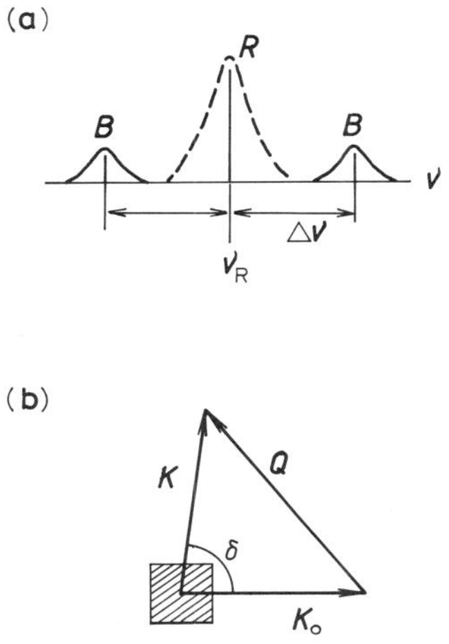

Figure 7.3. (a) A typical Brillouin spectrum, consisting of two Brillouin peaks (B) and a Rayleigh line (R). (b) A scattering geometry. δ: the scattering angle, which is normally 90°.

For a general scattering geometry at a scattering angle δ, as illustrated in Fig. 7.3(b),

$$|\boldsymbol{Q}| = 2|\boldsymbol{K}_0| \sin(\tfrac{1}{2}\delta),$$

we can use the relations

$$\Delta\omega = v|\boldsymbol{Q}| \qquad \text{and} \qquad |\boldsymbol{K}_0| = \frac{n}{c},$$

where the first relation is for acoustic phonons and v is the speed of sound waves, whereas the second one is for incident photons and n is the index of refraction. Combining these, the Brillouin shift can be expressed as

$$\frac{\Delta\omega}{\omega_0} = \pm\left(\frac{2nv}{c}\right)\sin(\tfrac{1}{2}\delta). \tag{7.12}$$

Here the order of v/c is 10^{-5}, and the Brillouin shift $\Delta\omega$ is typically $10^{-5}\omega_0$.

For spontaneous dielectric fluctuations in the critical region, we can repeat the same argument as above, assuming that $\delta\kappa(\boldsymbol{x}_0, t_0) \propto \sigma(\boldsymbol{x}_0 t_0)$. In this case, the dielectic fluctuation is described by the function $\delta\kappa(\pm\boldsymbol{q}, \mp\varpi)$, where $\boldsymbol{q}$ and ϖ signify the soft mode. Also, for such scattering from spontaneous fluctuations, the geometry should be made in such a way as $\boldsymbol{Q} \parallel \boldsymbol{q}$, depending on the symmetry change for the transition. Furthermore, to observe such scattering from the pseudospin mode, the soft-mode energy should be lower than the acoustic mode. In this case, being equivalent to the Brillouin intensity, the dielectric correlation is

$$\delta\kappa^*(\pm\boldsymbol{q}, t_0')\delta\kappa(\pm\boldsymbol{q}, t_0) \propto \frac{\Omega\gamma}{\{(\omega_0 + \varpi)^2 - \Omega^2\}^2 + \Omega^2\gamma^2} \exp\{i\Omega(t_0' - t_0)\}.$$

Such scatterings from optic modes at $\Omega = \omega_0 \pm \varpi$ are generally called Raman scatterings to distinguish them from Brillouin scattering from an acoustic mode, although observable with the Brillouin spectrometer.

As indicated by (7.12), the Brillouin shift depends on the scattering angle δ, while in contrast the Raman shift is temperature-dependent and observable at a specific scattering geometry Thus, the type of scattering can be identified by these features. From (7.9), both Raman and Brillouin spectra are essentially related to temporal fluctuations of the Fourier transform $\delta\kappa(\pm\boldsymbol{Q})$, providing therefore no information about spatial correlations of the pseudospin mode.

7.3. Dielectric Relaxation

In the critical region dielectric fluctuations are sinusoidal, making the crystal spatially inhomogeneous, when pinned by lattice defects. At noncritical temperatures, on the other hand, in the absence of such fluctuations dielectric results can be interpreted in terms of symmetry properties of the unit cell.

When a uniform oscillating field $\boldsymbol{D}_0 \exp(-i\omega t)$ is applied to a uniform crystal, it is logical to consider the dielectric response to the internal field $\boldsymbol{E}_0 \exp\{i(\boldsymbol{q}\cdot\boldsymbol{r} - \omega t)\}$, although the wavevector $\boldsymbol{q}$ is unspecified. In this case, we can deal with the Fourier transform of the response function $\chi_{\boldsymbol{q}}(\omega) = \chi(\omega)\exp(i\boldsymbol{q}\cdot\boldsymbol{r})$, which depends on the wavevector but only insignificantly at small $\boldsymbol{q}$. Accordingly, the susceptibility $\chi_{\boldsymbol{q}}(\omega)$ may be considered as a response to the applied uniform field $\boldsymbol{E}_0 \exp(-i\omega t)$. Such a linear response is a correct expression so long as there is no appreciable nonlinearity in a given system, signifying the behavior of order variables near T_c. We can approach the problem with such a linear response of the Fourier transform $\sigma_{\boldsymbol{q}}(t)$ at the wavevector $\boldsymbol{q}$, considering the nonlinearity as a perturbation.

In this approximation, the equation of motion of a polar pseudospin $\sigma_{\boldsymbol{q}}(t)$ is given by

$$m\left(\frac{\mathrm{d}^2\sigma_{\boldsymbol{q}}}{\mathrm{d}t^2} + \gamma\frac{\mathrm{d}\sigma_{\boldsymbol{q}}}{\mathrm{d}t} + \varpi^2\sigma_{\boldsymbol{q}}\right) = eE_0\exp(-i\omega t),$$

where m and e are the effective mass and charge of the pseudospin σ, and γ is the damping constant. Here, ϖ is the characteristic frequency of the condensate motion that is given by $\varpi^2 = k/m$ where k represents the restoring force constant. In such an oscillatory case, letting $\sigma_{\boldsymbol{q}} = \sigma_{\boldsymbol{q}0}\exp(-i\omega t)$ for the steady solution, we obtain the response function of a resonant type

$$\chi_q(\omega) \propto \frac{\sigma_{q0}}{E_0} = \frac{e/m}{[-\omega^2 - i\gamma\omega + \varpi^2]}, \tag{7.13}$$

which is maximum when $\omega = \varpi$.

On the other hand, if the kinetic energy for a large m does not follow the oscillating external field at a sufficiently fast rate, motion of the polar pseudospin is dictated by the relaxational damping force. In this case, the equation of motion is written as

$$\gamma \frac{d\sigma_q}{dt} + k\sigma_q = \frac{e}{m} E_0 \exp(-i\omega t), \tag{7.14}$$

where k is the restoring force constant, and the steady-state solution gives the response function

$$\chi_q(\omega) \propto \frac{e}{k - i\omega\gamma} = \frac{e/mk}{1 - i\omega\tau}. \tag{7.15}$$

The susceptibility of (7.15), known as the Debye relaxation, exhibits a relaxational decay with a characteristic time constant $\tau = \gamma/k$. Corresponding to $\chi_q(\omega)$, the dielectric function is given by

$$\varepsilon_q(\omega) = \varepsilon_0(1 + \alpha) + \frac{(e/mk)\varepsilon_0}{1 - i\omega\tau}, \tag{7.16}$$

where α is the ionic polarizability as defined in Chapter 4. It should be noticed that the LST relation for an oscillatory mode cannot be used for a relaxational motion, although an equivalent formula can be obtained from (7.13) for such a *relaxator*. It is noted that the complex function $\varepsilon_q(\omega)$ in (7.16) has a *pole* at $\omega = \omega_p$, where $i\omega_p\tau = 1$. Similar to an oscillatory mode, we can define specific values of $\varepsilon_q(\omega)$ at $0 = \infty$ and 0 by the equations

$$\varepsilon_q(\infty) = \varepsilon_0(1 + \alpha) \qquad \text{and} \qquad \varepsilon_q(0) = \varepsilon_q(\infty) + \left(\frac{e}{mk}\right)\varepsilon_0.$$

Letting $\varepsilon_q(\omega) = 0$, a specific frequency $\omega = \omega_l$ can also be defined as a solution, from which we derive the relation

$$i\omega_l\tau = \frac{\varepsilon_q(0)}{\varepsilon_q(\infty)}.$$

Comparing the frequency ω_l with the pole ω_p, we obtain the relation

$$\frac{\omega_l}{\omega_p} = \frac{\varepsilon_q(0)}{\varepsilon_q(\infty)}, \tag{7.17}$$

which corresponds to the LST relation for an oscillator mode. Applying (7.17) to a ferroelectric phase transition obeying the Curie–Weiss law $\varepsilon_q(0) \propto$

$(T - T_c)^{-1}$, we can obtain the temperature-dependent relaxation time

$$\tau \propto (T - T_c)^{-1}, \tag{7.18}$$

which is known as critical slowing-down, and utilized to characterize the relaxational mode near the transition temperature.

Using the dielectric constants at $\omega = 0$ and ∞, (7.16) can be written for the Debye relaxation in the following form:

$$\varepsilon(\omega) - \varepsilon(\infty) = \frac{\{\varepsilon(0) - \varepsilon(\infty)\}}{1 - i\omega\tau}, \tag{7.19}$$

in which the suffix $\boldsymbol{q}$ is omitted for brevity. In practice, for dielectric measurements, (7.14) can be expressed by real and imaginary parts of the complex dielectric function, i.e.,

$$\varepsilon(\omega) = \varepsilon'(\omega) + i\varepsilon''(\omega),$$

where

$$\varepsilon'(\omega) = \varepsilon(\infty) + \frac{\{\varepsilon(0) - \varepsilon(\infty)\}}{1 + \omega^2\tau^2} \quad \text{and} \quad \varepsilon''(\omega) = \frac{\{\varepsilon(0) - \varepsilon(\infty)\}\omega\tau}{1 + \omega^2\tau^2}.$$

Eliminating $\omega\tau$ from these parts, we obtain

$$[\varepsilon'(\omega) - \tfrac{1}{2}\{\varepsilon(0) + \varepsilon(\infty)\}]^2 + \varepsilon''(\omega)^2 = \tfrac{1}{4}\{\varepsilon(0) - \varepsilon(\infty)\}^2, \tag{7.20}$$

which is, in the ε'-ε'' plane, the equation of a semicircle with the radius $r = \frac{1}{2}\{\varepsilon(0) - \varepsilon(\infty)\}$ centered at the point $A(\frac{1}{2}\{\varepsilon(0) + \varepsilon(\infty)\}, 0)$ on the ε' axis, as shown in Fig. 7.4. The curve of (7.20) is known as the Cole–Cole semicircle for a Debye mode characterized by a single relaxation time τ, although in practice the center A is found slightly below the ε' axis. It is useful that the relaxation time τ can be determined from the maximum of $\varepsilon''(\omega)$ corresponding to the point P where $\omega_P\tau = 1$. Such a characteristic frequency $\omega_P = \tau^{-1}$ is often referred to as the *relaxation frequency*.

In Section 4.6, we discussed the dielectric response from a resonant soft mode that is coupled with a Debye relaxation mode. Considering a driving

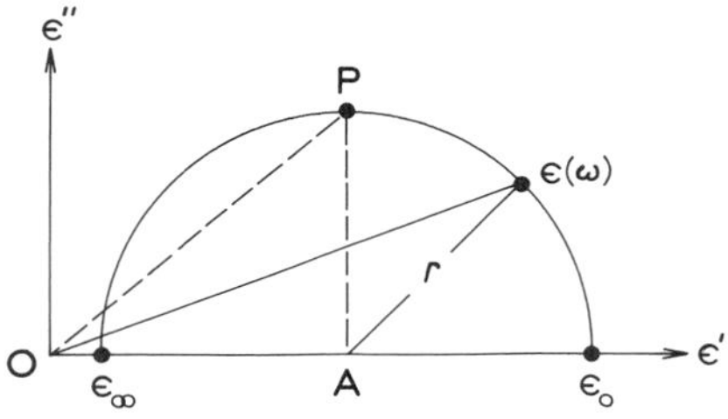

Figure 7.4. Idealized Cole–Cole diagram for dielectric measurements.

external field $\boldsymbol{E}_0 \exp(-i\omega t)$, the susceptibility is expressed by

$$\chi_q(\omega) = \left[\frac{\varpi^2 - \omega^2 - i\omega\gamma - \delta^2\omega\tau}{1 + i\omega\tau}\right]^{-1},$$

from which we derived the expression for the imaginary part composed of a typical resonant mode and a Debye relaxational mode, if $\gamma \ll \delta^2\tau$ and $\varpi \gg \tau^{-1}$. Although the soft mode by itself exhibits a relaxational character as T_c is approached, its relation to the Debye mode is clear in such a model for coupled modes when applied to the spectral analysis at very low frequencies.

The above expressions on dielectric responses are studied for the ferroelectric phase transition in TSCC crystals, which are summarized in the next section.

7.4. Dielectric Spectra in the Ferroelectric Phase Transition of TSCC

Resonant and relaxational dielectric responses were once considered to be exclusive, because the responsible mechanism was uncertain beyond the classical model of elementary dipoles. For example, order–disorder phase transitions exhibiting the Debye dielectric relaxation near T_c were regarded as signified by the absence of soft modes, while displacive systems were otherwise considered. In recent studies however, both characters were observed in the dielectric response from TSCC crystals, changing from a resonant type to a relaxation type, as T_c was closely approached.

For TSCC crystals there are a variety of dielectric results reported in the literature. Kozlov et al. [57] found in such dielectric spectra, measured at submillimeter-wave frequencies, that a typical soft mode remained underdamped at temperatures close to T_c, while Deguchi et al. [58] observed a relaxation mode at UHF frequencies showing a critical slowing-down behavior with the relaxation time obeying the Curie–Weiss law, in addition to another unidentified relaxation mode at lower frequencies. At microwave frequencies, Sawada and Horioka [38] showed that two types of dielectric response coexisted near the ferroelectric phase transition of TSCC at 163 K. Petzelt et al. [59] have attempted to generalize a model of a soft mode coupled with a relaxation mode for many other ferroelectric transitions, but it appeared to be difficult to do so only with their submillimeter-wave data. As clearly demonstrated by Sawada and Horioka, such high-frequency experiments should be extended to cover lower frequency ranges, while the central peak should be elucidated properly in relation to the intrinsic relaxation. Aside from the interpretation of central peaks, it is evident that the dielectric response near a phase transition is neither purely resonant nor relaxational but mixed in form, crossing from one type over to the other. For such results, the equation (4.27) was used for analysis. Such a coupling mech-

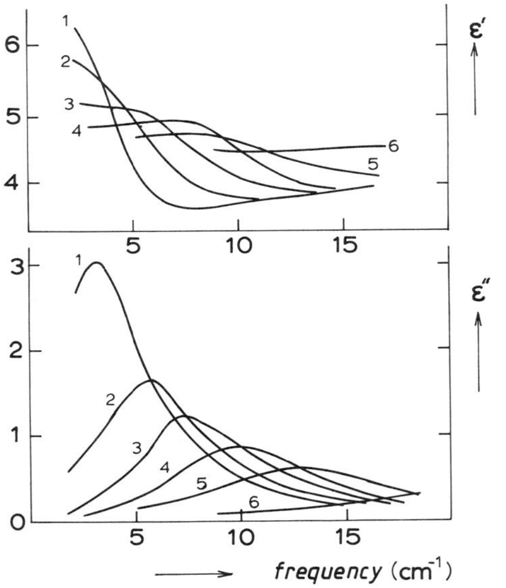

Figure 7.5. Real and imaginary parts of the dielectric response function $\varepsilon(\omega)$ in TSCC measured with submillimeter wave frequencies. (From A. A Kozlov, J. F. Scott, G. E. Feldkamp, and J. Petzelt, *Phys. Rev.* **B28**, 255 (1983).)

anism in the lattice mode was only recognized from the dielectric analysis, but is consistent with the condensate model theoretically.

Figure 7.5 shows the dielectric spectra at submillimeter-wave frequencies, obtained by Kozlov et al. [57], which exhibit a underdamped soft-mode peak when T_c was approached from above. The linewidth was temperature-dependent, becoming broader and overdamped, as T_c was approached, but was not close enough due to the low-frequency limit at about 50 GHz of their backward-wave oscillator. Figure 4.6(b) shows dielectric spectra of $\varepsilon'(0)$ from TSCC exhibiting a relaxational anomaly at high microwave frequencies, 35 GHz and 24 GHz, which was analyzed with (7.17) by Sawada and Horioka [38].

Pawlaczyk et al. [60] carried out dielectric dispersion measurements of TSCC in a wide range of frequencies from 6 GHz to 0.1 kHz in the critical region of the ferroelectric phase transition, confirming the earlier results of Deguchi et al. Their relaxation measurements were performed on high-quality samples characterized by a low internal bias field ($E_b \sim 15\ \mathrm{V\,cm^{-1}}$), and analyzed with Cole–Cole plots, as summarized in Figs. 7.6(a) and (b). There were basically two relaxational modes in the critical region, in addition to the overdamped soft mode, being signified by two relaxational frequencies 6 GHz to 1 GHz and 50 MHz to 5 MHz, which were called Mode I and Mode II by the authors, respectively. As the transition temperature T_c was closely approached from above, Mode I prevailed, where the Cole–Cole

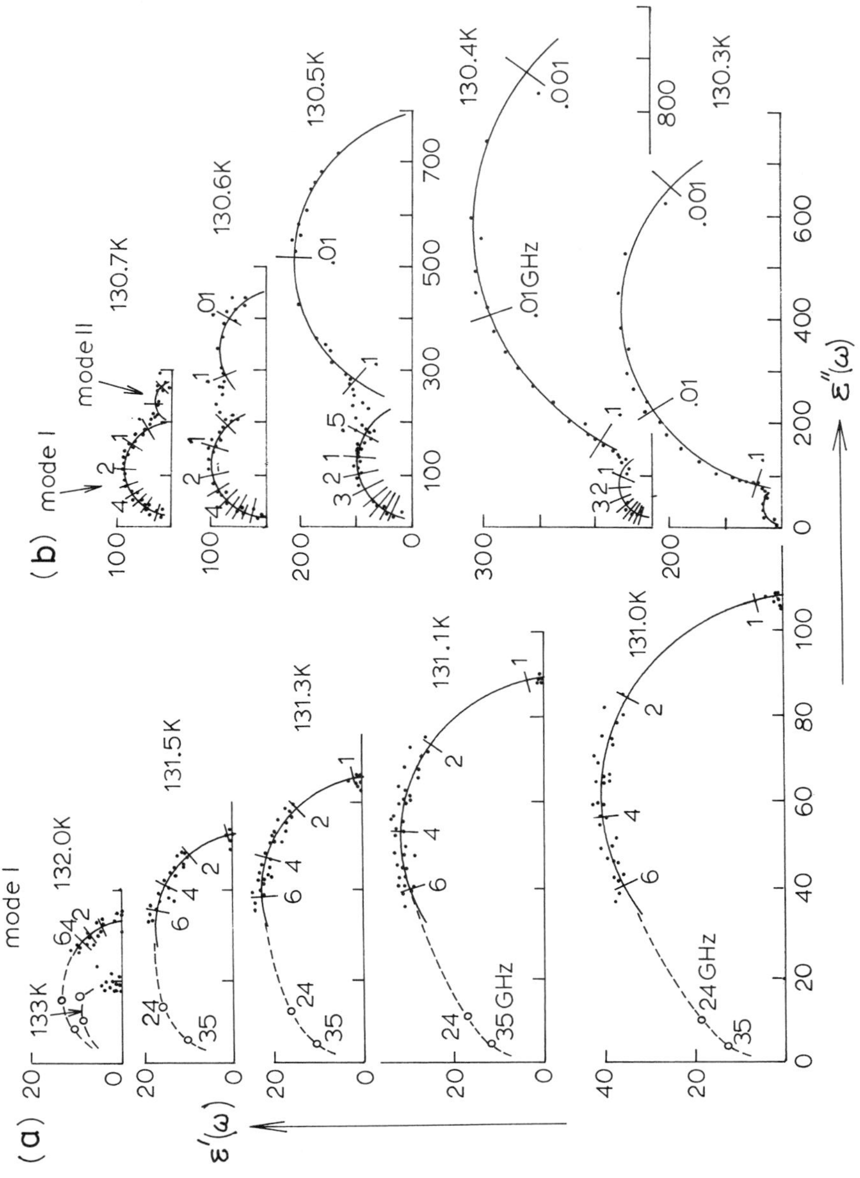

Figure 7.6. Cole–Cole plots of the dielectric response function in the critical region of the ferroelectric phase transition in TSCC. (From Cz. Pawlaczyk, H.-G. Unruh, and J. Petzelt, *Phys. Stat. Sol.* **(b)136**, 435 (1986); M. Fujimoto, Cz. Pawlaczyk, and H.-G. Unruh, *Phil. Mag.* **60**, 919 (1989).)

plots were hardly semicircular, as seen in Fig. 7.6(a). The plots included the Sawada–Horioka data at 35 GHz and 24 GHz from [38] to indicate a continuous transition from the oscillatory soft mode above T_c. The phase transition temperature was signified by Mode II that emerged after Mode I reached the maximum intensity, and determined as $T_c = 130.8$ K. Figure 7.6(b) shows the temperature change of Cole–Cole plots below T_c, where the growth of Mode II accompanies diminishing Mode I, while Mode II also diminishes after its maximum at around 130.4 K. They found that the intensity of Mode I was virtually independent of crystals, whereas the appearance of Mode II depended on the sample quality.

It is noticed from Fig. 7.6(b) that the right edge of Mode I coincides with the left edge of Mode II within experimental error. Therefore, observed double relaxations may be described by

$$\varepsilon(\omega) - \varepsilon(\infty) = \frac{S_1}{1 - i\omega\tau_1} + \frac{S_2}{1 - i\omega\tau_2}, \tag{7.21}$$

where

$$S_1 = \varepsilon_1(0) - \varepsilon_1(\infty), \qquad S_2 = \varepsilon_2(0) - \varepsilon_2(\infty),$$

and

$$\varepsilon_1(\infty) = \varepsilon_2(0). \tag{7.22}$$

Although introduced by Petersson [61] to deal with the contribution by lattice defects, equation (7.21) may be applied to the double relaxations comprising Mode I and Mode II, assuming that it represents dielectric responses from different states of the same collective model.

In the condensate model, these relaxation modes can be attributed to dielectric fluctuations in phase and amplitude modes. These modes are expressed as proportional to cos ϕ and sin ϕ, behaving as if stationary when pinned in crystals, and hence representing a thermally stable condensate at temperatures in the critical region. We may consider that the anharmonic quartic potential is responsible for such intrinsic pinning, when acting as a secular perturbation occurring independently of lattice defects. Therefore, these dielectric fluctuation modes should exhibit a relaxational response to the applied oscilating field. We can write equations of motion for these modes as

$$\gamma \frac{d\sigma_P}{dt} + k\sigma_P = \alpha_1 E_0 \cos \omega t \qquad \text{and} \qquad \gamma \frac{d\sigma_A}{dt} + k\sigma_A = \alpha_1 E_0 \sin \omega t,$$

which can be reexpressed in the following form using the complex variable defined as

$$\sigma = \sigma_P + i\sigma_A = \sigma_0 \exp(i\phi) = \sigma_q \exp(-i\omega t) \qquad \text{where} \quad \sigma_q = \sigma_0 \exp(iqx).$$

Namely,

$$\gamma \frac{d\sigma_q}{dt} + k\sigma_q = \alpha_1 E_0 \exp(-i\omega t),$$

where $\alpha_1 = e/m$ is the effective charge/mass ratio of the condensate. This equation has a steady-state solution $\sigma_q = \sigma_q(0)\exp(-i\omega t)$, where

$$\sigma_q(0) = \frac{\alpha_1 E_0}{k - i\omega\gamma}.$$

Defining the relaxation time $\tau_1 = \gamma/k$, the susceptibility of the condensate is expressed as

$$\chi_1 = \frac{\sigma_q(0)}{E_0} = \frac{S_1}{1 - i\omega\tau_1} \qquad \text{where} \quad S_1 = \frac{\alpha_1}{k},$$

which can be assigned to Mode I that represents the sinusoidal complex mode $\sigma_{qP} + i\sigma_{qA}$ of fluctuations. Hence, two components σ_{qP}, σ_{qA} cannot be distinguished in the dielectric response because of the same relaxation time τ_1. Figure 7.7(a) shows the relaxation time τ_1 of Mode I determined as a function of temperature in TSCC by Pawlaczyk et al. [44], exhibiting a critical slowing-down behavior as described by (7.14).

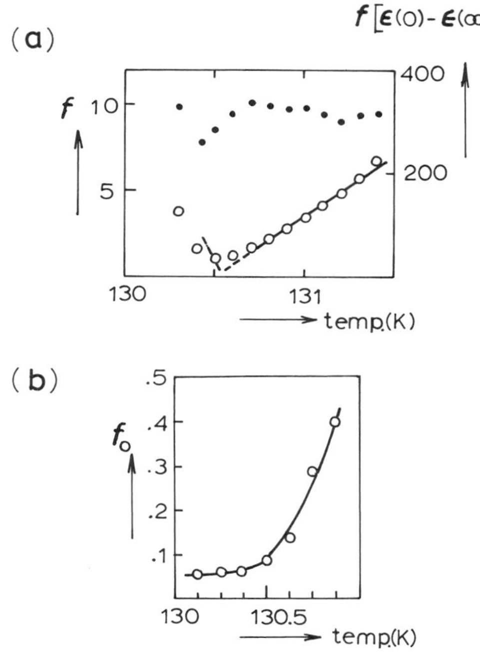

Figure 7.7. (a) Critical slowing down exhibited by the "relaxation mode I" in TSCC. (From Cz. Pawlacyk, H.-G. Unruh, and J. Petzelt, *Phys. Stat. Sol.* **(b)136**, 435 (1986).) (b) Pinning frequency g_0 of the phase mode versus temperature in TSCC determined from anomalous broadening of an Mn^{2+} line. (From M. Fujimoto, Cz. Pawlaczyk, and H.-G. Unruh, *Phil. Mag.* **60**, 919 (1989).)

In contrast, Mode II, expressed by the susceptibility

$$\chi_2 = \frac{S_2}{1 - i\omega\tau_2},$$

is characterized by a relaxation time τ_2 longer than τ_1, and so can be interpreted as related to an extended nonsinusoidal amplitude mode that is signified by a larger mass and stronger damping. Although it is not quite possible to make a positive identification, Mode II may be considered as representing pinned condensates by lattice defects by virtue of its sample dependence. This view is at least in qualitative agreement with interpretation as related to the domain wall, as speculated by many workers. While unidentifiable from dielectric results only, such an interpretation should be consistent with other observations of critical anomalies. In fact, Fujimoto et al. [45] interpreted EPR anomalies in TSCC as being compatible with the above dielectric relaxation results.

In Section 5.2 we discussed the dynamic equilibrium of a pinned condensate in a defect potential, ignoring the damping effect. However, in practice, the pinning frequency ω_0 is very low, and the kinetic energy of fluctuation is negligible so that the equation of motion can be written as relaxational, i.e.,

$$\frac{1}{\tau_1}\frac{d\sigma_q}{dt} + \omega_0{}^2\sigma_q = \alpha_2 E_0 \exp(-i\omega t),$$

where the second term on the left represents the restoring force by the pinning potential, and α_2 is the ratio of effective charge/mass of a pinned condensate, which is generally not identical to α_1 when considering the nonlinear nature of the condensate. According to this interpretation, the relaxation time for Mode II is given by

$$\tau_2 = \frac{1}{\omega_0{}^2\tau_1}, \tag{7.23}$$

which can be used to estimate the value of ω_0 from measured relaxation times τ_1 and τ_2 Figure 7.7(b) shows pinning frequencies $f_0 = \omega_0/2\pi$ determined with (7.23) at temperatures below T_c. The asymptotic value of about 50 MHz is a magnitude acceptable for domain-wall motion.

CHAPTER 8

The Spin Hamiltonian and Magnetic Resonance Spectroscopy

In this chapter, principles of magnetic resonance are briefly summarized for those who have no practical experience in this field of spectroscopy. I think that various definitions of spin-Hamiltonian parameters are generally complex and too specific for nonspecialists, and therefore in this chapter I discuss briefly their physical origins, which is considered as a preparatory introduction to what follows, where phase transitions are viewed as implied by magnetic resonance results. Those readers who are already familiar with the language of magnetic resonance can therefore skip this chapter to proceed directly to Chapter 9.

8.1. Introduction

From data for the energy and momentum exchange in neutron inelastic scattering experiments we cannot identify the active group responsible for a structural transformation in solids. While inferrable from its chemical composition in simple systems, active groups in complex crystals can be determined, in principle, in terms of constituent ions from the diffuse X-ray diffraction pattern in the critical region. By contrast, magnetic resonance experiments utilize nuclear and paramagnetic probes to obtain information about a local structural change in their surroundings. In a modulated crystal the spectra exhibit anomalous lineshapes due to spatial modification of the lattice. Sampling of condensates by such magnetic resonance probes is a unique technique in that the modulated structure could be visualized, while other methods deal basically with dynamical aspects of condensates. Providing complementary information to scattering and dielectric experiments, magnetic resonance sampling is indispensable for studies of order-variable condensates during a structural change.

Needless to say, the magnetic resonance method is primarily applicable to nonmagnetic systems, in which nuclear and paramagnetic probes with spins higher than $\frac{1}{2}$ are usable for detecting changes in the crystalline potential due to local distortion. On the other hand, magnetic probes with spins $\frac{1}{2}$ do not interact with nonmagnetic surroundings, providing no information about structural changes, unless the spectra are dominated by hyperfine structures arising from nuclear spins in the nearby lattice sites. It is significant, in any case, that these usable probes must be directly incorporated into the active group, otherwise the spectra are insensitive to local structural changes.

Also significant for magnetic resonance experiments is the timescale $t_0 \sim {\nu_L}^{-1}$ of the Larmor precession as compared with the characteristic frequency ϖ of fluctuations that is typically of the order of 10^{11} Hz $\sim 10^{10}$ Hz. In conventional magnetic resonance, the Larmor frequency ν_L is in the range of 35 GHz $\sim$ 5 GHz for paramagnetic probes, whereas 20 MHz $\sim$ 1 MHz for nuclear probes. Judging from the characteristic time of critical fluctuations $\tau = 2\pi/\varpi \sim 10^{-11} \sim 10^{-10}$ s, the condensate motion appears to be very slow or nearly static in the timescale of observation t_0, so that the spatial fluctuations are explicit in the lineshape. Moreover, comparative studies at the different Larmor frequencies are possible, yielding useful information as to slow dynamics of condensates. In spite of these practical advantages, it is a matter of chemical compatibility of magnetic probes when doping a given crystal without causing serious modification. In this context, it is usual practice to perform experiments using as many chemically accessible probes as possible. Nevertheless, while it is desirable to perform experiments at various measuring frequencies, nuclear and paramagnetic resonances are not competitive in most systems of interest, because the choice is limited in practice, and even suitable probes may not be found in some cases.

Combining with X-ray diffraction results, magnetic resonance studies can provide direct identification of the active group for a structural transformation in crystals. By sampling a modulated crystal, one can determine quantities related to basic features of the order-variable condensate, i.e., the amplitude σ_0 and the modulation form $f(\phi)$. In nuclear magnetic resonance, the spin-lattice relaxation time T_1 can be measured with accuracy, giving evidence for equilibrium between order variables and soft phonons that constitute a condensate.

Magnetic resonance is now a well-established method of the investigation for many problems in condensed matter, and many readers may already be familiar with its elemental principles. However, properties of paramagnetic ions in crystalline solids are normally described in terms of spin-Hamiltonian parameters that are a little too sophisticated for nonspecialists to grasp their meanings. Therefore, in this chapter I discuss various tensor parameters that may be modified in modulated crystals. Here, the theory is outlined to minimum necessities, but those readers who are interested in the detail are referred to the standard book by Abragam and Bleaney [62].

8.2. Magnetic Resonance and Relaxation

We consider nuclear and electronic magnetic moments that are carried by atomic nuclei and paramagnetic ions, respectively, in crystals. We assume that these moments are situated at fixed lattice sites. While for most applications diffusive motion of these carriers is ignored, here for simplicity these microscopic magnetic moments $\boldsymbol{\mu}$ are considered as free to rotate at their sites in nonmagnetic crystals. In the presence of a strong uniform magnetic field $\boldsymbol{B}_0$, the torque $\boldsymbol{\mu} \times \boldsymbol{B}_0$ acts on the moment $\boldsymbol{\mu}$, as described by the equation of motion

$$\frac{\mathrm{d}\boldsymbol{L}}{\mathrm{d}t} = \boldsymbol{\mu} \times \boldsymbol{B}_0, \tag{8.1}$$

where $\boldsymbol{L} = \gamma^{-1}\boldsymbol{\mu}$ is the angular momentum, and γ is the gyromagnetic ratio. Assuming $\boldsymbol{B}_0 = B_0\boldsymbol{e}_z$, where $\boldsymbol{e}_z$ is the unit vector along the z axis, the steady-state solution of (8.1) is given by

$$\mu_z = \mu \cos\theta = \text{const.} \tag{8.2a}$$

and

$$\mu_\perp = \mu \sin\theta \exp(i\varphi) \qquad \text{where} \quad \varphi = \gamma B_0 t. \tag{8.2b}$$

Here θ and φ are polar and azimuthal angles of $\boldsymbol{\mu}$ with respect to the z axis, as shown in Fig. 8.1(a). Equations (8.2a, b) describe the precessional motion of $\boldsymbol{\mu}$ around $\boldsymbol{B}_0$ at the angular frequency $\omega_L = \gamma B_0$, which is known as the Larmor precession.

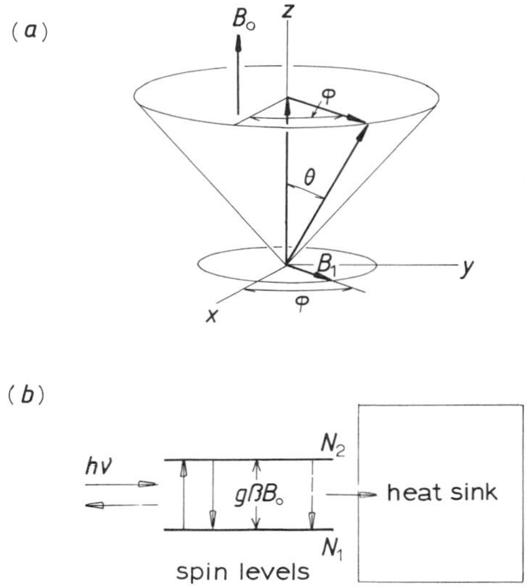

Figure 8.1. (a) Magnetic resonance of a classical spin $\boldsymbol{S}$. (b) Magnetic resonance of a quantum-mechanical spin $S = \frac{1}{2}$.

We consider a circularly rotating magnetic field $\boldsymbol{B}_1 = B_1(\boldsymbol{e}_1 \exp(i\omega t))$ applied perpendicularly to $\boldsymbol{B}_0$, i.e., $\boldsymbol{B}_1 \perp \boldsymbol{B}_0$ to observe the Larmor's precession, which is identified when $\boldsymbol{B}_1$ is synchronized with the motion of $\mu_\perp$. Namely, when the relation

$$\omega = \omega_L = \gamma B_0 \tag{8.3}$$

called the *magnetic resonance condition* is fulfilled, the high-frequency energy associated with $\boldsymbol{B}_1$ is absorbed by $\boldsymbol{\mu}$, resulting in a precession of a larger $\mu_\perp$ and smaller μ_z. It is noted that the magnetic resonance can be described in the rotating frame of reference at the frequency ω_L, in which the torque $\boldsymbol{\mu} \times \boldsymbol{B}_1$ rotates $\boldsymbol{\mu}$ to increase the angle θ.

In a macroscopic system of many identical $\boldsymbol{\mu}$, these moments will precess around $\boldsymbol{B}_0$ in random phases, which become coherent in phase on applying the rotating field $\boldsymbol{B}_1 \exp(i\omega t)$. In terms of macroscopic magnetization $\boldsymbol{M}$ composed of $\boldsymbol{\mu}$, $\langle M_\perp \rangle = 0$ and $M_z = \text{const.}$ for random phases, whereas $M_\perp$ is finite and significant if driven by $\boldsymbol{B}_1$. When $\boldsymbol{B}_1$ is switched on or off, the response of moments $\boldsymbol{\mu}$ is signified by phasing or dephasing, respectively, being characterized by a relaxation time T_2. Such a timescale depends on magnetic interactions among many $\boldsymbol{\mu}$ in the system, and is called the *spin–spin relaxation time.* Bloch [63] wrote the equation of motion for $M_\perp$ as

$$\frac{dM_\perp}{dt} = -\frac{M_\perp}{T_2} \tag{8.4a}$$

and

$$M_z = M_0 = \chi B_0, \tag{8.4b}$$

where χ is the magnetic susceptibility. In a macroscopic system, magnetic resonance does not occur sharply as given by the resonance frequency in (8.3), but is broadened around ω_L due to interactions with other magnetic moments $\boldsymbol{\mu}$ in the system. Such broadening arises not only from other $\boldsymbol{\mu}$'s of the same kind, but also includes magnetic interactions with different species. The Bloch equation (8.4) is applicable only to the former case of *homogeneous broadening*, but is also used for the approximate description of *inhomogeneously* broadened resonance lines due to interactions with different magnetic moments.

In the rotating field $\boldsymbol{B}_1$ the component M_z cannot be a constant of time, but relaxes to the static value M_0 after a certain time T_1 when $\boldsymbol{B}_1$ is switched off. For such a relaxation process, Bloch has given another equation

$$\frac{dM_z}{dt} = -\frac{M_z - M_0}{T_1}, \tag{8.5}$$

where T_1 is called the *spin-lattice relaxation time.* Comparing with rather loosely defined T_2, the time T_1 is well defined, signifying a characteristic time for the spins in a given system to establish thermal equilibrium. Microscopically, (8.5) is the rate equation for the population difference $n(T)$ between adjacent magnetic spin levels to become an equilibrium value $n_0(T)$ at a temperature T, i.e.,

$$\frac{dn}{dt} = -\frac{n - n_0}{T_1}.$$

Taking these two relaxation mechanisms into account, the macroscopic equation of motion for the magnetization vector $\boldsymbol{M}$ in the combined field of $\boldsymbol{B}_0$ and $\boldsymbol{B}_1 \exp(\pm i\omega t)$ is written as

$$\frac{dM_\pm}{dt} \pm i\gamma B_0 M_\pm + \frac{M_\pm}{T_2} = -i\gamma B_1 M_z \exp(\pm i\omega t) \tag{8.6a}$$

and

$$\frac{dM_z}{dt} + \frac{M_z - M_0}{T_1} = \tfrac{1}{2} i\gamma B_1 \{M_+ \exp(-i\omega t) + M_- \exp(i\omega t)\}. \tag{8.6b}$$

Equations (8.6a,b) are known as the Bloch equations, where $M_\pm$ represent the rotating component $M_\perp$ synchronizing with the rotating fields $B_1 \exp(\pm i\omega t)$. Although applicable to most cases, the Bloch equations are valid for cases where

$$\frac{1}{T_2} \ll |\gamma B_1|, \tag{8.6c}$$

which can be easily understood in the rotating frame of reference as described in the following. The relaxation time T_2 is essentially a measure of the local field fluctuation ΔB originating from spin–spin interactions in a given system. Therefore the magnitude of the transversal component $|\Delta B_\perp| = 1/\gamma T_2$ should be sufficiently smaller than B_1 for the synchronous motion of spins, otherwise the phase coherence in $M_\perp$ cannot be achieved when driven by the rotating field B_1. The value of $|\Delta B_\perp|$ is of the order of 10 G in typical diluted paramagnetic systems, giving rise to $T_2 \sim 10^{-8}$ s, whereas for a nuclear moment in nonmagnetic crystals $|\Delta B_\perp| \sim 1$ G and $T_2 \sim 10^{-4}$ s. Hence (8.6c) for a *slow passage* is generally met for nuclear spins, and also for most of the well-resolved electronic spectra.

The Bloch equations (8.6a, b) have a steady-state solution. Under the slow passage condition, we can consider that the dynamical process is described by the solution at all times at least approximately. In the steady state that is signified by $dM_z/dt = 0$, M_z accompanies transversal rotating components $M_\pm$, which are expressed in the laboratory frame of reference as

$$M_\pm = \frac{\gamma B_1 M_z \exp(\pm i\omega t)}{\omega + \gamma B_0 + i/T_2}.$$

Using these expressions in (8.6a,b), we can obtain the steady-state solutions

$$\frac{M_z}{M_0} = \frac{1 + (\omega - \omega_L)^2 T_2{}^2}{1 + (\omega - \omega_L)^2 T_2{}^2 + \gamma^2 B_1{}^2 T_1 T_2} \tag{8.7a}$$

and

$$\frac{M_\pm}{M_0} = \frac{\{(\omega - \omega_L)T_2 \pm i\}\gamma B_1 T_2 \exp(\pm i\omega t)}{1 + (\omega - \omega_L)^2 T_2{}^2 + \gamma^2 B_1{}^2 T_1 T_2}. \tag{8.7b}$$

Equation (8.7a) indicates that $M_z \approx M_0$ at resonance $\omega = \omega_L$ if $(\gamma B_1 T_1 T_2)^2 \ll 1$. Considering $M_z = M_0$, from (8.7a,b), the tilting angle θ of the vector $\boldsymbol{M}$ from $\boldsymbol{B}_0$ is given by

$$\tan\theta = \frac{M_\pm}{M_0} \approx \frac{\gamma B_1 T_2}{1 + (\omega - \omega_L)^2 T_2{}^2}.$$

Similar to the static magnetic susceptibility given by $M_0 = \chi_0 B_0$, the high-frequency susceptibility can be defined by means of the relation $M_\pm = \chi(\omega) B_1 \exp(\pm i\omega t)$, i.e.,

$$\frac{\chi(\omega)}{\chi_0} = \frac{\gamma B_1 T_2\{(\omega - \omega_L)T_2 + i\}}{1 + (\omega - \omega_L)^2 T_2{}^2 + (\gamma B_1 T_1 T_2)^2}.$$

The real and imaginary parts of the complex susceptibility $\chi(\omega) = \chi'(\omega) - i\chi''(\omega)$ are expressed as

$$\frac{\chi'(\omega)}{\chi_0} = \frac{\omega_L(\omega - \omega_L)}{(\omega - \omega_L)^2 + \delta\omega^2 + \gamma^2 B_1{}^2 T_1 \delta\omega}$$

and

$$\frac{\chi''(\omega)}{\chi_0} = \frac{\omega_L \delta\omega}{(\omega - \omega_L)^2 + \delta\omega^2 + \gamma^2 B_1{}^2 T_1 \delta\omega},$$

where $\delta\omega = 1/T_2$. The imaginary part becomes maximum at resonance, i.e., $\omega = \omega_L$, where

$$\frac{\chi''(\omega_L)}{\chi_0} \approx \frac{\omega_L}{\delta\omega},$$

if specifically $\gamma^2 B_1{}^2 T_1 \ll \delta\omega$. Hence at resonance $\chi''(\omega_L) \gg \chi_0$ for a sharp resonance, i.e., $\delta\omega \ll \omega_L$.

The magnetic resonance is interpreted as a phenomenon, where a system of magnetic moments in a uniform field $\boldsymbol{B}_0$ absorbs an energy from applied radiation via the magnetic interaction with the high-frequency field $\boldsymbol{B}_1$. When the absorbed energy by precessing magnetic moments is emitted or dissipated to the surroundings under the presence of the high-frequency radiation, the magnetic resonance can be observed as a steady process, in which the rate of energy absorption is described by

$$\left\langle \frac{dW}{dt} \right\rangle_t = \tfrac{1}{2}\omega_L \chi''(\omega_L) B_1{}^2 \quad \text{per one cycle.} \tag{8.8}$$

In a simple case, the magnetic resonance is related to two spin energy levels, which are in thermal equilibrium and populated by the Boltzmann statistics. Figure 8.1(b) shows a schematic diagram illustrating the principle of magnetic resonance. Consider two energy levels specified by the equilibrium population N_1 and N_2, where the level E_1 is lower than the level E_2. The magnetic resonance occurs when

$$\hbar\omega = E_2 - E_1 = \hbar\omega_{\mathrm{L}}$$

at a nonzero transition probability

$$w_{12} = \tfrac{1}{2}\pi\hbar^{-2}B_1{}^2|\mu_{12}|^2 f(\omega).$$

Here μ_{12} is an off-diagonal matrix element of the magnetic moment $\boldsymbol{\mu}$ between the two states, and $f(\omega)$ is the *shape function* normalized as $\int f(\omega)\,\mathrm{d}\omega = 1$, expressing resonance line broadening due to internal interactions. The lineshape is *Lorentzian* if broadened predominantly by a relaxational mechanism, whereas it is *Gaussian* if related to distributed local fields, representing homogeneously and inhomogeneously broadened lines, respectively.

It is obvious that a positive absorption takes place if $N_1 > N_2$, and in the high-temperature approximation, i.e., $\hbar\omega_{\mathrm{L}}/\mathrm{k_B}T \ll 1$, we can write

$$\begin{aligned}
\frac{\mathrm{d}W_{12}}{\mathrm{d}t} &= w_{12}\hbar\omega_{\mathrm{L}}(N_1 - N_2)\\
&= w_{12}\hbar\omega_{\mathrm{L}}N_1\left\{1 - \exp\left(-\frac{\hbar\omega_{\mathrm{L}}}{\mathrm{k_B}T}\right)\right\} \approx w_{12}\hbar\omega_{\mathrm{L}}N_1\left(\frac{\hbar\omega_{\mathrm{L}}}{\mathrm{k_B}T}\right)\\
&= N_1\left(\frac{\pi\omega_{\mathrm{L}}{}^2B_1{}^2}{\mathrm{k_B}T}\right)|\mu_{12}|^2 f(\omega_{\mathrm{L}}).
\end{aligned}$$

Since W in (8.8) is considered as equal to W_{12} in a two-level system, we obtain the expression

$$\chi''(\omega_{\mathrm{L}}) = \frac{N_1\pi\omega_{\mathrm{L}}|\mu_{12}|^2 f(\omega_{\mathrm{L}})}{\mathrm{k_B}T}, \tag{8.9}$$

so that the imaginary part $\chi''(\omega_{\mathrm{L}})$ at ω_{L} can be attributed to the energy absorption arising from the off-diagonal element μ_{12}.

The real part $\chi'(\omega_{\mathrm{L}})$, on the other hand, represents the dispersion of resonance frequencies in the vicinity of ω_{L}, while $\chi'(\omega)$ and $\chi''(\omega)$ are not independent but related by the Kramers–Krönig formula for a complex response function [64]. These parts of a complex susceptibility $\chi(\omega)$ can be measured independently, as discussed in the next section, although either part is sufficient in principle for a magnetic resonance measurement.

8.3. Magnetic Resonance Spectrometers

The magnetic resonance is signified experimentally by a maximum absorption of the radiation energy when the resonance condition $\omega = \gamma B_0$ is met in a uniform magnetic field B_0. Therefore, for a given magnetic moment specified by γ, the resonance can be detected in experiments either by scanning ω at a constant B_0 or by scanning B_0 at a given ω. Although just a technical matter, it is more practical to study magnetic resonance by the latter scheme for varying fields at a constant frequency, which is a commonly used method in practical spectrometers, using a conventional laboratory magnet producing a uniform field of the order of 15 kG. In such a field, nuclear resonance radio frequencies 1 MHz ~ 100 MHz are practical, whereas for paramagnetic resonance microwave frequencies in the range of 5 GHz ~ 40 GHz are commonly used.

A sample crystal is placed at the position of a resonator where the distributed high-frequency field B_1 gives a maximum strength. At radiofrequencies, such a resonator with no sample inside consists of a coil of inductance L_0 and an external capacitor C, so that its tuning frequency is given by $\omega = (L_0 C)^{-1/2}$, as illustrated in Fig. 8.2(a). At microwave frequencies, a sample is placed, as shown in Fig. 8.2(b), at a place of maximum B_1 inside a cavity resonator, which can also be described as an effective L_0-C resonator.

When a sample crystal is placed in a resonator, the inductance can be expressed as

$$L = (1 + \chi)L_0 = (1 + \chi')L_0 - i\chi'' L_0.$$

Hence, the impedance of such a resonator with a sample is

$$Z = R + i\omega L + (i\omega C)^{-1} = R + \omega\chi'' L_0 + i\omega(1 + \chi')L_0 - \frac{i}{\omega C}, \tag{8.10}$$

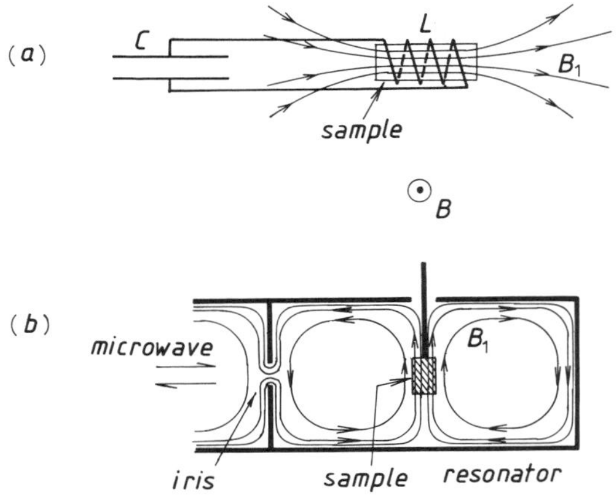

Figure 8.2. (a) A radio-frequency resonator consisting of a capacitor C and an inductor L with a sample crystal. (b) A microwave resonator. A sample is placed at a position (e.g., the center of the resonator), where the microwave field B_1 has the maximum amplitude. The iris is adjustable for the quality factor Q.

where R is the effective resistance, and the real part $R + \omega\chi'' L_0$ represents the total energy loss in the resonator. The term $\omega\chi'' L_0$ is the energy loss due to magnetic resonance absorption. Here, we have assumed that the whole volume V_m of the high-frequency magnetic field is occupied by a sample of the volume V_s, but in practice the sample occupies a part of the field where B_1 is effectively uniform with the maximum strength. In this context, the susceptibility χ should be replaced effectively by $(V_s/V_m)\chi$, where the volume ratio V_s/V_m is called the *filling factor*. Nevertheless, the basic principle for detecting magnetic resonance can be discussed simply by assuming $V_s = V_m$. For a typical resonator illustrated in Fig. 8.2, the energy loss is usually described in terms of the *quality factor* Q defined by

$$Q^{-1} = \frac{\text{total energy loss}}{\text{energy stored}} \quad \text{per cycle in a resonator.}$$

The electromagnetic energy stored per cycle is equal to $Q_0^{-1}\omega(\frac{1}{2}\mu_0 B_1{}^2 V_m)$, where Q_0 is the quality factor of the empty resonator, and according to (8.8) the energy dissipation for magnetic resonance at ω_L is given by $\{\frac{1}{2}\omega_L\chi''(\omega_L)B_1{}^2\}V_m$. Hence, the fractional loss of energy due to magnetic resonance is expressed as a fractional decrease of the quality factor

$$\frac{Q_0 - Q}{Q} = \mu_0{}^{-1}\chi''(\omega_L)Q_0,$$

indicating that magnetic resonance absorption $\chi''(\omega_L)$ is essentially proportional to $\Delta Q/Q$. In magnetic resonance experiments, a change in the quality factor ΔQ can be determined at a constant ω_L, yielding $\chi''(B)$ as a function of the field strength B.

As indicated by (8.10), the magnetic resonance absorption is observed as an additional energy dissipation, which is measured as a change in the real part of the complex impedance of the resonator. A typical magnetic resonance spectrometer is basically an *impedance bridge* designed for measuring the complex impedance of the sample resonator. Figure 8.3 shows a bridge that is commonly used for practical experiments. At the bridge, the complex impedance Z of the test resonator can be balanced with the reference impedance Z_0 either in phase or out of phase, so that the detector signal can be made proportional to either the real or imaginary part of Z. The complex Z given by (8.10) can be expressed as

$$Z = R + \omega_0 L_0 \chi'' + i\omega_0 L_0(1 + \chi')\left(\frac{\omega}{\omega_0} - \frac{\omega_0}{\omega}\right),$$

where $\omega_0 = (L_0 C)^{-1/2}$ is the resonance frequency. In the vicinity of the resonance, we obtain from Z that $\operatorname{Im} Z = 2L_0(1 + \chi')\Delta\omega$ for $\omega = \omega_0 + \Delta\omega$. Keeping the frequency ω_0 constant, such a variation as $\Delta\omega$ can be obtained by scanning the field B, i.e., $\Delta\omega = \gamma\Delta B$. Therefore for a magnetic resonance absorption we can write

$$\operatorname{Im} Z = 2L_0\{1 + \chi'(\Delta B)]\}\gamma\,\Delta B.$$

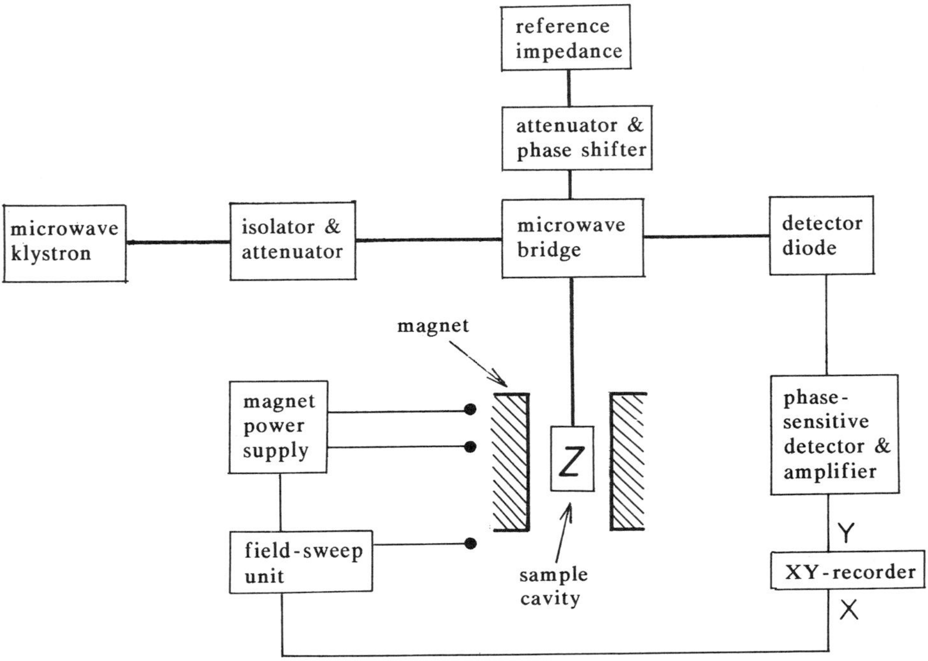

Figure 8.3. Block diagram of a microwave bridge spectrometer.

When the impedance bridge is balanced for the reactance Im $Z = 0$ at resonance, in the vicinity of $B = B_0$, a signal that is directly related to the magnetic dispersion $\chi'(B)$ can be obtained. If, on the other hand, the bridge is adjusted for Re Z,

$$\mathrm{Re}\, Z = R + \omega_0 L_0 \chi''(\Delta B),$$

and the absorption $\chi''(\Delta B)$ is directly proportional to Δ Re Z. For the detail of practical spectrometers, interested readers are referred to standard reference books of magnetic resonance spectroscopy.

8.4. The Spin Hamiltonian

Magnetic probe ions for studying structural phase transitions should preferably be of a spin larger than $\frac{1}{2}$, which however requires a very complex spectral analysis. In this section we outline the basic method for magnetic resonance analysis of such probes. Depending on the magnitude of resonance fields, the spectral analysis has to be performed with a higher-order accuracy. However, as will be discussed in Chapter 9, the type of lattice modulation can be determined in the first-order accuracy from changes in spin-Hamiltonian parameters deduced directly from the observed spectra. Leaving the accuracy problem to individual spectra, we discuss here the method of spin Hamiltonian in normal crystals, prior to introducing modified magnetic resonance parameters for modulated structures.

Unpaired electrons in a paramagnetic probe in a crystal can be described primarily by the total orbital and spin angular moments, $\boldsymbol{L}$ and $\boldsymbol{S}$, which are then perturbed by a *crystalline potential* at the probe site. It is noted that such a description of electronic states of an ion known as the Russel–Saunders coupling scheme is adequate for transition ions of the iron group that are most commonly used as magnetic resonance probes.

The crystalline potential is expressed normally by an anisotropic power series in the coordinates x, y, z with respect to the probe center, which is consistent with the local symmetry. For example, in an orthorhombic crystal the potential is expressed in the lowest order as a quadratic form

$$V(x, y, z) = Ax^2 + By^2 + Cz^2,$$

where the relation

$$A + B + C = 0$$

is required for these coefficients because such a static potential should obey the Laplace equation $\Delta V = 0$. Specifically for a uniaxial potential, $A = B$ and the above relation can be written as $C = -2A$, resulting in the potential

$$V(x, y, z) = A(x^2 + y^2 - 2z^2) = A(r^2 - 3z^2)$$

where $r^2 = x^2 + y^2 + z^2$, represents a crystal field of tetragonal or trigonal symmetry, which is generally axial symmetry along the z direction.

Another typical example is the quartic potential

$$V(x, y, z) = D(x^4 + y^4 + z^4),$$

which represents a crystal field of cubic symmetry in the lowest order. It is noted that in cubic symmetry the coefficients of a quadratic potential should be equal, i.e., $A = B = C = 0$, and hence such quartic terms prevail in the crystalline potential at the lowest order. Such a quartic potential represents a small anisotropic distortion from an otherwise spherical ionic potential of the same order, i.e., $V_0 = \alpha r^4$. It is clear that the difference potential $V(x, y, z) - V_0$ satisfies the Laplace equation if considering $\alpha = -\frac{3}{5}D$ in this case. It is noted that the coordinate axes x, y, z should be consistent with the symmetry axes a, b, c, if there is only one probe accommodated in a unit cell, otherwise unrelated to the crystal symmetry. In the latter case, there should be more than one position for the probe in a unit cell in order for the macroscopic crystal symmetry not to be violated. The theory of crystalline potentials was founded on Bethe's work in 1929 [65], which has since been refined for applications to many optical and magnetic problems in solids. For magnetic applications, Abragam and Pryce [66] have laid the theoretical foundation for the method of spin Hamiltonian to deal with paramagnetic ions in their ground states, which provides the basic formulation of magnetic resonance analysis.

In a uniaxial crystalline field, the oribital angular momentum $\boldsymbol{L}$ of the ion is quantized primarily along the unique z axis, and hence $\int \psi_0{}^* L_z \psi_0 \, dv = 0$, although the ground state can be further perturbed via the off-diagonal element $\int \psi_\varepsilon{}^* L_\perp \psi_0 \, dv$ in the second-order approximation. In a quadratic crystalline potential for $A > 0$, the energy separation between the excited and the ground states is normally larger than $k_B T$, resulting in the magnetic moment due primarily to the spin angular momentum $\boldsymbol{S}$ only. In this case, the ground state energy is independent of the spin direction, being degenerate for $\pm S_z$. In this case, the orbital angular momentum is said *quenched* by the crystalline potential, exhibiting the ground state characterized by degeneracy known as *Kramers' doublet.*

For such a spin doublet we have to further consider the spin-orbit coupling $\lambda \boldsymbol{L} \cdot \boldsymbol{S}$, which is of the order of $100 \text{ cm}^{-1} \sim 800 \text{ cm}^{-1}$ for iron-group elements, and responsible for significant interactions with the lattice. While weaker than the crystalline potential, the ground state ψ_0 of an ion in the crystalline potential can be significantly perturbed by the excited state ψ_ε via the off-diagonal matrix element $\int \psi_0{}^* L_\perp \psi_\varepsilon \, dv$. Therefore, the electron spin $\boldsymbol{S}$ cannot be *free* from the crystalline potential, and the Zeeman energy of the ion in an applied uniform field $\boldsymbol{B}$ is expressed as

$$\mathscr{H}_Z = -\boldsymbol{\mu}_e \cdot \boldsymbol{B},$$

where

$$\boldsymbol{\mu}_e = -\beta \int \psi'^*(\boldsymbol{L} + g_e \boldsymbol{S})\psi' \, dv.$$

Here,

$$\psi' = \psi_0 + \lambda \sum_\varepsilon \left[\frac{\int \psi_\varepsilon{}^* L_\perp \psi_0 \, dv}{\Delta\varepsilon}\right] \psi_\varepsilon$$

is the perturbed wave function of the ground state, $\beta = e\hbar/2m_e c = 0.927 \times 10^{-20}$ emu (the Bohr magneton), $g_e = 2.0023$ the Landé factor of a free electron, and $\Delta\varepsilon$ represents energy gaps between the state ψ_ε and the ground state ψ_0.

When the applied field is parallel to the z axis, $\int \psi_\varepsilon{}^* L_\perp \psi_0 \, dv = 2 \exp(t\Delta\varepsilon/\hbar)$ for a transition $\Delta L = 0 \to 1$. Therefore, only the z component of $\boldsymbol{\mu}_e$ is stationary in this approximation, and the Zeeman energy is expressed as

$$\mathscr{H}_Z = \beta g_z S_z B \qquad \text{where} \quad g_z = g_e\left(1 + \frac{2\lambda}{\Delta\varepsilon}\right). \tag{8.11a}$$

For the $\boldsymbol{B} \| x$ or y axis, we consider that $L_\perp = L_y + iL_z$ or $L_z + iL_x$, respectively, and obtain

$$\mathscr{H}_Z = \beta g_x S_x B \qquad \text{or} \qquad = \beta g_y S_y B,$$

where

$$g_x = g_y = g_e\left(1 - \frac{\lambda}{\Delta\varepsilon}\right). \tag{8.11b}$$

When the magnetic field is applied in an arbitrary direction, $\boldsymbol{B} = B\boldsymbol{n}$ where $\boldsymbol{n}$ is the unit vector in the direction of $\boldsymbol{B}$, we can consider that g_x, g_y, and g_z constitute the principal values of a tensor $\mathbf{g}$, and the Zeeman energy is expressed in the form

$$\mathscr{H}_Z = \beta \boldsymbol{S}\cdot\mathbf{g}\cdot\boldsymbol{B}. \tag{8.12}$$

Equation (8.12) can be interpreted either that the effective magnetic moment $-\beta\boldsymbol{S}\cdot\mathbf{g}$ precesses around the field $\boldsymbol{B}$, or that the spin $\boldsymbol{S}$ precesses around the effective field $\mathbf{g}\cdot\boldsymbol{B}$. Since the spin $\boldsymbol{S}$ is normally quantized in a strong field approximation, the latter scheme is more convenient than the former to express eigenstates of the Hamiltonian $\mathscr{H}_Z$ in terms of the spin components. Such a Hamiltonian including other interactions related to the spin variable $\boldsymbol{S}$ is known as the spin Hamiltonian.

Considering $B' = |\mathbf{g}\cdot\boldsymbol{B}|$ as the effective magnetic field that acts on the spin $\boldsymbol{S}$, we have a relation

$$B' = g_n B \qquad \text{where} \quad g_n{}^2 = \boldsymbol{n}\cdot\mathbf{g}^\dagger\mathbf{g}\cdot\boldsymbol{n} = \boldsymbol{n}\cdot\mathbf{g}^2\cdot\boldsymbol{n}. \tag{8.13}$$

Here, the tensor $\mathbf{g}^\dagger$ is the transposed matrix g_{ji} from $\mathbf{g} = (g_{ij})$, but $\mathbf{g}^\dagger = \mathbf{g}$ for the symmetrical tensor $g_{ij} = g_{ji}$, therefore $\mathbf{g}^\dagger\mathbf{g} = \mathbf{g}^2$. The eigenvalues of the Zeeman spin Hamiltonian $\mathscr{H}_Z$ are expressed as

$$\mathscr{H}_Z{}^{(1)} = \beta g_n \boldsymbol{S}\cdot\boldsymbol{B} = g_n \beta B M_S, \tag{8.14}$$

when the direction of $\boldsymbol{B}$ is specified by the direction cosines $\boldsymbol{n} = n_i$ ($i = a, b, c$), and M_S is the magnetic quantum number with regard to this quantization direction. Equation (8.13) indicates that $g_n{}^2$ is expressed as a quadratic form with respect to n_i

$$g_n{}^2 = \sum_{ij}\left(\sum_k g_{ik}g_{kj}\right)n_i n_j.$$

In magnetic resonance spectroscopy, it is routine practice to determine the principal axes (x, y, z) from measured g_n in spectra observed in all directions of $\boldsymbol{B}$, which can be obtained by a coordinate transformation. With respect to the principal axes,

$$g_n{}^2 = g_x{}^2 n_x{}^2 + g_y{}^2 n_y{}^2 + g_z{}^2 n_z{}^2$$

and

$$n_x{}^2 + n_y{}^2 + n_z{}^2 = 1.$$

The first vector $\boldsymbol{n}$ in $\boldsymbol{n}\cdot\mathbf{g}^2\cdot\boldsymbol{n}$ of (8.13) acting on the matrix $\mathbf{g}^2$ from the right is expressed as a one-row matrix, while the second $\boldsymbol{n}$ multiplying from the left is a column matrix, according to the matrix multiplication rule. In this context, it is convenient to write these vectors as a *bra* and *ket* notations, i.e.,

$\langle n|$ and $|n\rangle$. Using these notations,

$$g_n{}^2 = \langle n|\mathbf{g}^2|n\rangle. \tag{8.13a}$$

An important feature of the $\mathbf{g}$ tensor is that the trace is related to its isotropic value g_e, i.e.,

$$\text{trace } \mathbf{g} = g_x + g_y + g_z = 3g_e, \tag{8.15}$$

as verified from (8.11a,b), playing a significant role in the spin Hamiltonian. Usually, a shift of the g factor from $g_e = 2.0023$ of the free electron or from $g = 2.0036$ of an organic free radical called DPPH (α–α' diphenyl picryl hydrazil) as a practical standard is measured. In the above case, the shift tensor $\Delta\mathbf{g} = \mathbf{g} - g_e\mathbf{e}$, where $\mathbf{e}$ is the unit matrix, is determined from observed spectra. It is noticed that such a shift tensor is traceless, i.e.,

$$\text{trace } \Delta\mathbf{g} = 0, \tag{8.15a}$$

which is a useful relation for spectral analysis.

8.5. The Fine Structure

In Section 8.4 the spin-orbit coupling was considered as a significant perturbation to the ground state of an ion in a crystalline field. Although vanishing in the first order, the coupling perturbs the magnetic property of an ion in a crystalline potential in the second order, as seen from the anisotropic g shift $\Delta\mathbf{g}$. In addition, in an anisotropic crystal potential the charge cloud of an ion is significantly deformed from the spherical symmetry, for which the spin-orbit coupling is also responsible. The energy associated with such ionic deformation appears as the *fine structure* energy in magnetic resonance spectra.

The spin-orbit coupling energy can be expressed as

$$\mathscr{H}_{SL} = \lambda(L_x S_x + L_y S_y + L_z S_z),$$

where the indexes x, y, and z designate the principal axes of the crystalline potential, and the constant λ is assumed to remain primarily isotropic as in the isolated ion. The coupling $\mathscr{H}_{SL}$ gives a second-order perturbation energy due to nonvanishing off-diagonal matrix elements of $\boldsymbol{L}$ between the ground state ψ_0 and the excited state ψ_ε, which is expressed as

$$E_{SL}{}^{(2)} = \frac{\lambda^2}{\Delta\varepsilon}\sum_{ij} S_i S_j \left\{\left(\int \psi_0{}^* L_i \psi_\varepsilon \, dv\right)\left(\int \psi_\varepsilon{}^* L_j \psi_0 \, dv\right)\right\}.$$

It is noted that $E_{SL}{}^{(2)}$ is in a quadratic form with respect to the spin components S_x, S_y, and S_z, when considering

$$D_{ij} = \frac{\lambda^2}{\Delta\varepsilon}\left\{\left(\int \psi_0{}^* L_i \psi_\varepsilon \, dv\right)\left(\int \psi_\varepsilon{}^* L_j \psi_0 \, dv\right)\right\} \tag{8.16a}$$

as the coefficients. Therefore, the energy $E_{SL}{}^{(2)}$ can be expressed as

$$E_{SL}{}^{(2)} = \sum_{ij} S_i D_{ij} S_j = \langle S|\mathbf{D}|S\rangle. \tag{8.16b}$$

The fine structure tensor $\mathbf{D}$ defined by (8.16a) is symmetrical, i.e., $D_{ij} = D_{ji}$, and also traceless as can be seen from its definition. Namely,

$$\text{trace } \mathbf{D} = \sum_i D_{ii} = \frac{\lambda^2}{\Delta\varepsilon} \sum_i \langle 0|L_i|\varepsilon\rangle\langle\varepsilon|L_i^*|0\rangle = \frac{\lambda^2}{\Delta\varepsilon}\left\langle 0\left|\sum_i L_i L_i^*\right|0\right\rangle = 0. \tag{8.17a}$$

In classical language, the $\mathbf{D}$ tensor represents a distorted ionic charge from a spherical distribution, reflecting symmetry of the crystalline potential that is orthorhombic in this approximation.

The fine-structure tensor $\mathbf{D}$ can be determined from observed magnetic resonance spectra with respect to the symmetry axes a, b, and c of a given crystal, which is then transformed into the principal form that is specified by the principal values D_x, D_y, D_z and their direction cosines of the principal axes x, y, z with respect to the symmetry axes a, b, c. Since the tensor is traceless, in the principal frame of reference

$$\text{trace } \mathbf{D} = D_x + D_y + D_z = 0. \tag{8.17b}$$

Accordingly, if, for example, $D_z > 0$, $D_x + D_y = -D_z < 0$, indicating that the distorted charge distribution can be characterized by an electric quadrupole moment, as schematically illustrated in Fig. 8.4(a). It is noted that such a fine structure energy is significant for an ion whose spin S is higher than $\frac{1}{2}$, otherwise it vanishes entirely for $S = \frac{1}{2}$. This can be seen by assuming that the $\mathbf{D}$ tensor is uniaxial, and

$$\begin{aligned}\langle S|\mathbf{D}|S\rangle &= D_x(S_x{}^2 + S_y{}^2) + D_z S_z{}^2 = -\tfrac{1}{2}D_z(S^2 - S_z{}^2) + D_z S_z{}^2 \\ &= \tfrac{1}{2}D_z\{3S_z{}^2 - S(S+1)\},\end{aligned}$$

which is zero for $S = \frac{1}{2}$.

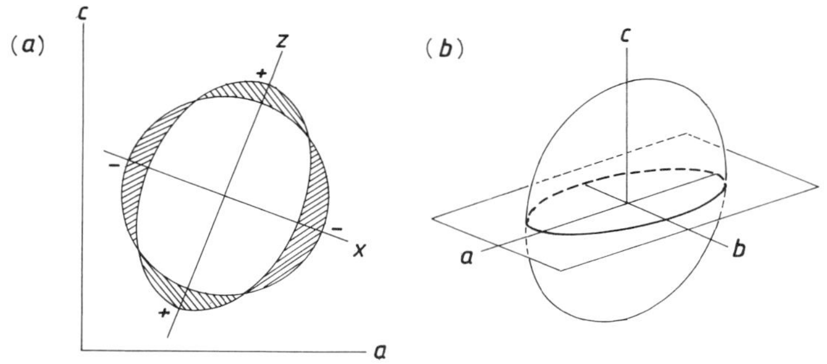

Figure 8.4. (a) An ellipsoidal deformation of the electronic charge arising from a crystal potential, resulting in a fine structure interaction $\langle S|\mathbf{D}|S\rangle$. (b) An elliptic cross-section between the ellipsoid of a $\mathbf{D}$ tensor and a plane of observation in which the applied field $\mathbf{B}_0$ is rotated.

In the foregoing argument we consider that both **g** and **D** tensors originate from the second-order perturbation of a spin-orbit coupling in a crystalline potential. Therefore, their symmetries should be identical to that of the crystalline potential. However, generally these tensors are mixed in magnetic resonance spectra, and the **D** tensor needs to be calculated in relation to the **g** tensor that can be determined separately in high-field approximation. In a given uniform magnetic field $\boldsymbol{B}$, the spin Hamiltonian is

$$\mathscr{H} = \mathscr{H}_Z + \mathscr{H}_F = \beta\langle S|\mathbf{g}|B\rangle + \langle S|\mathbf{D}|S\rangle. \tag{8.18}$$

In a strong field approximation, the Zeeman energy should be considerably larger than the fine structure energy. In this case, the spin $\boldsymbol{S}$ is quantized with respect to the effective field $\mathbf{g}\cdot\boldsymbol{B}$, and hence the spin variables $\langle S|$ in $\mathscr{H}_Z$ and $\mathscr{H}_F$ are replaced by $g_n{}^{-1}\langle S|\mathbf{g} = \langle S_n|$ and $\langle S| = \langle S_n| = S_n\langle n|$, respectively, in the first-order perturbation. Thus, for $\mathscr{H}_Z$ we obtain (8.14) and for $\mathscr{H}_F$

$$E_F{}^{(1)} = \langle S_n|\mathbf{D}|S_n\rangle = \langle n|\mathbf{D}|n\rangle M_S{}^2 = D_n M_S{}^2, \tag{8.19}$$

where the coefficient $D_n = \langle n|\mathbf{D}|n\rangle$ is a quadratic with respect to the direction cosines (n_x, n_y, n_z) of the vector $\langle n|$, i.e., for the matrix $\mathbf{D} = (D_{ij})$

$$\langle n|\mathbf{D}|n\rangle = \sum_{ij} D_{ij} n_i n_j. \tag{8.20}$$

For the spin Hamiltonian of (8.18), the magnetic resonance at a given frequency ω is given by the condition

$$\hbar\omega = g_n\beta B M_S + D_n M_S{}^2, \tag{8.21}$$

to which the selection rules $\Delta M_S = \pm 1$ apply for magnetic dipole transitions.

If on the other hand, the fine structure energy is greater than the Zeeman energy, the spin $\langle S|$ should be first quantized along the direction of the principal axis for the largest value of D. In such a weak-field approximation, the energy eigenvalues should be worked out to higher-order corrections, considering the Zeeman energy as a perturbation to $\mathscr{H}_F$. In this case, the result is too complex to deal with structural modification in modulated crystals, except for specific directions of the applied field, in which a profile of modulation scheme can be revealed. We therefore do not attempt to discuss the intermediate or weak-field cases in detail, leaving the analysis to individual cases of interest. Spectra that can be analyzed in a strong field are somewhat specific, but give useful insight into the three-dimensional order-variable vector in a given crystal. Nevertheless, low-field analyses are required for some problems, as discussed briefly in the next section, in addition to the Zeeman studies of nuclear quadrupole resonance and of triplet excited states in molecular probes.

8.6. Hyperfine Interactions and Forbidden Transitions

The magnetic interaction between an electronic magnetic moment $\boldsymbol{\mu}_e$ and a nuclear magnetic moment $\boldsymbol{\mu}_n$ located nearby is generally called the *hyperfine interaction.* Such a nuclear moment can be of the nucleus of the ion, or of an ion in the ligand structure. These magnetic moments can interact with each other as in a classical dipole–dipole interaction as well as in the quantum-mechanical *contact* mechanism. The classical interaction is anisotropic, whereas the contact interaction, known as the Fermi interaction, is isotropic. The hyperfine interaction is generally involved in these two mechanisms, and expressed by

$$\mathscr{H}_{HF} = \left\{\frac{\boldsymbol{\mu}_e \cdot \boldsymbol{\mu}_n - 3(\boldsymbol{\mu}_e \cdot \boldsymbol{r})(\boldsymbol{\mu}_n \cdot \boldsymbol{r})}{r^{-2}}\right\} r^{-3} + \frac{8\pi}{3}|\psi(0)|^2 \boldsymbol{\mu}_e \cdot \boldsymbol{\mu}_n,$$

where $\boldsymbol{r}$ is the distance between $\boldsymbol{\mu}_e$ and $\boldsymbol{\mu}_n$, and $|\psi(0)|^2$ is the density of the electronic charge at the position of the nucleus, $\boldsymbol{r} = 0$. Here $\boldsymbol{\mu}_e = -\beta \mathbf{g} \cdot \boldsymbol{S}$ and $\boldsymbol{\mu}_n = \gamma \boldsymbol{I}$, where γ is the gyromagnetic ratio of the nuclear moment of spin $\boldsymbol{I}$. Using the direction cosines (l, m, n) of the vector $\boldsymbol{r}$, $\mathscr{H}_{HF}$ can be expressed as

$$\mathscr{H}_{HF} = \langle \mu_e | \mathbf{A}_d | \mu_n \rangle + \frac{8\pi}{3}|\psi(0)|^2 \langle \mu_e \mid \mu_n \rangle,$$

where the tensor

$$\mathbf{A}_d = r^{-3}\begin{bmatrix} 1 - 3l^2 & -3lm & -3nl \\ -3ml & 1 - 3m^2 & -3mn \\ -3ln & -3nm & 1 - 3n^2 \end{bmatrix}$$

represents the dipole–dipole interaction. Since $l^2 + m^2 + n^2 = 1$, it is clear that trace $\mathbf{A}_d = 0$. Combining the two contributions into one, the hyperfine interaction energy is expressed as

$$\mathscr{H}_{HF} = \langle \mu_e | \mathbf{A} | \mu_n \rangle \qquad \text{where} \quad \mathbf{A} = \mathbf{A}_d + \frac{8\pi}{9}|\psi(0)|^2 \mathbf{E} \tag{8.22}$$

and $\mathbf{E}$ is a unit matrix, and we have the trace relation for the tensor $\mathbf{A}$

$$\text{trace } \mathbf{A} = \frac{8\pi}{9}|\psi(0)|^2. \tag{8.23}$$

The hyperfine interaction tensor $\mathbf{A}$ for the central nucleus can normally be coaxial with the tensors $\mathbf{g}$ and $\mathbf{D}$, but not necessarily so with a ligand nucleus. Using the expressions $\langle \mu_e | = -\beta \langle S | \mathbf{g}$ and $\langle \mu_n | = \gamma \langle I |$ in (8.22),

$$\mathscr{H}_{HF} = -\beta\gamma \langle S | \mathbf{g} \cdot \mathbf{A} | I \rangle. \tag{8.24}$$

This expression, together with the nuclear Zeeman energy $\langle B | \gamma \boldsymbol{I} \rangle$, can be interpreted in such a way that the nuclear moment $\gamma \boldsymbol{I}$ precesses at its position

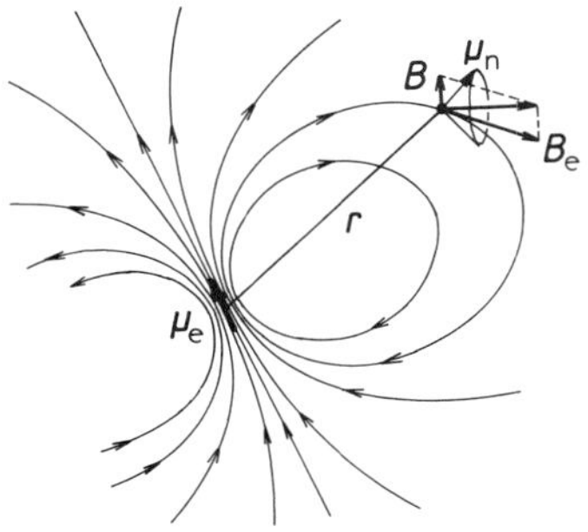

Figure 8.5. Magnetic dipole interaction between an electronic moment $\boldsymbol{\mu}_e$ and a nuclear moment $\boldsymbol{\mu}_n$ at a distance $\boldsymbol{r}$. The magnetic field $\boldsymbol{B}_e$ due to $\boldsymbol{\mu}_e$ and the uniform applied field $\boldsymbol{B}_0$ are combined to interact with $\boldsymbol{\mu}_n$.

around the effective magnetic field due to the electron spin, $\langle B_e| = \beta S_n \langle n|\mathbf{g}\cdot\mathbf{A}$, plus the external field $\langle B|$. Therefore, the nuclear spin-quantum number M_I' should be referred to the combined field $\langle B| + \langle B_e|$, as illustrated in Fig. 8.5. However, generally, the nuclear Zeeman energy is negligibly small in a magnetic field for electronic resonance experiments, and

$$E_{\rm HF}{}^{(1)} + E_{NZ} = -\gamma\langle B_e + B|I\rangle \approx -\gamma|B_e|M_I'.$$

Here,

$$|B_e| = \beta M_S \langle n|\mathbf{g}\cdot\mathbf{A}\cdot\mathbf{A}^\dagger\cdot\mathbf{g}^\dagger|n\rangle^{1/2}.$$

Writing

$$g_n{}^2K_n{}^2 = \gamma^2\langle n|\mathbf{g}^2\cdot\mathbf{A}^2|n\rangle, \tag{8.25a}$$

the hyperfine energy in the first-order approximation can be expressed as

$$E_{\rm HF}{}^{(1)} = g_n\beta K_n M_S M_I'. \tag{8.25b}$$

For practical applications of (8.25a,b), it is convenient to include γ in the expression of the tensor $\mathbf{A}$. On the other hand, we shall keep the constant β as in the above expression for $E_{\rm HF}{}^{(1)}$, since, in practice, the constant K_n is measured in units of the applied magnetic field. In other words, $E_{\rm HF}{}^{(1)}/g_n\beta = K_n M_S M_I'$ is in units of Gauss in practical magnetic resonance spectra.

For an isotropic $\mathbf{g}$ tensor, (8.25a) is specially written as

$$K_n{}^2 = \langle n|\mathbf{A}^2|n\rangle,$$

and splittings measured from the spectra in all directions determine the tensor $\mathbf{A}$. In general, however, $\mathbf{g}$ and $\mathbf{A}$ tensors are not coaxial, and so the coefficients K_n (in units of B) in (8.25b) and g_n in (8.14) are determined from magnetic resonance spectra independently, and the tensor $\mathbf{A}$ can then be calculated. A practical example for such calculation is shown in Section 8.6.

In conventional magnetic resonance experiments, a magnetic hyperfine energy is usually smaller than Zeeman and fine-structure energies, and so (8.25a) in the first-order approximation is generally sufficiently accurate to deal with observed spectra. For such an analysis, the selection rules for resonance transitions are given by

$$\Delta M_S = \pm 1 \qquad \text{and} \qquad \Delta M_I' = 0, \tag{8.26}$$

which are called *allowed transitions.*

In the foregoing we did not consider the fine structure term for simplicity, which is however significant for a spin S larger than $\frac{1}{2}$, in which case the spin Hamiltonian should be expressed as

$$\mathscr{H} = g_n\beta\langle S_n|B\rangle + \langle S_n|\mathbf{D}|S_n\rangle + g_n\beta\gamma\langle S_n|\mathbf{A}|I\rangle.$$

As remarked, it is not easy to obtain the corresponding energy value, but we realize that in the second-order approximation the above selection rules for electronic transitions can be accompanied by nuclear-spin transitions $\Delta M_I' = \pm 1$. To simplify the argument, we assume that both $\mathbf{g}$ and $\mathbf{A}$ are isotropic, and write only $\mathbf{A}$ in place of $g_n\beta\gamma\mathbf{A}$. Further, for a uniaxial $\mathbf{D}$ tensor the above spin Hamiltonian can be expressed as

$$\begin{aligned}\mathscr{H} = g\beta BS_z &+ \tfrac{1}{2}D\left\{\frac{S_z^2 - S(S+1)}{3}\right\}(3\cos^2\theta - 1)\\ &+ D(S_zS_x + S_xS_z)\cos\theta\sin\theta + \tfrac{1}{4}D(S_+^2 + S_-^2)\sin^2\theta\\ &+ A\boldsymbol{S}\cdot\boldsymbol{I} - \gamma\boldsymbol{B}\cdot\boldsymbol{I},\end{aligned}$$

where θ is the angle between S_z and the unique direction of the $\mathbf{D}$ tensor. It is noted that when $\boldsymbol{B}$ is applied parallel to either the z or x axes, i.e., $\theta = 0$ or $\frac{1}{2}\pi$,

$$\begin{aligned}E^{(1)}(M_S, M_I) &= g\beta BM_S + DM_S^2 + AM_SM_I - \gamma BM_I \qquad \text{or}\\ &= g\beta BM_S - \tfrac{1}{2}DM_S^2 + AM_SM_I - \gamma BM_I,\end{aligned}$$

respectively, magnetic resonance transitions are all allowed, i.e., $\Delta M_S = \pm 1$ and $\Delta M_I = 0$.

If, on the other hand, $0 < \theta < \frac{1}{2}\pi$, an off-diagonal product such as $D\cos\theta\sin\theta S_zS_\pm \times AS_-I_\pm$ make additional second-order contributions $E^{(2)}(M_S, M_I)$ that are proportional to $S_z^2(I_+ + I_-)$, giving rise to *forbidden transitions*

$$\Delta M_S = \pm 1 \qquad \text{and} \qquad \Delta M_I = \pm 1. \tag{8.27}$$

According to Abragam and Bleaney [62], intensities of these forbidden lines relative to allowed ones are given by

$$\left(\frac{3D\sin 2\theta}{4g\beta B}\right)^2\left\{\frac{1 + S(S+1)}{3M_S(M_S+1)}\right\}^2\{I(I+1) - M_I^2 + M_I\}.$$

They have also shown that cross-products $DS_+^2(AS_-I_+)^2$ and $DS_-^2(AS_+I_-)^2$ can give another set of forbidden transitions $\Delta M_I = \pm 2$, and the combined hyperfine structure arising from the second-order perturbation of these terms is described by

$$AM_SM_I - \gamma BM_I + \left(\frac{A^2}{2g\beta B}\right)[M_I\{M_S^2 - S(S+1)\} - M_S\{M_I^2 - I(I+1)\}]$$
$$+ \left(\frac{D\sin 2\theta}{4g\beta B}\right)^2 (2AM_IM_S)[\{M_S^2 - S(S+1)\} - M_S^2]$$
$$+ \left(\frac{D\sin^2\theta}{4g\beta B}\right)^2 (2AM_SM_I)\{2M_S^2 + 1 - 2S(S+1)\}.$$

Such forbidden transitions appear in spectra as a function of the angle θ, varying from a maximum intensity when $\theta = 45°$ to zero on the principal axes.

8.7. Tensor Analysis for Spin-Hamiltonian Parameters

Considering the spin $\langle S|$ is primarily quantized along the external field $B\langle n|$, energy eigenvalues of the spin Hamiltonian $\mathscr{H} = \mathscr{H}_Z + \mathscr{H}_F + \mathscr{H}_{HF}$ in the first-order accuracy are expressed in terms of $\langle S_n| = M_S\langle n|$ as

$$E(M_S, M_I') = g_n\beta BM_S + D_nM_S^2 + K_nM_SM_I',$$

where

$$g_n^2 = \langle n|\mathbf{g}^2|n\rangle, \qquad K_n^2 = \langle n|\mathbf{A}^2|n\rangle, \qquad \text{and} \qquad D_n = \langle n|\mathbf{D}|n\rangle.$$

It is noted that trace $\mathbf{g} = g_e/3$ and trace $\mathbf{A} = (8\pi/3)|\psi(0)|^2 = A_0/3$ are positive and isotropic, whereas trace $\mathbf{D} = 0$. Geometrically, these quadratic forms represent ellipsoids, which are regarded as deformed from spheres of radii g_e^2, A_0^2 for the first two, and so for the $\mathbf{D}$ tensor if a large sphere of an arbitrary radius D_0 is added, as illustrated in two dimensions in Fig. 8.5(a).

Figures 8.6 and 8.7 show typical paramagnetic resonance spectra of VO^{2+} probes in the normal phase of BCCD, and Mn^{2+} spectra in normal TSCC and BCCD crystals, respectively. Here, the VO^{2+} spectra exhibit eight hyperfine lines due to the ^{51}V nucleus, which are practically equally spaced but with unequal linewidths for transitions at different nuclear quantum numbers. The difference between the spectra at different microwave frequencies is due to a large $g_a - g_e$, making the resonance field at 35 GHz significantly higher than at 9.2 GHz. In the Mn^{2+} spectra values of g_n and K_n are isotropic, but the fine-structure parameters D_n are very different in these crystals, i.e., normal in TSCC whereas unusually large in BCCD.

In principle, all of these parameters g_n, K_n, and D_n can be determined from spectra observed for any direction $\langle n|$ of the applied field, and hence the plots

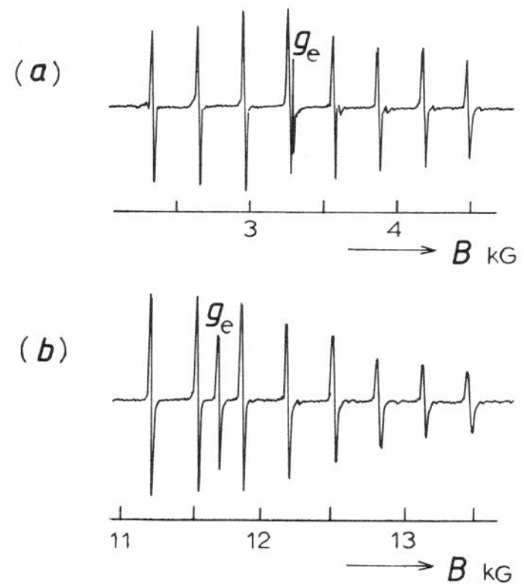

Figure 8.6. Typical EPR spectra characterized by an anisotropic g factor and hyperfine splittings. Shown here are $^{51}VO^{2+}$ spectra in the normal phase of BCCD crystals for $B_0 \| a$. (a) at 9.2 GHz; and (b) at 35 GHz.

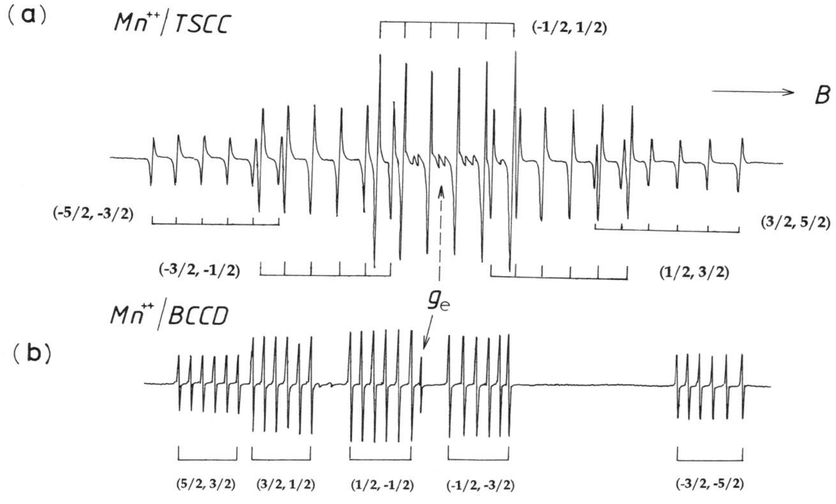

Figure 8.7. (a) A representative $^{55}Mn^{2+}$ spectrum in the normal phase of TSCC, which are dominated by an anisotropic fine structure and an isotropic hyperfine interaction with a ^{55}Mn-nucleus. (b) A $^{55}Mn^{2+}$ spectrum from BCCD in the normal phase, which is dominated by a large anisotropic **D** tensor. All the lines are due to allowed magnetic transitions.

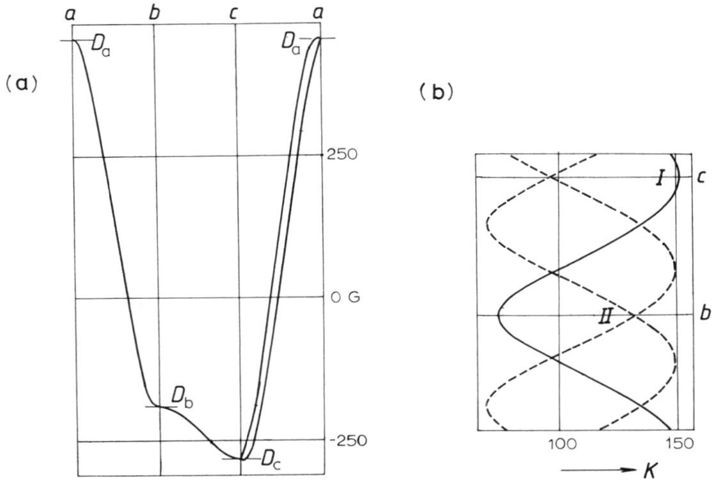

Figure 8.8. (a) Angular variations of the fine-structure splitting D_n in Mn^{2+} spectra in the normal TSCC. (b) Angular variations of the hyperfine splitting K_n in VO^{2+} spectra in the *bc* plane of normal TSCC, showing three magnetically independent orientations in the unit cell. The solid curve is consistent with the orthorhombic symmetry of the crystal, while a pair of broken lines are symmetrical with respect to the *b* and *c* axes, and hence the two lines together retain the crystal symmetry.

of $g_n{}^2$, $K_n{}^2$, and D_n against the angle of rotation of the sample crystal should exhibit $\cos^2$ (or $\sin^2$) curves, as shown in Fig. 8.8. Geometrically, as illustrated by Fig. 8.4(b), such a periodic curve represents radii of the cross-sectional ellipse when the ellipsoid is cut by a plane with the normal vector $\langle n|$. In practice, we rotate the crystal around three orthogonal symmetry axes *a*, *b*, and *c*, and from a set of three elliptic curves we can determine all the coefficients of the quadratic form representing each one of these spin-Hamiltonian parameters. For example, for a direction $\langle n| = (n_1, n_2, n_3)$

$$\begin{aligned} g_n{}^2 = {} & g_{11}{}^2 n_1{}^2 + g_{12}{}^2 n_1 n_2 + g_{13}{}^2 n_1 n_3 \\ & + g_{12}{}^2 n_1 n_2 + g_{22} n_2{}^2 + g_{23}{}^2 n_2 n_3 \\ & + g_{13}{}^2 n_1 n_3 + g_{23} n_2 n_3 + g_{33}{}^2 n_3{}^2, \end{aligned}$$

where the matrix

$$\begin{pmatrix} g_{11}{}^2 & g_{12}{}^2 & g_{13}{}^2 \\ g_{12}{}^2 & g_{22}{}^2 & g_{23}{}^2 \\ g_{13}{}^2 & g_{23}{}^2 & g_{33}{}^2 \end{pmatrix}$$

can be determined completely. Then this matrix is diagonalized as

$$\begin{pmatrix} g_x{}^2 & 0 & 0 \\ 0 & g_y{}^2 & 0 \\ 0 & 0 & g_z{}^2 \end{pmatrix}$$

by a linear coordinate transformation

$$(n_{x1}=l_{x1}a+m_{x1}b+n_{x1}c,\quad n_{x2}=l_{x2}a+m_{x2}b+n_{x2}c,\quad n_{x3}=l_{x3}a+m_{x3}b+n_{x3}c),$$

$$(n_{y1}=l_{y1}a+m_{y1}b+n_{y1}c,\quad n_{y2}=l_{y2}a+m_{y2}b+n_{y2}c,\quad n_{y3}=l_{y3}a+m_{y3}b+n_{y3}c),$$

$$(n_{z1}=l_{z1}a+m_{z1}b+n_{z1}c,\quad n_{z2}=l_{z2}a+m_{z2}b+n_{z3}c,\quad n_{z3}=l_{z3}a+m_{z3}b+n_{z3}c),$$

each representing the principal axes x, y, z for the diagonalized $g_z{}^2$, $g_y{}^2$, $g_z{}^2$, or g_x, g_y, g_z.

The significance of diagonalized parameter values can normally be attributed to the symmetry of the crystalline potential deforming the charge distribution of the probe ion. If, for example, judging from the unique principal value, as often is the case, the corresponding principal axis can be regarded as representing the symmetry axis of the potential. The above calculation for obtaining principal values and directions can be carried out by a routine numerical analysis of magnetic resonance spectra.

CHAPTER 9

Magnetic Resonance Sampling and Nuclear Spin Relaxation Studies of Pseudospin Condensates

9.1. Paramagnetic Spins in a Modulated Crystal

In a normal crystal phase governed by translational symmetry, magnetic probes distributed in the lattice are all equivalent, exhibiting an identical spectrum, whereas in a modulated phase where unit cells are not exactly identical, magnetic resonance lines are inhomogeneously broadened, exhibiting an anomalous shape reflecting the specific type of lattice modulation.

When *doping* with a small quantity of magnetic impurities, it is assumed that these probes are randomly distributed in the lattice, thereby showing a weak *diluted* paramagnetism. For studies of phase transitions, such probes must be incorporated in the active groups either interstitially or in substitution for constituent ions, otherwise insensitive to a structural change of the crystal. In such a doped crystal, however, we should be aware of the fact that properties of a crystal are always modified to some extent by doping, and it is therefore essential to make sure if such a modification is insignificant for the pseudospin dynamics. On the other hand, when the active group contains an approriate magnetic nucleus, its nuclear magnetic resonance spectrum can also be informative about the structural change. Nevertheless, because of a larger extent of the electronic wavefunction that encompasses the surroundings, paramagnetic probes are generally more sensitive to a structural change than nuclear spins. Nuclear probes are only weakly coupled with the local electric-field gradient that interacts with their nuclear quadrupole moments.

In practice, ions of the transition-group elements are commonly used as paramagnetic probes, whenever exhibiting well-defined spectra. It is realized that such probes are associated with charge defects that compensate excess impurity charges in the lattice, often making the situation unnecessarily more complicated than in undoped crystals. Irradiated crystals can also be studied, since damaged constituents called *free radicals* are generally paramagnetic. However, we should be cautious about the modified transition mechanism in irradiated crystals to see if they are compatible with the corresponding unirradiated crystals. In any event, properties of doped or irradiated crystals should be carefully examined to see if their spectra can provide relevant information to unmodified crystals.

In a nonmagnetic crystal, the magnetic property of a probe ion is perturbed by the crystal potential through the spin-orbit coupling, as discussed in Chapter 8. In consequence, in a modulated phase the spin vector $\boldsymbol{S}'$ is quantized in a direction different from the spin $\boldsymbol{S}$ in the corresponding unmodulated phase, depending on its site in the crystal. We assume that the quantization axis for the spin $\boldsymbol{S}'$ can be transformed from $\boldsymbol{S}$ by a matrix $\mathbf{a}$, i.e.,

$$\boldsymbol{S}' = \mathbf{a} \cdot \boldsymbol{S} \qquad \text{where} \quad \mathbf{a} = \mathbf{1} + \langle \sigma | \mathbf{e} \tag{9.1}$$

is the transformation matrix between these spin axes. Here $\mathbf{1} = (\delta_{ij})$ is the unit matrix with elements given by Kronecker's delta, $\mathbf{e}$ is the basic deformation matrix for an active group, and $\sigma = \sigma_0 f(\phi)$ is the order variable emerging from zero at T_c. By (9.1) we mean that the spatial variation of σ is *sampled* by the spins $\boldsymbol{S}'$ that are randomly located along the condensate $\sigma(\phi)$. It is important that at a sufficiently low concentration of magnetic probes the pseudospin condensate is modified only to a negligibe extent.

The deformation matrix $\mathbf{e} = (\mathrm{e}_{ij})$ can be considered as composed of symmetrical and antisymmetrical parts, i.e.,

$$(\mathrm{e}_{ij}) = \tfrac{1}{2}(\mathrm{e}_{ij} + \mathrm{e}_{ji}) + \tfrac{1}{2}(\mathrm{e}_{ij} - \mathrm{e}_{ji}),$$

where the first term represents structural deformation, and the second matrix describes a *rigid-body rotation* of an active group.

For a small symmetrical matrix (e_{ij}), the active group is deformed with no appreciable volume change, the strain tensor is characterized by

$$\mathrm{e}_{ij} = \mathrm{e}_{ji} \qquad \text{and} \qquad \text{trace}\, \mathbf{e} = \mathrm{e}_{11} + \mathrm{e}_{22} + \mathrm{e}_{33} = 0, \tag{9.2}$$

and one of these axes represents a unique direction $\boldsymbol{e}$ in a binary system. On the other hand, for a rigid-body rotation the antisymmetric elements are given by

$$\mathrm{e}_{ij} = -\mathrm{e}_{ji} \qquad \text{and} \qquad \mathrm{e}_{11} = \mathrm{e}_{22} = \mathrm{e}_{33} = 0, \tag{9.3}$$

where the vector defined by $\boldsymbol{e} = (\mathrm{e}_{23}, \mathrm{e}_{31}, \mathrm{e}_{12})$ represents the axis of rotation. With such a rotating model, the binary pseudospin may be characterized by two directions given by the vector $\boldsymbol{e}$ and its inversion $-\boldsymbol{e}$. For example, as shown in Subsection 9.3.2, in the first incommensurate phase of BCCD crystals between 164 K and 127 K, the librational motion of the active group can be described by the amplitude proportional to $\cos\phi$ around the axis $\boldsymbol{e}$ determined by $\phi = 0$ and π, so that the vectors $\pm\sigma\boldsymbol{e}$ are equivalent to the matrix $\pm\sigma\mathbf{e}$. On the other hand, in TSCC, the pseudospin behaves like a classical vector $(\cos\phi, \sin\phi)$, where ϕ is the angle from the a axis in the *ac* plane. In this case, considering the diagonal elements, $e_{11} = 1$, $e_{22} = 1$, $e_{33} = -1$, and zero off-diagonal elements for the matrix **e**, a vector such as $\sigma_0(\boldsymbol{i} \cos \phi - \boldsymbol{k} \sin \phi)$ is equivalent to the matrix $\langle \sigma | \mathbf{e}$ in (9.1). Equations (9.2) and (9.3) can also be used to describe the distorted active group $BO_6{}^{2-}$ for structural changes in perovskite crystals.

9.2. The Spin Hamiltonian in Modulated Crystals

Assuming (9.1) for magnetic spins sampling a pseudospin condensate, we can write the spin Hamiltonian for modified spins $\langle S' |$ as

$$\mathcal{H}' = \beta \langle S' | \mathbf{g} | B \rangle + \langle S' | \mathbf{D} | S' \rangle + \langle S' | \mathbf{A} | I \rangle$$

$$= \mathcal{H} + \sigma\beta \langle S | \mathbf{e}^{\dagger} \cdot \mathbf{g} | B \rangle \tag{9.4a}$$

$$+ \sigma \langle S | \mathbf{e}^{\dagger} \cdot \mathbf{D} + \mathbf{D} \cdot \mathbf{e} | S \rangle + \sigma^2 \langle S | \mathbf{e}^{\dagger} \cdot \mathbf{D} \cdot \mathbf{e} | S \rangle \tag{9.4b}$$

$$+ \sigma \langle S | \mathbf{e}^{\dagger} \cdot \mathbf{A} | I \rangle, \tag{9.4c}$$

where

$$\mathcal{H} = \beta \langle S | \mathbf{g} | B \rangle + \langle S | \mathbf{D} | S \rangle + \langle S | \mathbf{A} | I \rangle$$

is the spin Hamiltonian in unmodulated crystals. Here the nuclear spin $\langle I |$ is not directly related to any lattice variables, so long as the quadrupole interaction can be ignored. In this case, we can consider that nuclear spins are not influenced primarily by lattice modulation. It is noted that those anisotropic product tensors **g**, **D**, and **A**, multiplied by the deformation matrix **e**, are not generally zero in (9.4a, b, c), and considered as responsible for modulated spectra. In practice, observed magnetic resonance spectra can be analyzed to obtain these tensor parameters modified by **e**, from which the modulation effect can be evaluated.

9.2.1. The g Tensor Anomaly

In the first-order approximation, the Zeeman energy of a paramagnetic ion is expressed in terms of the stationary spin component S_n of the spin $\langle S |$ precessing around the static magnetic field $\langle B | = B \langle n |$. In a modulated crystal, the modified magnetic moment $\beta \langle S | \mathbf{g}'$ can be regarded as quantized along the field $\langle B |$, in which the steady component is given by $g_n' \beta \langle S_n |$, where $\langle S_n | = M \langle n |$. Here M is the magnetic quantum number, and g_n' represents the modified g_n factor. In this first-order approximation, the Zeeman energies are expressed as

$$E_Z{}^{(1)} = g_n' \beta B M,$$

where the modified g factor is determined from

$$g_{\mathrm{n}}'^{2} = \langle n|\mathbf{g}'^{2}|n\rangle = \langle n|\mathbf{a}^{\dagger}\cdot\mathbf{g}^{2}\cdot\mathbf{a}|n\rangle$$
$$= g_{\mathrm{n}}^{2} + \sigma\langle n|\mathbf{e}^{\dagger}\cdot\mathbf{g}^{2} + \mathbf{g}^{2}\cdot\mathbf{e}|n\rangle + \sigma^{2}\langle n|\mathbf{e}^{\dagger}\cdot\mathbf{g}^{2}\cdot\mathbf{e}|n\rangle, \tag{9.5}$$

and

$$g_{\mathrm{n}}^{2} = \langle n|\mathbf{g}^{2}|n\rangle$$

gives the unmodified g factor. Equation (9.5) indicates that the modulation effect is described by the two terms proportional to σ and σ^2, where the factors $\langle n|\mathbf{e}^{\dagger}\cdot\mathbf{g}^{2} + \mathbf{g}^{2}\cdot\mathbf{e}|n\rangle$ and $\langle n|\mathbf{e}^{\dagger}\cdot\mathbf{g}^{2}\cdot\mathbf{e}|n\rangle$ are not generally zero. The first factor can however be zero only if the tensor $\mathbf{e}$ is traceless and the $\mathbf{g}$ tensor is isotropic. The second factor is negligibly small for small distortion, unless the g factor is unusually large. Thus, in most cases the modulation anomaly arises from the first term, and

$$g_{\mathrm{n}}'^{2} - g_{\mathrm{n}}^{2} \approx \langle n|\mathbf{e}^{\dagger}\cdot\mathbf{g}^{2} + \mathbf{g}^{2}\cdot\mathbf{e}|n\rangle\sigma(\phi).$$

A binary ordering between $\sigma = \pm\sigma_0$ can be characterized by

$$\Delta g_{\mathrm{n}} = g_{\mathrm{n}}'(+\sigma_0) - g_{\mathrm{n}}'(-\sigma_0) \qquad \text{and} \qquad g_{\mathrm{n}} = \tfrac{1}{2}\{g_{\mathrm{n}}'(+\sigma_0) + g_{\mathrm{n}}'(-\sigma_0)\},$$

and we can write relations

$$\Delta g_{\mathrm{n}} = c_{\mathrm{n}}\cos\phi \qquad \text{and} \qquad c_{\mathrm{n}}\sin\phi \tag{9.6}$$

for fluctuations in phase and amplitude modes, respectively, with an abbreviation

$$c_{\mathrm{n}} = \frac{\sigma_0\langle n|\mathbf{e}^{\dagger}\cdot\mathbf{g}^{2} + \mathbf{g}^{2}\cdot\mathbf{e}|n\rangle}{2g_{n}}.$$

In practice, the fluctuations are so slow at temperatures close to T_{c} that their spatial variation can be observed in Δg_{n} averaged over the timescale t_0 of magnetic resonance experiments. Such a timescale is related to the Larmor period $2\pi/\omega_{\mathrm{L}}$, which is a time required to sample a condensate during at least one complete spin precession. Mathematically, it can be verified that averages of $\cos\phi$ and $\sin\phi$ will not vanish if $\omega t_0 \leq 1$, as discussed in Section 6.5 for neutron scattering intensities. Writing $\phi = \phi_{\mathrm{s}} - \omega t$ for now, where $\phi_{\mathrm{s}} = qx + \phi_0$,

$$\langle\cos\phi\rangle_{\mathrm{t}} = t_0^{-1}\int_{t}^{t+t_0}\cos(\phi_{\mathrm{s}} - \omega t)\,\mathrm{d}t = \frac{\sin(\omega t_0)}{(\omega t_0)}\cos\{\phi_{\mathrm{s}} - \omega(t + \tfrac{1}{2}t_0)\}$$

and

$$\langle\sin\phi\rangle_{\mathrm{t}} = t_0^{-1}\int_{t}^{t+t_0}\sin(\phi_{\mathrm{s}} - \omega t)\,\mathrm{d}t = \frac{\sin(\omega t_0)}{(\omega t_0)}\sin\{\phi_{\mathrm{s}} - \omega(t + \tfrac{1}{2}t_0)\}$$

Here the relation

$$\lim_{\omega t_0\to 0}\frac{\sin(\omega t_0)}{(\omega t_0)} = 1$$

signifies that the amplitude of σ is generally reduced from σ_0 under observation at a timescale t_0, but remains at the order of 1 if $2\pi/\omega \geq t_0$. In this context, for such observation we can use the redefined amplitude and phase as $\sigma_0 \sim \sigma_0 \sin(\omega t_0)/(\omega t_0)$ and $\phi \sim \phi_s - \omega(t + \frac{1}{2}t_0)$, respectively, in (9.6) for observed g anomalies.

In the critical region of a phase transition, the Zeeman energy is thus modulated as described above, resulting in distributed magnetic resonance frequencies. Arising from fluctuations in amplitude and phase modes, the anomaly should consist of two resonances, where the broadenings are described by

$$\mathrm{h}\Delta\nu_n(\mathrm{P}) = \Delta g_\mathrm{n}(\mathrm{P})\beta B = \sigma_0 c_\mathrm{n}\beta B \cos\phi \quad \text{for the phase mode,} \tag{9.7a}$$

and

$$\mathrm{h}\Delta\nu_\mathrm{n}(\mathrm{A}) = \sigma_0 c_\mathrm{n}\beta B \sin\phi \quad \text{for the amplitude mode,} \tag{9.7b}$$

where $0 \leq \phi \leq 2\pi$, as previously defined. Writing, for brevity, $\nu_1 = \sigma_0 c_\mathrm{n}\beta B/\mathrm{h}$, frequency distributions in these modes are expressed by $\Delta\nu_\mathrm{n}(\mathrm{P}) = \nu_1 \cos\phi_\mathrm{P}$ and $\Delta\nu_\mathrm{n}(\mathrm{A}) = \nu_1 \sin\phi_\mathrm{A}$, and the corresponding lineshapes are given by

$$f(\phi_\mathrm{P})\,\mathrm{d}\phi_\mathrm{P} = \frac{f(\phi_\mathrm{P})\,\mathrm{d}\Delta\nu_\mathrm{n}(\mathrm{P})}{|\mathrm{d}\Delta\nu_\mathrm{n}(\mathrm{P})/\mathrm{d}\phi_\mathrm{P}|} = \frac{f(\mathrm{P})\,\mathrm{d}\Delta\nu_\mathrm{n}(\mathrm{P})}{|\nu_1 \sin\phi_\mathrm{A}|}$$

and

$$f(\phi_\mathrm{A})\,\mathrm{d}\phi_\mathrm{A} = \frac{f(\mathrm{A})\,\mathrm{d}\Delta\nu_\mathrm{n}(\mathrm{A})}{|\nu_1 \cos\phi_\mathrm{A}|},$$

respectively. These expressions describe the densities of resonance frequency distribution due to fluctuations in amplitude and phase modes when the phases ϕ_P and ϕ_A are replaced by the frequency variations $\Delta\nu_\mathrm{n}(\mathrm{P})$ and $\Delta\nu_\mathrm{n}(\mathrm{A})$. Namely, by letting $\Delta\nu_\mathrm{n}(\mathrm{P}) = \Delta\nu_\mathrm{n}(\mathrm{A}) = \Delta\nu_\mathrm{n}$

$$f(\phi_\mathrm{P})\,d\phi_\mathrm{P} = \frac{F(\Delta\nu_\mathrm{n}]\,\mathrm{d}\Delta\nu_\mathrm{n}}{[\nu_1{}^2 - \Delta\nu_\mathrm{n}{}^2]^{1/2}}, \tag{9.8a}$$

which has singularities at $\Delta\nu_\mathrm{n} = \pm\nu_1$, corresponding to $\phi_\mathrm{P} = 0$ and π, whereas

$$f(\phi_\mathrm{A})\,\mathrm{d}\phi_\mathrm{A} = \frac{F(\Delta\nu_\mathrm{n})\,\mathrm{d}\Delta\nu_\mathrm{n}}{\Delta\nu_\mathrm{n}} \tag{9.8b}$$

whose singularity is at $\phi_\mathrm{A} = \pm\frac{1}{2}\pi$. Being identical to Fig. 6.6(a), these densities are illustrated again in Fig. 9.1(a) where the corresponding magnetic resonance lineshapes in field derivatives are shown for comparison. It is noted that the anomaly is explicit for the phase mode, as characterized by distributed resonance frequencies between two edge frequencies $\pm\nu_1$, whereas the amplitude mode shows only a featureless single line and is not informative about lattice modulation. Nevertheless, to substantiate the condensate mech-

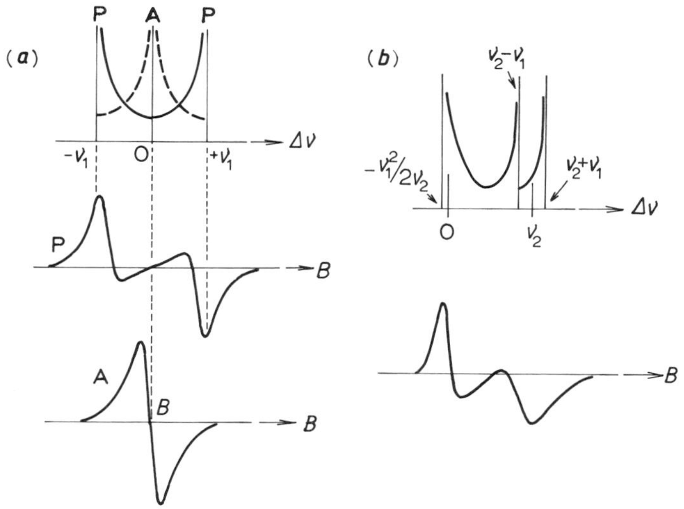

Figure 9.1. (a) Anomalous broadening of a magnetic resonance line for P and A modes. (b) Anomalous broadening of a magnetic resonance for $\Delta\nu = \nu_1 \cos\phi + \nu_2 \cos^2\phi$ $(\nu_1 > \nu_2)$. Magnetic resonance spectra are normally displayed as derivatives with respect to the field.

anism, it is important experimentally that such a single line be identified as arising from the amplitude mode. As discussed in Subsection 9.3.1, the polar amplitude mode in the ferroelectric phase transition of TSCC crystals was so identified, as it was converted to a phase mode by applying a weak electric field.

When such anomalous lines as in Fig. 9.1(a) are observed, the edge frequencies can be estimated in acceptable accuracy. The edge separation is expressed as

$$2\nu_1 = \frac{(\Delta g_n)_{\rm edge}\beta B}{\rm h},$$

where

$$(\Delta g_n)_{\rm edge} = \frac{\sigma_o}{2g_n}\frac{\sin(\omega t_o)}{(\omega t_o)}\langle n|\mathbf{e}^\dagger\cdot\mathbf{g}^2 + \mathbf{g}^2\cdot\mathbf{e}|n\rangle \tag{9.9}$$

and here g_n is the average g value over the distribution as given by the value at the center between the two edges. The separation $(\Delta g_n)_{\rm edge}$ is anisotropic, depending on the direction $\langle n|$ of the static field, hence providing useful information about a spatially modulated structure as shown later in the example of VO^{2+} anomalies in the incommensurate phase of BCCD crystals.

9.2.2. The Hyperfine Anomaly

In Section 8.6 we discussed that the hyperfine interaction is described by the Hamiltonian

$$\mathcal{H}_{\text{HF}} = \beta\gamma\langle S|\mathbf{g}\cdot\mathbf{A}|I\rangle,$$

whose first-order energies in a uniform field $\langle B|$ are given in the high-field approximation as

$$E_{\text{HF}}{}^{(1)} = g_{\text{n}}\beta K_{\text{n}} M_{\text{S}} M_I{}'.$$

Here, the hyperfine splitting K_{n} is generally expressed as

$$K_{\text{n}}{}^2 = \frac{\langle n|\mathbf{g}^2\cdot\mathbf{A}^2|n\rangle}{g_{\text{n}}{}^2},$$

which is in units of (field)^2, and the constant γ is included in the tensor $\mathbf{A}$. Writing $\mathbf{C} = (\mathbf{g}/g_{\text{n}})\cdot\mathbf{A}$ for brevity,

$$K_{\text{n}} = \langle n|\mathbf{C}^2|n\rangle^{1/2}, \tag{9.10}$$

which can be directly determined from spectra for all directions $\langle n|$.

In a modulated crystal, the hyperfine interaction is modified by the transformation matrix $\mathbf{a}$ tn the same manner as the $\mathbf{g}$ tensor. Namely, the hyperfine splitting is modulated as

$$\begin{aligned} K_{\text{n}}'^2 &= \langle n|\mathbf{a}^\dagger\cdot\mathbf{C}^2\cdot\mathbf{a}|n\rangle \\ &= \langle n|\mathbf{C}^2|n\rangle + \sigma\langle n|\mathbf{e}^\dagger\cdot\mathbf{C}^2 + \mathbf{C}^2\cdot\mathbf{e}|n\rangle + \sigma^2\langle n|\mathbf{e}^\dagger\cdot\mathbf{C}^2\cdot\mathbf{e}|n\rangle. \end{aligned}$$

Considering that the last term is negligible for the small deformation matrix $\mathbf{e}$,

$$K_{\text{n}}'^2 - K_{\text{n}}{}^2 = \langle n|\mathbf{e}^\dagger\cdot\mathbf{C}^2 + \mathbf{C}^2\cdot\mathbf{e}|n\rangle\sigma(\phi).$$

Therefore, similar to the g anomaly, hyperfine lines exhibit anomalous shapes characterized by the edge separation

$$(\Delta K_{\text{n}})_{\text{edge}} = \frac{\sigma_0\langle n|\mathbf{e}^\dagger\cdot\mathbf{C}^2 + \mathbf{C}^2\cdot\mathbf{e}|n\rangle}{2K_{\text{n}}}. \tag{9.11}$$

It is clear from (9.11) that the line is not broadened, if the tensor $\mathbf{C}$ is isotropic. In Mn^{2+} complexes in TSSC and BCCD crystals, for example, the g- and ^{55}Mn-hyperfine tensors are isotropic, and therefore not responsible for the observed anomalies. On the other hand, in VO^{2+} complexes in these crystals the g- and ^{51}V-hyperfine tensors are anisotropic, and the observed anomalies can be interpreted by (9.9) and (9.11), as will be discussed in Section 9.3.

9.2.3. The Fine Structure Anomaly

Magnetic resonance sampling of a modulated structure can also be performed on the fine-structure energy. However, the fine-structure energy is quadratic with respect to spin components, and hence treated differently from g and hyperfine tensors that are linearly related to the spin $\langle S|$.

The fine-structure energy $\langle S|\mathbf{D}|S\rangle$ is significant for a spin larger than $\frac{1}{2}$, expressing a charge distribution deformed from spherical symmetry in the crystalline potential. In some cases, such a deformation energy can be larger than the Zeeman energy in a conventional magnetic field $B\langle n|$, but in many practical cases we can consider that Zeeman levels are determined by the spin component $\langle S_n|$ in a given field $B\langle n|$, and the magnetic transitions are modified by the first-order fine structure energy. We therefore assume that in a modulated crystal the spin $\langle S|$ is modified as

$$\langle S_{\mathrm{n}}'| = \langle S_{\mathrm{n}}|\mathbf{a}^{\dagger},$$

and, consequently, the fine-structure energy is expressed as

$$\begin{aligned}\mathscr{H}_{\mathrm{F}}' &= \langle S_{\mathrm{n}}'|\mathbf{D}|S_{\mathrm{n}}'\rangle = \langle S_{\mathrm{n}}|\mathbf{a}^{\dagger}\cdot\mathbf{D}\cdot\mathbf{a}|S_n\rangle \\ &= \langle S_{\mathrm{n}}|\mathbf{D}|S_{\mathrm{n}}\rangle + \sigma\langle S_{\mathrm{n}}|\mathbf{e}^{\dagger}\cdot\mathbf{D} + \mathbf{D}\cdot\mathbf{e}|S_{\mathrm{n}}\rangle + \sigma^2\langle S_{\mathrm{n}}|\mathbf{e}^{\dagger}\cdot\mathbf{D}\cdot\mathbf{e}|S_{\mathrm{n}}\rangle. \quad (9.12)\end{aligned}$$

Here, judging from the order of magnitude of $\mathbf{D}$ in the usual cases, these correctional terms are not at all negligible, and first-order energies of $\mathscr{H}_{\mathrm{F}}'$ are written as

$$E_{\mathrm{F}}'^{(1)} = E_{\mathrm{F}}^{(1)} + (a_{\mathrm{n}}\sigma + b_{\mathrm{n}}\sigma^2)M_S^2,$$

where

$$a_{\mathrm{n}} = \langle n|\mathbf{e}^{\dagger}\cdot\mathbf{D} + \mathbf{D}\cdot\mathbf{e}|n\rangle \qquad \text{and} \qquad b_{\mathrm{n}} = \langle n|\mathbf{e}^{\dagger}\cdot\mathbf{D}\cdot\mathbf{e}|n\rangle.$$

Both coefficients a_{n} and b_{n} depend on the direction $\langle n|$ of the applied field, and are responsible for anomalous spectra arising from the fine structure.

Energy eigenvalues of the Hamiltonian $\mathscr{H}_{\mathrm{Z}}' + \mathscr{H}_{\mathrm{F}}'$ in a modulated system are generally written as

$$E'(M_{\mathrm{S}})^{(1)} = g_{\mathrm{n}}\beta B M_{\mathrm{S}} + (D_{\mathrm{n}} + a_{\mathrm{n}}\sigma + b_{\mathrm{n}}\sigma^2)M_{\mathrm{S}}^2.$$

An allowed magnetic resonance absorption of a microwave photon hν between M_S and $M_S + 1$ can therefore be expressed as

$$\mathrm{h}\nu = E'(M_S + 1) - E'(M_S) = g_{\mathrm{n}}\beta B + (D_{\mathrm{n}} + a_{\mathrm{n}}\sigma + b_{\mathrm{n}}\sigma^2)(2M_S + 1).$$

If, for example, $S = \frac{5}{2}$ as in Mn^{2+} or Fe^{3+} ions in a normal phase, for example, the fine structure constant D_{n} can be directly obtained from the allowed transitions other than $(-\frac{1}{2}\leftrightarrow\frac{1}{2})$ for which the resonance condition is simply $\mathrm{h}\nu = g_{\mathrm{n}}\beta B_0$, where B_0 is the resonance field and there is no contribution from the fine-structure terms. On the other hand, for the transition $(\frac{3}{2}\leftrightarrow\frac{5}{2})$, for instance, D_{n} can be directly evaluated from the resonance field, $B_0 + 4D_{\mathrm{n}}$. In the modulated phase, such resonance lines other than at B_0 accompany broadening $\Delta\nu_{\mathrm{n}}$, i.e.,

$$\mathrm{h}\Delta\nu_{\mathrm{n}} = 4(a_{\mathrm{n}}\sigma + b_{\mathrm{n}}\sigma^2) \qquad \text{for} \quad M_S = \tfrac{3}{2}.$$

When the fluctuations are in an amplitude mode $\sigma = \sigma_0 \cos\phi$, the broadening is described as

$$\Delta\nu_n = \nu_1(n)\cos\phi + \nu_2(n)\cos^2\phi, \tag{9.13}$$

where $\nu_1(n) = 4\sigma_0 a_n/h$ and $\nu_2(n) = 4\sigma_0{}^2 b_n/h$. In this case, the density of the resonance frequency distribution is given by

$$\left(\frac{d\phi}{d\Delta\nu_n}\right)^{-1} = \frac{1}{[(\nu_1{}^2/2\nu_2 + \Delta\nu_n)\{\nu_1{}^2 - (\nu_2 - \Delta\nu_n)^2\}]^{1/2}}, \tag{9.14}$$

which is characterized by three singular frequencies at $\nu_2 \pm \nu_1$ and $-\nu_1{}^2/2\nu_2$ as illustrated in Fig. 9.1(b), where the corresponding lineshape is sketched for $\nu_2 > \nu_1$. If $\nu_1 > \nu_2$, on the other hand, the lineshape is similar to Fig. 9.1(a), due dominantly to linear fluctuations. Examples of such anomalous lines in practical systems are shown in the next section.

9.3. Structural Phase Transitions as Observed by Paramagnetic Resonance Spectra

Although signified by anomalous line broadening, it is generally difficult to obtain full information about the nature of order variables responsible for the structural change from magnetic resonance spectra. Magnetic resonance spectra of probes with $S > \frac{1}{2}$ are often too complex to deduce the detail of σ, unless allowed transitions can be identified in all directions of $\langle B|$. For $S = \frac{1}{2}$, on the other hand, it is necessary for the **g** and **A** tensors to be anisotropic to detect the structural change. Phase transitions in TSCC and BCCD crystals, among others, were thoroughly studied systems, where the modulated structures are adequately interpreted in the light of the condensate model. Although still difficult in some aspects, we summarize in this section the magnetic resonance results from these crystals, providing a useful guide for further investigations in other systems as well.

9.3.1. The Ferroelectric Phase Transition in TSCC Crystals

(i) *Symmetry of Active Complexes*

TSCC crystals can be doped with a small quantity of iron-group elements such as VO^{2+}, Cr^{3+}, Mn^{2+}, and Fe^{3+} in about one per $10^5 \sim 10^6$ Ca^{2+} ions, providing spectra with a sufficient intensity for comparative studies on the ferroelectric phase transition at 163 K [67]. However, even at such a low impurity density, properties of doped crystals may be significantly modified from undoped crystals, in which case the experimental results have to be carefully examined for compatible information with the ordering mechanism in the hosting lattice.

The active group for the ferroelectric phase transition in TSCC is the Ca^{2+} (sarcosine)$_6$ complex, where the central Ca^{2+} ion can be replaced by an impurity probe. The complex structure in the normal phase is in approximately trigonal symmetry along the *a* axis (Fig. 3.4(b)), while it can be further deformed by an impurity ion when substituted for the Ca^{2+} ion. It is known empirically that such a distortion appears to retain the local trigonal symmetry, as evidenced by angular variations of the hyperfine splitting in VO^{2+} spectra, indicating that one principal axis is nearly in parallel with the *a* axis of the orthorhombic lattice (Fig. 3.4(a)). In the *bc* plane, as shown in Fig. 8.8(b), angular variations of the ^{51}V-hyperfine splitting in the VO^{2+} spectra were nearly identical in three independent orientations, suggesting that the corresponding strains in the lattice can be considered as nearly equivalent. However, these component spectra were not equivalent in the *ab* and *ca* planes, and so classified into types I and II. Here, the VO^{2+} complex of type I is characterized by one of the principal axes perpendicular to the mirror plane (010), while the spectra of type II consist of two magnetically inequivalent components that are related by reflection on the mirror plane.

Cr^{3+} and Fe^{3+} impurities also exhibited these two spectra corresponding to types I and II of the VO^{2+} complexes, although the type I spectrum was not clearly visible and overlapped with the spectra of type II, as shown in the angular variation of Fe^{3+} spectra in Fig. 9.2. In contrast, Mn^{2+} spectra in TSCC consisted only of the type I component, which is advantageous over the other probes showing spectra of two types, in order to deal with breaking mirror symmetery in the ferroelectric phase transition. Taking such spectral differences in different probes for granted, we can make use of specific features of these probes to analyze modulated structures.

(ii) *Symmetry Breaking at the Ferroelectric Phase Transition*

In the magnetic resonance spectra of Mn^{2+}, Cr^{3+}, and Fe^{3+} impurities in TSCC, the ferroelectric phase transition is signified by anomalous broadening at temperatures near T_c. On the other hand, macroscopically, the phase transition is characterized by the loss of mirror symmetry. Representing the mirror plane, the Mn^{2+} spectra of type I are therefore suitable for studying symmetry breaking at T_c.

Figure 9.3 shows anomalous broadening of a representative Mn^{2+} line observed for a representative direction of $\langle B|$ in the *bc* plane when the temperature was varied through T_c in the critical region. Although exhibiting different lineshapes at different microwave frequencies, the broadened line consists generally of a single line at the center and a symmetrical pair of peaks. Except for the single line, the anisotropic broadening in the *bc* plane behaved as if the paraelectric line started to split into two anomalous peaks at T_c, whereas in the *ab* plane the broadening was smaller and obscured by the single line at the center. In the mirror plane in contrast, the spectra showed absolutely no change at T_c, indicating that the pseudospin vector is

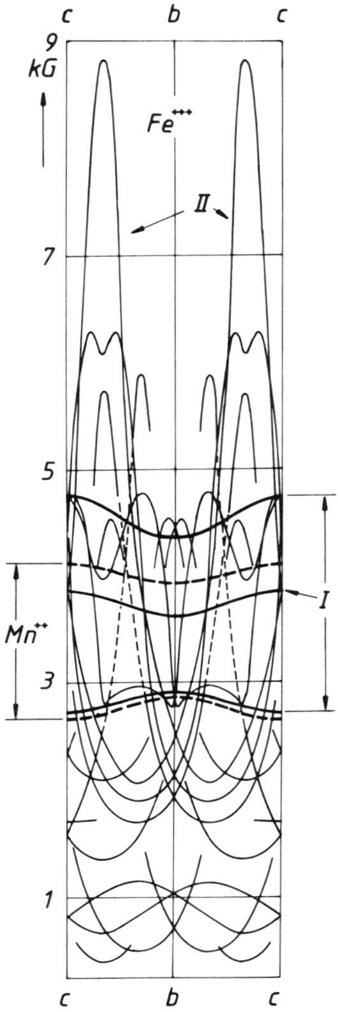

Figure 9.2. Fe^{3+} spectra of type II in the *bc* plane of TSCC crystals. Mn^{2+} are unavoidably included in the sample crystal. Reonance lines of type I are unresolved, but confined to the region marked I.

confined to the *ac* plane. Figure 9.4 shows angular variations of the measured splitting D_n in all directions of the static field at 108 K. Also noticed is that there is no such broadening along the symmetry axes *a*, *b*, and *c*. Leaving the single line to later discussions, the anomalous broadening in the *bc* plane is clearly due to distributed resonances as described by the pseudospin $\sigma = \sigma_0 \cos \phi$. While there is no reason to ignore quadratic fluctuations σ^2, the

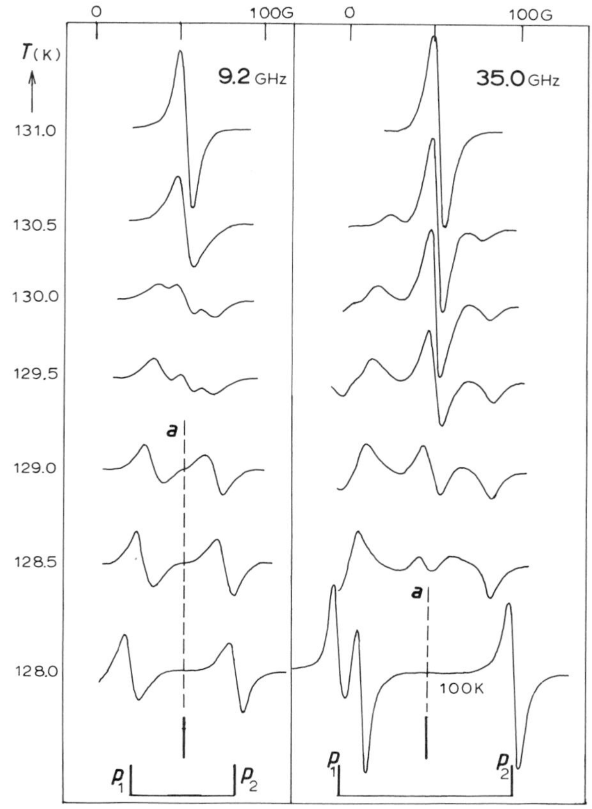

Figure 9.3. Critical anomalies near T_c in TSCC observed from line broadening of the Mn^{2+} spectra in the *bc* plane. $B_0 \| (90°, 45°, 45°)$.

observed anomaly can be expressed by

$$\Delta D_n = A_a \sigma_0 \cos \phi \qquad \text{and} \qquad A_b = A_c = 0. \tag{9.15}$$

To explain this experimental result, we can assume that for each Mn^{2+} probe two magnetically inequivalent "states," 1 and 2, below T_c emerged from a single line above T_c as a result of slow fluctuations in phase mode between σ and $-\sigma$ related by inversion (Fig. 9.4). In this interpretation, these states 1 and 2 represent such binary states $\pm\sigma$ that are represented by

$$(\Delta D_n)_1 = a_n \sigma_0 \cos \phi + b_n \sigma_0{}^2 \cos^2 \phi$$

and

$$(\Delta D_n)_2 = a_n \sigma_0 \cos \phi' + b_n \sigma_0{}^2 \cos^2 \phi'.$$

On the other hand, "fine-structure splittings" in the spectra are expressed as

$$\Delta D_n = (\Delta D_n)_1 - (\Delta D_n)_2. \tag{9.15a}$$

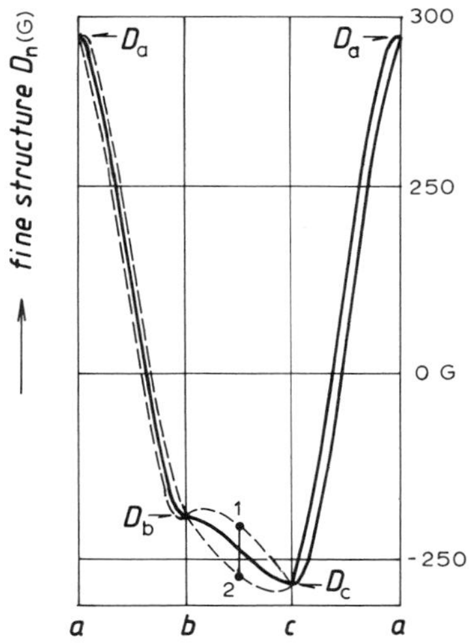

Figure 9.4. Angular variations of fine-structure splittings in Mn^{2+} spectra in TSCC below T_c. The broadening in Fig. 9.3 was measured as the separation between 1 and 2 in the *bc* plane.

In order for (9.15a) to be consistent with (9.15), it is sufficient to propose the phase relation

$$\cos\phi = -\cos\phi' \qquad \text{or} \qquad \phi' = \pi - \phi$$

for $(\Delta D_n)_1$ and $(\Delta D_n)_2$, while the relation $A_n = 2a_n$ holds for the magnitude, and the terms for ${\sigma_0}^2$ are canceled in (9.15a). In this analysis (9.15a) clearly represents two distributed resonances, where pseudospins $\sigma_1 = \sigma\cos\phi$ and $\sigma_2 = \sigma\cos\phi'$ representing states 1 and 2 are related by reflection on the mirror plane, i.e., $\sigma_2 = -\sigma_1$. Thus, observed Mn^{2+} anomalies in the *bc* plane exhibit clear-cut evidence for symmetry breaking at the ferroelectric phase transition of TSCC. It is noticed that under ordinary circumstances, such states 1 and 2 represent two independent Mn^{2+} sites in the unit cell, while in contrast, in crystals undergoing binary phase transitions such fluctuations in phase mode play a role leading the crystal to a two-domain structure with opposite polarizations.

(iii) *The Long-Range Order in Ferroelectric TSCC*

While short-range correlations among pseudospins prevail in the critical region, long-range correlations become increasingly significant with decreasing temperature. As discussed in Chapter 4, in the mean-field approximation such long-range correlations can be described in terms of the average local field with growing intensity, which represents an internal electric field in ferroelectric phases. Evidenced by a diffused phase transition under an applied electric field in moderate strength, polar condensates are considered to be under such an internal electric field with a temperature-dependent strength arising from increasing long-range correlations.

In Section 3.6 we discussed the mean-field argument for distant correlations, and derived

$$\lambda_{\mathrm{m}}\sigma_{\mathrm{m}} = F_{\mathrm{m}}, \tag{3.23}$$

where $F_{\mathrm{m}} = \sum_{\mathrm{n}} J_{\mathrm{mn}}\sigma_{\mathrm{n}}$ represents the local field at site m, arising from all the near and distant σ_{n}. While such a field F_{m} or the corresponding pseudosin σ_{m} is not directly measurable, their Fourier transforms F_q and σ_q can be sampled by a spin-Hamiltonian parameter of paramagnetic probes. In this context, the local field F_q related to the amplitude σ_0 is represented by the edge separation in the anomalous line broadening for the phase mode $\sigma_0 \cos \phi$. On the other hand, for a single line represented by the amplitude mode $\sigma_{\mathrm{p}} \sin \phi$, the sampling frequency has the maximum density of distribution at the unmodulated frequency, similar to what are given by (6.20a, b), so that the distributed field F_q is observed as if averaged out.

Although derived for an isolated condensate, the field F_q in the Fourier mode σ_q of (3.23) may be contributed, in principle, by all correlations inside the mode as well as from outside interactions. For the latter, dipolar interactions with neighboring condensates, as well as the interaction with an external electric field E, are particularly significant in a ferroelectric phase. Paying attention to the fact that an external uniform field E is not symmetrical, whereas the internal field is symmetrical with respect to $\boldsymbol{q}$, we can consider the edge frequency proportional to the amplitude σ_0 in terms of the internal and external fields, i.e.,

$$\pm\nu_1 \propto \pm\sigma_0 \propto \pm F_q + E.$$

In this treatment, in the presence of E the edge frequencies for the phase mode become asymmetrical as given by $(F_q + E, -F_q + E)$, whereas the amplitude mode is characterized by $F_q = 0$ when $E = 0$, but broadened in the presence of nonzero E with an edge separation $(E, -E)$. This is consistent with what was discussed in Chapter 5, i.e., a condensate in amplitude mode can be in equilibrium with an applied field E after shifting the phase ϕ from 0 to $\pm\frac{1}{2}\pi$. Thus, the amplitude mode stabilized by an external field E can be distinguished from a paramagnetic line by an anomaly showing E-dependent edge separation.

Figure 9.5 shows the effect of a weak applied field E to an anomalous Mn^{2+} line at a temperature very close to T_c, where the two modes are clearly resolved when no field is applied [21]. At $E = 0.6\ \mathrm{kV\ cm^{-1}}$ the amplitude mode *a* split and changed to a doublet (p_1', p_2') with increasing separation, which moved toward the doublet of the phase mode (p_1, p_2) to join in eventually as E was increased to $2.5\ \mathrm{kV\ cm^{-1}}$. The edge separation in the phase mode became wider with decreasing temperature, indicating the growing internal spontaneous field F_q.

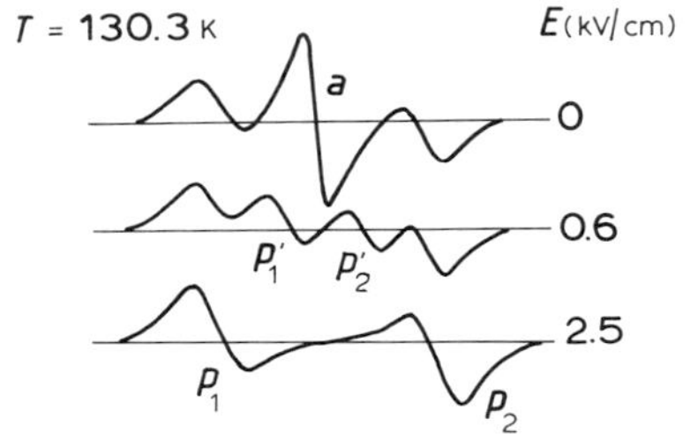

Figure 9.5. Effects of a weak applied field to an anomalous Mn^{2+} line at 130.3 K < T_c. The line marked a for $E = 0$ split into a pair (p_1', p_2') as E was increased to 0.6 kV cm^{-1}, which eventually joined in (p_1, p_2) when $E = 2.5$ kV cm^{-1}.

In the presence of distant correlations, the pseudospin function $\sigma_P(\phi)$ cannot be sinusoidal, showing increasing nonlinearity that is expressed by the Jacobi elliptic function in a one-dimensional chain as discussed in Chapter 5. From (5.13)

$$\sigma_P(\phi) = 2^{1/2}\sigma_{P0}\frac{\kappa}{(1+\kappa^2)^{1/2}}\,\mathrm{sn}\left(\frac{\phi}{1+\kappa^2}\right). \tag{9.16}$$

Here, the nonlinearity in correlated pseudospins is signified by the modulus $0 < \kappa \leq 1$, for which the long-range mean field is essentially responsible, depending on the temperature. When sampled by magnetic resonance probes as described by (9.15), the anomalous lineshape as related to a nonlinear $\sigma_P(\phi)$ can be numerically simulated by (9.16). The curves in Fig. 9.6 drawn by broken lines between peaks show such simulated results for $\kappa = 0.4$ and 0.6, which appear to represent observed lines from Mn^{2+} in TSCC at $T_c - T =$ 0.4 K and 1.4 K, respectively. The single line a due to fluctuations in the amplitude mode decreases the intensity on lowering temperature, because of

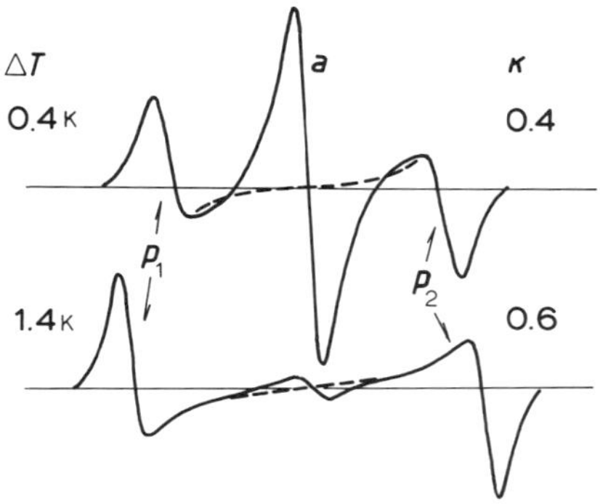

Figure 9.6. Numerical simulation of an EPR anomaly for $\kappa = 0.4$ and 0.6, which were fitted observed anomalies at $\Delta T = 0.4$ K and 1.4 K, respectively.

its conversion to the phase mode by the increasing internal field F_q. It is noted that such a nonlinear pseudospin mode $\sigma_P(\phi)$ below T_c is still slowly fluctuating, but decoupled from the phonon mode (A_1) at higher frequencies.

When sampling the nonlinear pseudospin mode σ_P below T_c, the magnetic resonance frequency varies as

$$h\Delta\nu_n = A_n\sigma_P(\phi) \qquad \text{or} \qquad \Delta\nu_n = \nu_{1n}\,\mathrm{sn}\left(\frac{\mu\phi}{2^{1/2}}\right),$$

where $\nu_{1n} = A_n\lambda\sigma_0/h$. With this we can obtain formula for the edge frequencies and separation of such a nonsinuoidal distribution. To derive the expression for the density of magnetic resonance frequencies, it is convenient to use an angle θ that is related to the phase ϕ as $\sin\theta = \mathrm{sn}(\mu\phi/2^{1/2})$ following the argument in Section 5.3. In this case, $\Delta\nu_n/\nu_{1n} = \sin\theta$, and the density $d\phi/d(\Delta\nu_n)$ can be obtained by differentiating the elliptic integral of (5.14), i.e.,

$$\frac{\mu}{2^{1/2}}\frac{d\phi}{d\theta} = (1 - \kappa^2\sin^2\theta)^{-1/2},$$

which is

$$\frac{\mu}{2^{1/2}}\frac{d\phi}{d\Delta\nu_n} = \frac{1}{\nu_{1n}(1 - \sin^2\theta)^{1/2}(1 - \kappa^2\sin^2\theta)^{1/2}}.$$

Hence

$$\frac{d\phi}{d\Delta\nu_n} = \frac{\lambda\nu_{1n}}{2^{1/2}(\nu_{1n}{}^2 - \Delta\nu_n{}^2)^{1/2}\{(\nu_{1n}/\kappa)^2 - \Delta\nu_n{}^2\}^{1/2}}.$$

Since $(\nu_{1n}/\kappa) > \nu_{1n}$ for $0 < \kappa < 1$, we can only consider $|\Delta\nu_n| < \nu_{1n}$ for the density to be real. Accordingly, the edge frequencies for a elliptic distribution are also given by $\pm\nu_{1n}$, which is consistent with a sinusoidal case. In the limit $\kappa = 1$, the density is given by

$$\left(\frac{d\phi}{d\Delta\nu_n}\right)_{\kappa\to 1} = \frac{\lambda\nu_{1n}}{2(\nu_{1n}{}^2 - \Delta\nu_n{}^2)},$$

where the density of the frequency distribution between the two edges is less than the sinusoidal case, but not exactly zero. Experimentally, such a doublet as observed at 100 K in Mn^{2+} spectra (Fig. 9.3) appears to have a lineshape compatible with this view.

(iv) *Edge Separation and the Frequency of Critical Fluctuations*

As seen from Fig. 9.3, anomalous lineshapes below T_c are appreciably different when observed by magnetic resonances at different Larmor frequencies. It is evidently due to the fact that the frequency ω is so slow that critical fluctuations are not averaged out completely, depending on the timescale of observation $t_0 \sim 2\pi/\omega_L$. In Subsection 9.2.1, we have already discussed such a time effect in relation to the average sinusoidal variation at a frequency ω over a short period t_0, which does not vanish if $\omega t_0 \leq 1$. The basic result is

$$\langle \cos \phi \rangle_t = t_0^{-1} \int_t^{t+t_0} \cos(qx - \omega t + \phi_0)\, dt = \alpha(t_0) \cos(qx - \omega \underline{t} + \phi_0),$$

where

$$\alpha(t_0) = \frac{\sin(\omega t_0)}{(\omega t_0)} \quad \text{and} \quad \underline{t} = t + \tfrac{1}{2} t_0. \tag{9.17}$$

Considering for the pseudospin the rescaled amplitude σ_0 and time t as given by (9.17), we can express the edge frequency for the line $\frac{3}{2} \leftrightarrow \frac{5}{2}$, for example, as

$$\nu_1(n, t_0) = \frac{4\alpha(t_0)\sigma_0 a_n}{h}. \tag{9.18}$$

Using this relation, the ratio between edge separations measured at 9.2 GHz and 35 GHz is

$$\frac{\nu_1\{n, t_0(9.2)\}}{\nu_1\{n, t_0(35)\}} = \frac{\alpha(t_0(9.2))}{\alpha(t_0(35))},$$

where $t_0(9.2)$ and $t_0(35)$ are 6.8 and 1.8×10^{-10} s. Therefore, from the ratio of these edge frequencies measured at the same temperature below T_c, the critical frequency ω can be estimated, which were 6.2 GHz and 4.8 ± 0.3 GHz at 0.2 K and 0.4 K below T_c, respectively. These frequencies should be compared with the terminal frequency of the B_{2u} soft mode as T_c was approached from above, which was of the order of 20 GHz. Although lacking the detail in the range $0 \leq T_c - T \leq 0.2$ K, we consider that these estimated frequencies provide evidence for a rapid decrease of ω in the critical region due to the increasing long-range field F_q. It is considered that the ferroelectric order in TSCC can be established in the limit of $\omega \to 0$, however, the wavevector q may not necessarily be zero in this limit. If F_q remains constant at a finite q, the crystal may exhibit a modulated phase.

In the paraelectric phase of TSCC, VO^{2+} impurity ions exhibit spectra of types I and II, as shown by the angular plot of ^{51}V-hyperfine splittings in Fig. 8.8(b). Although nearly trigonal in the *bc* plane, it was recognized that the type I complex is characterized by one of the principal axes lying perpendicular to the mirror plane *ac*, while the type II complex is not oriented as involved in symmetry planes.

At temperatures below T_c, VO^{2+} spectra of both types split into two as shown in Fig. 9.7(a), except that the type I spectra in the *ac* plane exhibited no splitting at all directions. Corresponding to the Mn^{2+} spectra in the mirror plane that was unchanged at the transition, the VO^{2+} spectra of type I also showed no change in *b* plane, indicating that the ^{51}V tensor changed only in the perpendicular direction, for which $^{51}K_b$ is proportional to the component σ_b. Although consistent in this regard with the Mn^{2+} results, the transition anomalies in ^{51}V-hyperfine splittings were hardly noticeable, pre-

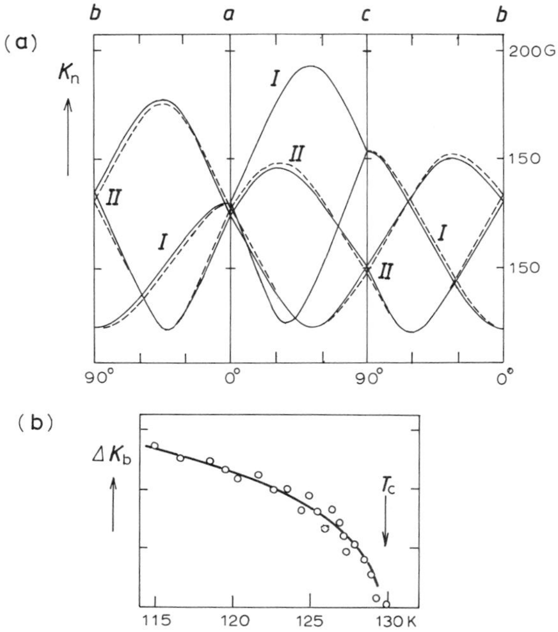

Figure 9.7. (a) Angular variations of the ^{51}V-hyperfine splitting in VO^{2+} spectra from TSCC at 130 K. Notice that type I spectra did not split in the mirror *ac* plane, while type II split in all directions. (b) Temperature-dependence of the splitting ΔK_b, representing the order parameter below T_c.

sumably because of the electric field of dipolar VO^{2+} ions, which is apparently responsible for the absence of the amplitude mode in diffuse changes of spectra at the threshold of the critical region. Nevertheless, the hyperfine splitting $^{51}K_b$ indicated approximately a parabolic temperature-dependence as shown in Fig. 9.7(b), and we can consider that $\Delta K_b = K_b(T) - K_b(T_c) \propto \sigma_b$ below T_c, where the critical behavior was not explicit.

9.3.2. Structural Phase Transitions in BCCD

Orthorhombic crystals of (betaine)-$CaCl_2 2H_2O$ known as BCCD, where the betaine is an organic amino acid, $(CH_3)_3NCH_2COOH$, exhibit successive structural phase transitions under atmospheric pressure between the normal phase above 164 K and the ferroelectric phase below 45 K. Rother et al. [68] first observed dielectric anomalies at these transitions, suggesting that a polar mechanism is involved in these stepwise structural changes. Brill and Ehses [69] carried out an X-ray study on BCCD crystals, and found that in the

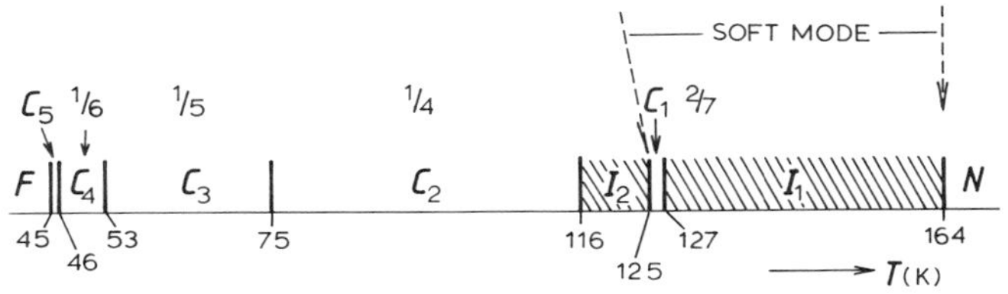

Figure 9.8. Successive phase transitions in BCCD crystals: I_1 and I_2 are incommensurate, being led by soft modes when the transition threshold is approached, whereas C_1, C_2, C_3, and C_4 are commensurate, for which corresponding fractional parameters δ are indicated.

range between 164 K and 45 K there are seven modulated phases that are characterized by the wavevector $Q_c = \delta(T)c^*$, where c^* is the reciprocal lattice constant for the c axis. Figure 9.8 illustrates the series of phase transitions in BCCD crystals under normal atmospheric pressure, where values of the parameter $\delta(T)$ are indicated, as determined by the X-ray results of [69]. Using a submillimeter technique, Volkov et al. [70] observed underdamped soft modes near the threshold temperatures of these incommensurate phases, which were later identified by Ao and Schaack [71] as of B_{2u} symmetry from their infrared measurements. The molecular arrangement in BCCD at room temperature was determined by Brill et al. [72], who reported the presence of appreciable librational fluctuations of Ca-(betaine)$_2$ complexes in the b plane. Figure 9.9 shows the unit-cell structure sketched after their X-ray results.

The Ca-(betaine)$_2$ complex in BCCD crystals is asymmetrical with regard

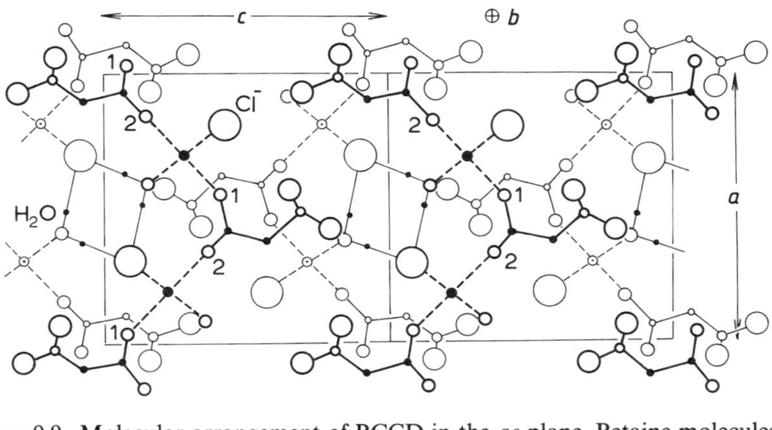

Figure 9.9. Molecular arrangement of BCCD in the ac plane. Betaine molecules are planar lying in the ac plane. Each Ca^{2+} ion is coordinated by O(1) and O(2) of the ligand betaines, and two Cl^- and two OH_2, where the two hydrogens are out of the ac plane.

to the coordination of two –COOH groups of ligand betaine molecules, i.e., O(1) of one betaine and O(2) of the other. If such an asymmetrical coordination can fluctuate between two such structures that are related by inversion in the b plane, these complexes should be heavily correlated in the lattice, being in collective motion during the structural change. Evidenced by VO^{2+} sampling results, the active groups are predominantly in librational motion between these inverted structures, and which can be described as binary states ± 1 of a pseudospin σ.

Placing such a pseudospin variable at each Ca^{2+} site, we can consider the pseudospin lattice for ordering in BCCD crystals, which is similar to Fig. 3.5 illustrated for the bc plane in TSCC. We can therefore consider short-range correlations similar to TSCC, although in BCCD the correlations between adjacent mirror planes (J_d) and those in the mirror plane (J_c) are not exactly on the bc plane, judging from the diagram in Fig. 9.9. However, assuming that these significant correlations are in the bc plane, we can simplify the calculation of correlation energies for the normal-to-incommensurate phase transition ($N \to I_1$). Although unknown, out-of-plane correlations associated with J_d may be significant for subsequent transitions below the phase I_1. Referring to Subsection 3.5.2, with such a model, we can predict the presence of a modulated mode that is characterized by an irrational wavevector k_c given by $\cos(k_c c) = -J_d/J_c$, where $|-J_d/J_c| < 1$. In general, these correlations are temperature-dependent, and written as

$$k_c = \{1 - \delta(T)\}c^* \qquad \text{and} \qquad Q_c = \delta(T)c^*. \tag{9.19}$$

(i) *VO^{2+} Spectra*

Diatomic VO^{2+} probes were found useful in identifying the order variable for the first incommensurate phase transition, although looking at the transition mechanism as distorted by their electric dipole moments. As in TSCC, critical fluctuations in the amplitude mode were suppressed by the polar field of VO^{2+}, and in the phase mode the transition appeared to be diffuse. In the VO^{2+} spectra the first incommensurate phase I_1 was clearly distinguished, whereas subsequent serial structural changes were entirely unnoticeable. Despite that, the fluctuation in the incommensurate phase I_1 was revealed in detail, providing a very instructive example of magnetic resonance sampling. In contrast to the narrow critical region in TSCC, anomalies in the VO^{2+} spectra with well-defined edge splittings were observed at all temperatures in the incommensurate phase I_1.

In Fig. 9.10 are shown representative VO^{2+} spectra observed at temperatures above and below the critical temperature $T_i = 164$ K, using different microwave frequencies 9.2 GHz and 35 GHz. A considerable difference between these spectra at different frequencies signifies the characteristic time of the fluctuations that is comparable with or longer than the timescale of these microwave observations. It is also noticed that intensities and widths of eight hyperfine lines exhibit, at all temperatures, a marked variation from one line

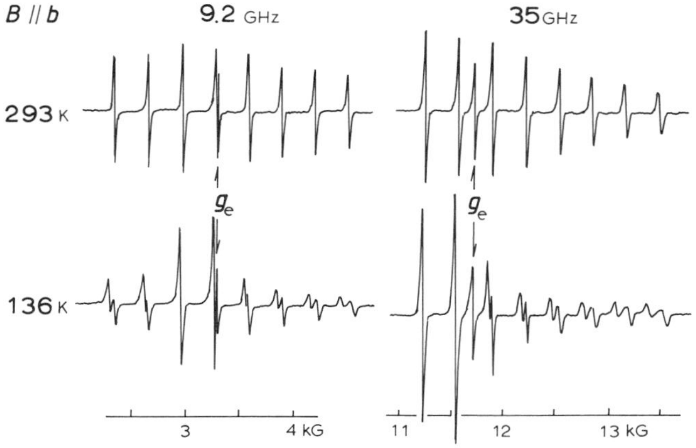

Figure 9.10. Representative VO^{2+} spectra in BCCD for $B_0 \| b$, observed at 9.2 GHz and 35 GHz in the normal and the first incommensurate phases.

to another, depending on the nuclear spin quantum number M_I. Such a spectrum signifies the presence of librational fluctuations in crystals as consistent with the X-ray results [72], although it was uncertain if the mode above T_i is the same as was suggested by Brill et al. Below T_i the resonance lineshape turned into an anomalous shape characterized by a symmetrical edge separation, thereby suggesting that fluctuations in the same mode as above T_i are slowed down considerably.

In the normal phase, the **g** and **A** tensors of VO^{2+} complexes were found to be coaxial with well-defined principal values [73], [74]. Crystallographically, there are four independent unique axes in the unit cell that are symmetrically located with respect to the b axis. Accordingly, as observed in symmetry planes (100), (010), and (001), spectra are signified in general by two magnetically inequivalent complexes, while for $B \| b$ all four VO^{2+} complexes showed exactly identical spectra, as shown in Fig. 9.10.

Energy levels of a VO^{2+} ion in a uniform field $\langle B| = B\langle n|$ are expressed in the first order of the high-field approximation as

$$E(M_S, M_I) = g_n \beta B M_S + g_n \beta K_n M_S M_I.$$

The magnetic resonance condition for allowed electronic transitions $M_S \rightarrow M_S + 1$, and M_I unchanged, is

$$\mathrm{h}\nu = E(M_S + 1, M_I) - E(M_S, M_I) = g_n \beta B + g_n \beta K_n M_I,$$

where $M_I = \frac{7}{2}, \frac{5}{2}, \ldots, -\frac{7}{2}$ are eight nuclear states of ^{51}V. For usual magnetic resonance experiments where the frequency ν is kept constant, resonance fields $B(M_I)$ for allowed transitions for these nuclear states M_I are given by

$$B(M_I) = B_0 + K_n M_I, \tag{9.20}$$

where $B_0 = \mathrm{h}\nu / g_n \beta$ is the resonance field without the hyperfine interaction.

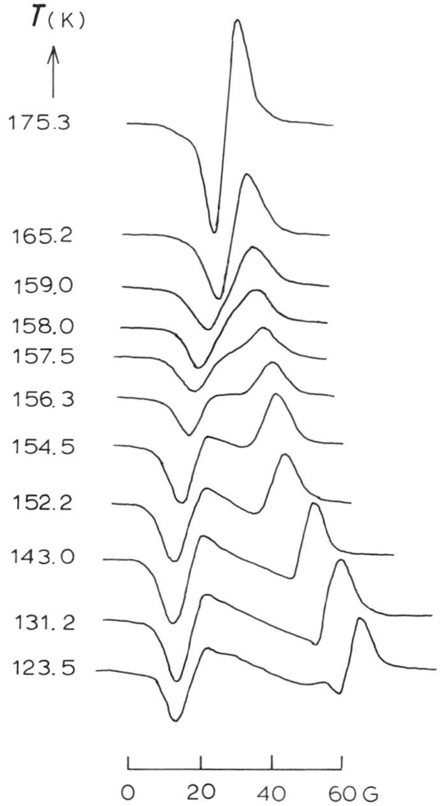

Figure 9.11. A change in the lineshape of a VO^{2+} spectra in the phase I_1 of BCCD.

The line broadening below T_i is typical for fluctuations in phase mode $\sigma = \sigma_0 \cos\phi$ as shown in Fig. 9.11 for a representative hyperfine line, although dependent on the value of M_I. In general, the broadening in VO^{2+} spectra arises from variations in g_n and K_n. From (9.20)

$$\Delta B(M_S) = \Delta B_0 + M_I \Delta K_n, \tag{9.21a}$$

where ΔB_0 is the broadening due to fluctuations in g_n, i.e.,

$$\Delta B_0 = \frac{h\nu}{\beta}\frac{(-\Delta g_n)}{g_n{}^2} = -B_0\frac{\Delta g_n}{g_n},$$

while those contributed by fluctuations in hyperfine structure are given by $M_I \Delta K_I$. Equation (9.21a) indicates that broadening in each resonance line depends on the nuclear spin quantum number M_I, being competitive with ΔB_0 arising from g anisotropy. Namely,

$$\{\Delta B(M_I)\}_{\text{edge}} = -\left(\frac{B_0}{g_n}\right)(\Delta g_n)_{\text{edge}} + M_I(\Delta K_n)_{\text{edge}} \tag{9.21b}$$

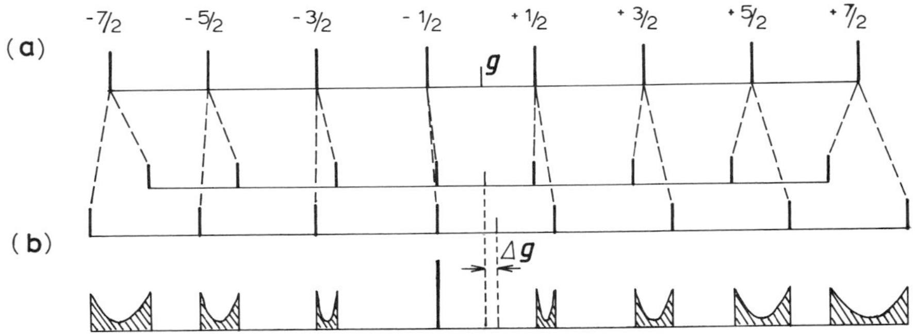

Figure 9.12. Interpretation of the anomalous VO^{2+} spectra from BCCD. The anomaly is due to Δg_n and ΔK_n.

gives the broadening for each hyperfine line for M_I as measured by the edge separation. Figure 9.12 shows, in a stick diagram, how such M_I-dependent anomalies arise from anisotropic **g** and **A** tensors in a representative VO^{2+} spectrum, where $\Delta g_n < 0$ is assumed.

Figure 9.13 shows a comparison of angular variations of K_n, $(\Delta g_n)_{edge}$, and $(\Delta K_n)_{edge}$ measured from the VO^{2+} spectra at 130 K. It is interesting to notice that these anisotropic edge separations indicate changing sign from plus to minus at directions for zero anomalies, which are clearly identified in these curves. As marked by vertical arrows in Fig. 9.13, such directions for zero $(\Delta g_n)_{edge}$ agreed with those for zero $(\Delta K_n)_{edge}$ within an accuracy of $\pm 2°$, which we consider to represent the direction of the librational axis. The librational axis for Ca^{2+}(betaine)$_2$ complexes can be represented by the unique axes of the **g** and **A** tensors of VO^{2+}(betaine)$_2$ complexes. Table 9.1 shows a comparison of these directions determined experimentally.

Rogers and Pake [75] observed such intensity anomalies in VO^{2+} spectra from randomly tumbling vanadyl radicals in liquids, and explained the M_I-dependent linewidths in terms of fluctuating g and hyperfine anisotropies in the diatomic VO^{2+}. While in modulated crystals such broadening is attributed to nonzero averages of $\langle \Delta g \rangle$ and $\langle \Delta K \rangle$ in slow fluctuations, for rapidly tumbling VO^{2+} molecules in liquids the root-mean-squared averages $\langle \Delta g \rangle^{1/2}$ and $\langle \Delta K^2 \rangle^{1/2}$ are essential for M_I-dependent broadening. Nevertheless, these cases share the common physical ground for observed anomalies.

As shown in Fig. 9.10, VO^{2+} spectra were markedly different when observed at 9.2 GHz and 35 GHz, due to line broadening that depends on the timescale of observation. Sampling results of a sinusoidal condensate near T_i during a timescale t_0 can be expressed by the basic formula

$$\langle \sigma \rangle_t = \sigma_0 \frac{\sin(\omega t_0)}{(\omega t_0)} \cos\phi \qquad \text{where} \quad 0 \le \phi \le 2\pi.$$

As in Mn^{2+} anomalies in TSCC, we can estimate the critical frequency ω from separations measured at 9.2 GHz and 35 GHz, which was of the order

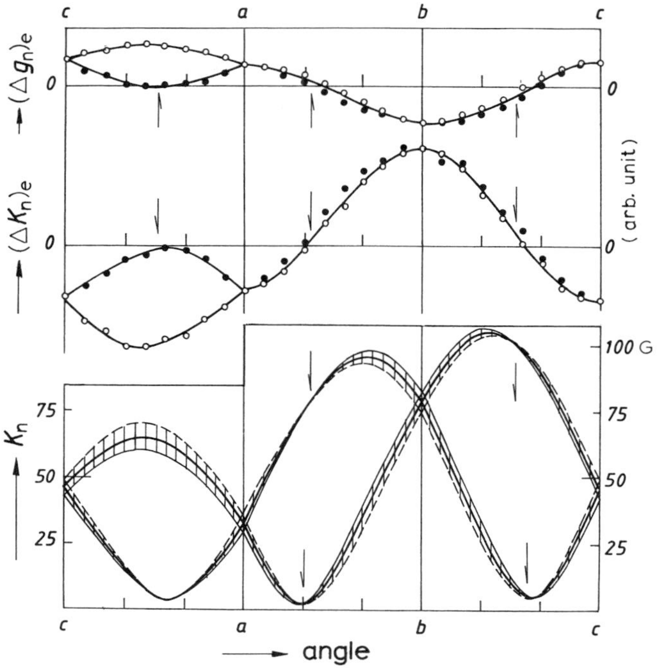

Figure 9.13. Angular variations of K_n, $(\Delta K_n)_{edge}$, and $(\Delta g_n)_{edge}$ in VO^{2+} spectra from BCCD at 130 K.

of a few GHz at temperatures close to T_i. At 136 K in contrast, the edge separations in *soliton-like* anomalies were virtually identical at these frequencies, where the modulation was quasi-static.

(ii) *Transition Anomalies in Mn^{2+} Spectra*

Mn^{2+} spectra in BCCD crystals are dominated by a large fine structure, while ^{55}Mn-hyperfine interaction is isotropic. Consequently, it is not only

Table 9.1. **g** and ^{51}V Hyperfine Tensors in VO^{2+} Ions in BCCD Crystals.

Librational Axes		(0.60, ±0.38, ±0.71)	(−0.60, ±0.38, ±0.71)
Principal values		Direction cosines	
g	$K(G)$		
2.0315	135	(0.559, ∓0.581, ±0.592)	(−0.559, ∓0.581, ±0.592)
1.9538	75	(−0.511, ∓0.624, −0.591)	(0.511, ∓0.624, −0.591)
1.9623	22	(−0.831, ∓0.324, ±0.453)	(0.831, ∓0.324, ±0.453)

difficult to carry out a precision analysis with high-field approximation, but a large number of forbidden transitions make the analysis almost impossible in most directions of the applied field. In BCCD crystals, Mn^{2+} complexes have two orientations in the unit cell, while VO^{2+} complexes arise from four symmetrical orientations with respect to the b axis. Yet, the Mn^{2+} spectra in BCCD are far more complex than the VO^{2+} spectra, although most phases in the transition series were identified faithfully in the Mn^{2+} spectra.

While it was not possible to perform a full analysis of the Mn^{2+} spectra, the principal directions of the **D** tensor were easily identifiable, where the behavior of pseudospins can be studied from distinct allowed transition lines that exhibit anomalous lineshapes in the critical region. Figure 9.14 shows the Mn^{2+} spectrum where B is applied exactly parallel to a principal direction of one of the oriented complexes. This particular direction is signified by allowed transitions of one of the two complexes, while the other exhibits many forbidden lines in the area marked "site 2." In the spectrum of Fig. 8.8(b), these two complexes have a common principal direction in parallel with the c axis, hence giving all allowed transitions. Transition anomalies in allowed transition lines at critical temperatures provide significant information about the phase ϕ and the magnitude σ_0 of the pseudospin vector σ, but its dimensionality could be obtained only when the spectra were analyzed in all directions. On the other hand, Mn^{2+} impurities are supposed to be better probes than dipolar VO^{2+}, in a sense that the transition dynamics is less perturbed by nonpolar probes. In this context, the serial structural changes in BCCD crystals are "faithfully" sampled by Mn^{2+} ions, in spite of the spectra that are too complex to analyze.

Central Transitions $(-\frac{1}{2} \leftrightarrow +\frac{1}{2})$

Allowed transitions for $M_S = -\frac{1}{2} \leftrightarrow +\frac{1}{2}$ at the center of Mn^{2+} spectra are unrelated to the fine structure energy, hence providing no information about

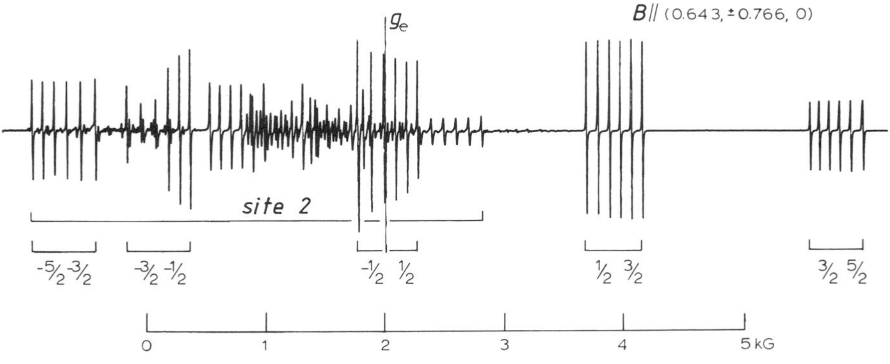

Figure 9.14. An Mn^{2+} spectrum from BCCD at room temperature. B_0 was applied in the ab plane, in the direction 45° from the axes. The spectrum from site 1 consisted of all allowed transitions, whereas the one at site 2 was not well resolved.

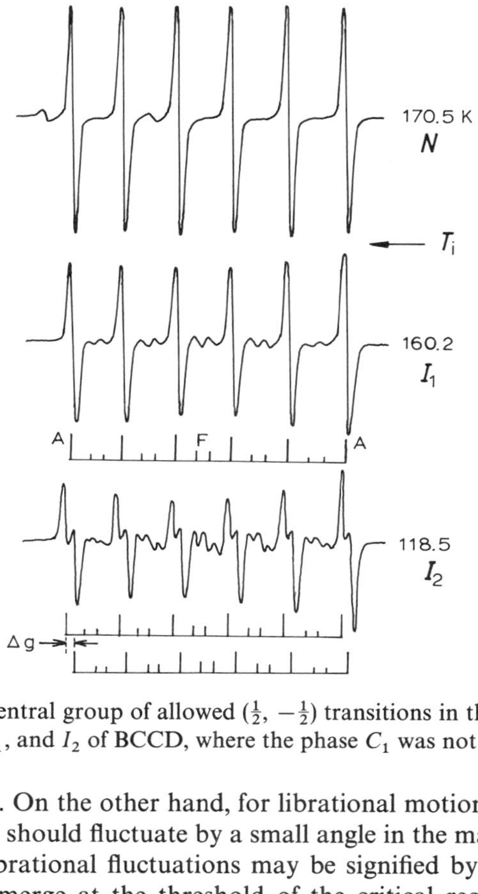

Figure 9.15. The central group of allowed $(\frac{1}{2}, -\frac{1}{2})$ transitions in the Mn^{2+} spectrum in the phases I_1, C_1, and I_2 of BCCD, where the phase C_1 was not distinguishable.

the pseudospin σ. On the other hand, for librational motion of the complex, the principal axis should fluctuate by a small angle in the magnetic field, and therefore such librational fluctuations may be signified by weak forbidden lines that may emerge at the threshold of the critical region. Figure 9.15 shows a temperature change of the central group of Mn^{2+} spectra composed of six lines for ^{55}Mn nuclear states ($I = \frac{5}{2}$). The incommensurate phase I_1 is clearly characterized by forbidden lines between allowed lines. According to Abragam and Bleaney [62], intensities of the forbidden lines increase sensitively with the librational angle (see Section 8.6), resulting in increasing intensity observed with decreasing temperature. However, the transition to the commensurate phase C_1 was not detected in such central transitions that are unrelated to σ. In contrast, the second incommensurate phase I_2 was clearly distinguishable by an additional splitting in both allowed and forbidden lines. Although unidentified, such a change in the Mn^{2+} spectra should be attributed to a mechanism leading to a g anisotropy resulting from a structural distortion associated with the second soft mode at 125 K.

Low-Field Transitions $(-\frac{5}{2} \leftrightarrow -\frac{3}{2})$ *for* $B \| c$

In Fig. 9.16 are shown the change of six hyperfine lines for the allowed transitions $(-\frac{5}{2} \leftrightarrow -\frac{3}{2})$ in the Mn^{2+} spectrum for $B \| c$, when observed at

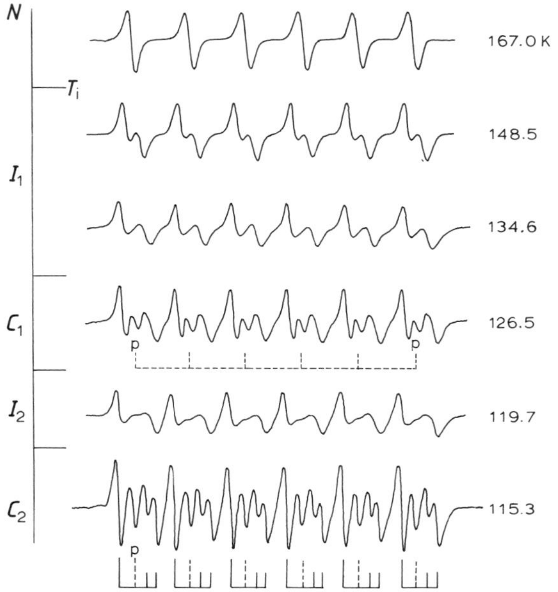

Figure 9.16. The low-field group of allowed ^{55}Mn transitions $(-\frac{3}{2}, -\frac{5}{2})$ in successive phases of BCCD.

various temperatures below T_i. It is noticed that in the range between room temperature and 100 K each phase of the series of structural changes is signified by a characteristic lineshape, whereby identifying serial phase transitions in BCCD.

In the incommensurate phases I_1 and I_2, the anomalous lineshape is typically given by the curve in Fig. 9.1(b) due to fluctuations contributed by terms proportional to σ as well as to σ^2 as in (9.13) for $\nu_2 > \nu_1$, where the phase ϕ is continuous in the range $0 \leq \phi \leq 2\pi$. According to the analysis of the $(\frac{1}{2} \leftrightarrow -\frac{1}{2})$ transitions in this section, phase I_2 could be characterized by a small g anomaly Δg proportional to σ, which is however practically insignificant for the lineshape dominated by a larger fine-structure fluctuation.

The commensurate locked-in phases C_1 and C_2 are characterized by discrete phases determined as

$$\phi = Q_c x \quad \text{where} \quad Q_c = \delta(T)c^*, \qquad x = nc \quad (n\text{: integer})$$

and $\delta(C_1) = \frac{2}{7}$ and $\delta(C_2) = \frac{1}{4}$, according to Brill and Ehses [69]. Therefore, there should be *discommensuration phases* $\pi/7$, $2\pi/7$, $3\pi/7$, $5\pi/7$ in the locked-in phase C_1 and $\pi/4$ in the phase C_2 between two edges $(0, \pi)$ of the amplitude mode. In addition, there should be a single line (p) of the phase mode, since there is no appreciable long-range order in these phases of BCCD. In the phase C_1, the center line in each hyperfine component may represent the phase mode, where peaks of discommensuration appear to be unresolved. In the phase C_2, the line p may coincide with the only discommensuration line, which is considered as evidenced by its intensity stronger than in C_1. While the anomalies from measurements at 9.2 GHz were discussed in the above, the lineshape at 35 GHz were very different from those in Fig. 9.16. As reported in [73], the difference is emphasized by lines of the phase mode with a stronger intensity at all temperatures, making it difficult to identify discommensuration peaks in locked-in phases. Owing to a nearly four times longer timescale than at 35 GHz. the phase mode was observed with somewhat suppressed intensity in the narrower critical region in measurements at 9.2 GHz.

Although discommensuration lines in locked-in phases were difficult to identify with Mn^{2+} probes, incommensurate phases were clearly distinguishable by the continuous phase ϕ. Nevertheless, at temperatures lower than 100 K, Mn^{2+} spectra in BCCD were just too complex to deduce further information about structural changes. On the other hand, it is known from dielectric measurements that the transition scheme in BCCD crystals is considerably modified under hydrostatic and uniaxial pressure, which may constitute a worthy subject for further studies.

9.4. Nuclear Quadrupole Relaxation in Incommensurate Phases

Nuclear magnetic probes can, in principle, be used for sampling order-variable condensates for structural phase transitions, since the nucleus with spin $I > \frac{1}{2}$ can interact with the lattice via the quadrupole energy. The quardrupole Hamiltonian is usually expressed as $\mathscr{H}_Q = \langle I|\mathbf{QT}|I\rangle$, where $\mathbf{Q}$ is the nuclear quadrupole moment tensor and $\mathbf{T} = \nabla E$ is the electric field gradient tensor at the nuclear site. In a modulated crystal, elements of the tensor $\mathbf{T}$ composed of second derivatives of the lattice potential [76] can be modified as $\mathbf{T}' = \mathbf{a}^\dagger\mathbf{T}\mathbf{a}$, if the lattice displacement is considered to change as $\boldsymbol{u}_{\boldsymbol{q}} \to \boldsymbol{u}_{\boldsymbol{q}} + \delta\boldsymbol{u}_{\boldsymbol{q}} = \mathbf{a}\boldsymbol{u}_{\boldsymbol{q}}$. Here the transformation matrix is defined by $\mathbf{a} = \mathbf{1} + \sigma\mathbf{e}$, where $\delta\boldsymbol{u}_{\boldsymbol{q}} = \sigma\mathbf{e}\boldsymbol{u}_{\boldsymbol{q}}$, and the modified quardrupole Hamiltonian can then be written as

$$\mathscr{H}_Q' = \mathscr{H}_Q + \sigma\langle I|\mathbf{e}^\dagger\mathbf{QT} + \mathbf{QTe}|I\rangle + \sigma^2\langle I|\mathbf{e}^\dagger\mathbf{QTe}|I\rangle,$$

where the term proportional to σ^2 is small, and normally ignored for an approximate analysis. The tensor $\mathbf{T}$ represents the anisotropic local field, and hence is independent of the spin $\langle I|$. Neglecting the last quadratic term, an additional energy for quadrupole modulation is given as

$$\mathscr{H}_Q' - \mathscr{H}_Q = \Delta\mathscr{H}_Q = \sigma\langle I|\mathbf{e}^\dagger\mathbf{Q} + \mathbf{Qe}|I\rangle\mathbf{T} = \sigma\langle I|\mathbf{Q}'|I\rangle\mathbf{T}, \qquad (9.22)$$

where $\mathbf{Q}' = \mathbf{e}^\dagger\mathbf{Q} + \mathbf{Q}\mathrm{e} = (\mathbf{e}^\dagger + \mathbf{e})\mathbf{Q}$ is the effective quadrupole tensor in the modulated lattice. In this treatment, we see that the nuclear spin $|I\rangle$ may be considered as if modulated, i.e., $|I'\rangle = \mathbf{a}|I\rangle$, although the tensor $\mathbf{T}$ is actually modulated in the lattice.

In the presence of a uniform magnetic field $|B\rangle = B|n\rangle$, the nuclear Zeeman energy for a spin $I > \frac{1}{2}$ can also be modulated as

$$\mathscr{H}_Z' - \mathscr{H}_Z = \Delta\gamma_n B \qquad \text{where} \quad \Delta\gamma_n = \gamma\langle n|\mathbf{e}^2|n\rangle\sigma,$$

and the corresponding magnetic resonance line is broadened when the frequency is distributed by $\Delta\gamma_n$ just as the anomalous g_n factor. Therefore, as in a paramagnetic resonance, we can write a similar resonance formula for a nuclear magnetic resonance perturbed by a quadrupole interaction in a modulated crystal. For the fluctuating σ in the critical region, such a nuclear resonance frequency is broadened as

$$\Delta\nu_n(\mathrm{P}) = \nu_1 \cos\phi \qquad \text{and} \qquad \Delta\nu_n(\mathrm{A}) = \nu_1 \sin\phi, \tag{9.23a}$$

where

$$\nu_1 = \frac{\gamma\sigma_0\langle n|\mathbf{e}^2|n\rangle}{\mathrm{h}} + \frac{\sigma_0[\langle M_I|\mathbf{Q}'\mathbf{T}|M_I\rangle - \langle M_I'|\mathbf{Q}'\mathbf{T}|M_I'\rangle]}{\mathrm{h}}$$

is the edge frequency in the first-order accuracy, and $0 \le \phi \le 2\pi$.

Magnetic resonance anomalies are primarily due to the first-order term $\langle M_I|\mathbf{e}^\dagger\mathbf{QTe}|M_I\rangle$, but may also be contributed to by the second-order term that is effectively proportional to σ^2. In this case, (9.23a) should be revised as

$$\Delta\nu_n = \nu_1 \cos\phi + \nu_2 \cos^2\phi, \tag{9.23b}$$

as already discussed for fine structure anomalies in a paramagnetic resonance spectrum.

Equation (9.22) is a quadratic form with respect to principal axes x, y, and z of the traceless tensor $\mathbf{Q}'$. Taking the z axis as the direction for quantization, nuclear transitions $\Delta m = 0, \pm 1, \pm 2$ between nuclear spin states can be conveniently described with components I_z and $I_\pm = I_x \pm iI_y$. The tensor $\mathbf{T}$ is a function of time t, and the elements in this (x, y, z) system constitute coefficients of the quadratic form of (9.22). Therefore, in the first order

$$\langle I|\Delta\mathscr{H}'|I\rangle = \sigma \sum_{\Delta m} T_{\Delta m}(t)Q_{\Delta m}, \tag{9.24}$$

where

$$T_0 = T_{zz}, \qquad T_{\pm 1} = T_{xz} \pm iT_{yz}, \qquad T_{\pm 2} = \frac{T_{xx} - T_{yy}}{2} \pm iT_{xy},$$

$$Q_0 = A(3I_z^2 - I^2), \qquad Q_{\pm 1} = A(I_zI_\pm + I_\pm I_z), \qquad Q_{\pm 2} = I_\pm^2,$$

and

$$A = \frac{e^2Q}{4I(2I-1)}.$$

Here, the parameter Q is known as the nuclear quadrupole moment.

Blinc et al. [77], [78] carried out extensive NMR studies on the incommensurate phase transitions in Rb_2ZnBr_4 and Rb_2ZnCl_4 crystals, and discovered anomalous ^{87}Rb lines for $(\frac{1}{2} \rightarrow -\frac{1}{2})$ transitions related to two modes of fluctuations below T_i, which resembled anomalous Mn^{2+} EPR lines at the ferroelectric phase transition of TSCC. While originally called "phase fluctuations" and "floating modulation" by these authors, such anomalies are undoubtedly due to pseudospin fluctuations in phase and amplitude modes expressed by (9.23a).

Assuming that in a pseudospin condensate nuclear energies absorbed due to transitions Δm are transferred to the soft mode through the critical coupling, sinusoidal fluctuations described approximately by (9.23a) should be responsible for the quadrupole relaxation, which should be different from random fluctuations caused by the lattice as a whole. Nevertheless. the energy transfer is indicated by the spin-lattice relaxation time T_1 when the nuclear magnetic resonance is observed for transitions either by $\Delta m = \pm 1$ or ± 2.

At this point, we should also consider that in critical regions and incommensurate phases fluctuating order variables are in amplitude and phase modes, contributing to T_1 independently. Hence,

$$\frac{1}{T_1} = \frac{1}{T_{1A}} + \frac{1}{T_{1P}}, \tag{9.25}$$

where

$$\frac{1}{T_{1A}} \propto \sigma_A{}^2 \sum_{\Delta m} \tau^{-1} \int_0^\tau T_{\Delta m}(A, 0)\, T_{\Delta m}(A, t) \exp(i\omega t)\, dt$$

and

$$\frac{1}{T_{1P}} \propto \sigma_P{}^2 \sum_{\Delta m} \tau^{-1} \int_0^\tau T_{\Delta m}(P, 0)\, T_{\Delta m}(P, t) \exp(i\omega t)\, dt.$$

Writing $J_{\Delta m}(A)$ and $J_{\Delta m}(P)$, these integrals are correlation functions with a characteristic time τ, which are maximum at $\omega = |\Delta m|\omega_I$ where $\omega_I = \gamma B$ is the nuclear Larmor frequency. Accordingly, (9.25) can be written as

$$\frac{1}{T_1} \propto \left\{\sum_{\Delta m} J_{\Delta m}(P)\right\} \sigma_0{}^2 \cos^2\phi + \left\{\sum_{\Delta m} J_{\Delta m}(A)\right\} \sigma_0{}^2 \sin^2\phi. \tag{9.26}$$

As seen from the derivation, magnitudes of these correlation functions depend on the densities of condensates in phase and amplitude modes that are proportional to $\cos^2\phi$ and $\sin^2\phi$, respectively, and to functions of the temperature. Equation (9.26) is the equation originally derived by Blinc et al. [78], who analyzed with this formula the ^{14}N-spin-lattice relaxation rate in the incommensurate phase of $\{N(CH_3)_4\}_2ZnCl_4$ crystals at 14° C. Figure 9.17(a) and (b) show results of Blinc et al.'s ^{14}N nuclear magnetic resonance studies, where the Zeeman energy is perturbed by the quadrupole interaction. Here, the resonance frequency is closely as distributed described by (9.22) for

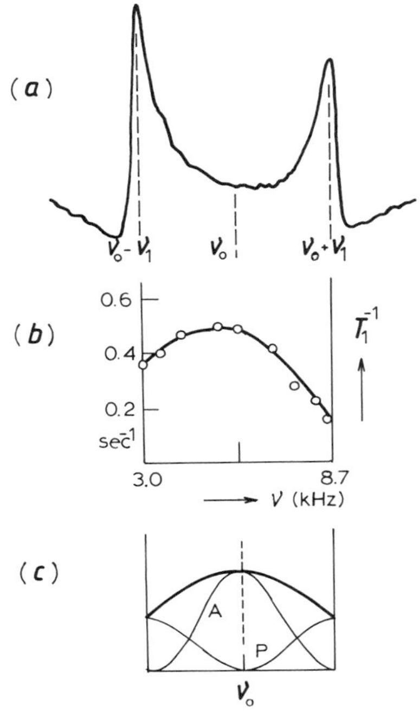

Figure 9.17. (a) An anomalous ^{14}N NMR line from the incommensurate phase of a $[N(CH_3)_4]_2ZnCl_4$ crystal. (From S. Zumer and R. Blinc, *J. Phys.* **C14**, 465 (1981).) (b) Observed T_1^{-1}, corresponding to (a). (c) Calculated contributions from the amplitude mode (A) and from the phase mode (P), where the coefficients in (9.26) was assumed to be equal. The thick lines for the net contribution are somewhat similar to the observed T_1^{-1}.

a predominant P mode, and the corresponding $1/T_1$ curve indicates an approximate distribution given by $\cos^2 \phi$ in this range. According to the Blinc et al., the finite T_1 at the edges $\phi = 0$ and π may be attributed to fluctuations in the A mode, although the slight asymmetry in these curves may be explained by (9.23b) for $\nu_1 \gg \nu_2$.

The above example gives significant evidence for the condensate model, where a specific excitation of the lattice may hold collective pseudospins in stable motion. Blinc [80] published an extensive review of nuclear relaxation results from various systems undergoing structural phase transitions, to which interested readers are strongly referred.

In magnetic resonance spectra from a modulated crystal, it is generally not easy to identify the line due to the amplitude mode, simply because of missing spectral details. For example, a condensate that is released from the pinning potential is free to propagate through the lattice, and such a "depinned" condensate exhibits a motionally narrowed spectrum that is not distinguish-

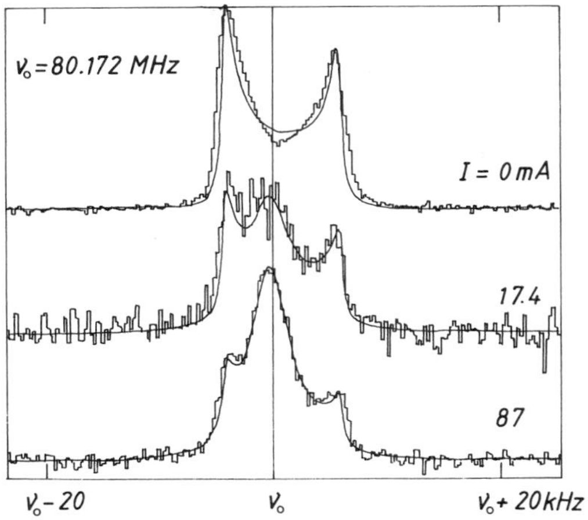

Figure 9.18. Electric current "di-pinning" observed in a ^{87}Rb NMR line from the charge-density-wave (CDW) state of $Rb_{0.3}MoO_3$. The center line appeared with increasing current in the NMR spectrum at 80.17 MHz. (From P. Segransan, A. Jánossy, C. Berther, J. Mercus, and P. Buraud, *Phys. Rev. Lett.* **56**, 1654 (1986).)

able from the phase mode. Segransan et al. [81] carried out an ^{87}Rb-NMR experiment in the CDW-state of $Rb_{0.3}MoO_3$, showing that the modulated resonance line of pinned CDW condensates can be converted to free-running condensates by increasing electrical current through the conductor. Figure 9.18 shows their experimental results, which is similar to coexisting fluctuations in P and A modes in Rb_2ZnCl_4 and related crystals.

CHAPTER 10

Structural Phase Transitions in Miscellaneous Systems

In Chapter 9, transition anomalies in oxide perovskites, TSCC, and BCCD crystals were described in some detail as examples to facilitate the concept of order-variable condensates. In this chapter, we look at other systems of interest on the basis of the knowledge acquired from these examples to see how much of the transition mechanism can be elucidated. While principles for pseudospin condensates discussed in Part One provide a useful guideline, phase transitions in crystalline systems are of such an enormous variety that no systematic description is possible at the present stage. We therefore focus our attention on structural changes in representative systems that were studied intensively, to conclude our discussion.

10.1. Cell-Doubling Transitions in Oxide Perovskites

Crystals of the oxide–perovskite family designated by the chemical formula ABO_3, where A = K, Sr, Ba, Ca, ..., and B = Ta, Ti, Al, Pb, ..., are rich in types of structural changes that are all attributed to the collective motion of deformed or displaced octahedral units BO_6^{2-} in the lattice, providing a prototype of the active group. As already described, the ferroelectric phase transition in $BaTiO_3$ is related to uniaxial deformation of the TiO_6^{2-} octahedron along one of the C_4 symmetry axes, whereas in the cell-doubling transition of $SrTiO_3$ cooperative rotation of these TiO_6^{2-} units around a C_4 axis is responsible for the structural change at $T_c = 105$ K. Substituting paramagnetic impurities for ions inside active octahedra, their magnetic spectra allow us to sample the condensate mode for information about their critical behavior, if the condensate is not significantly perturbed by impurities. Since such a model can be adapted to structural transformations in other systems, it may be worth summarizing the essential results from oxide perovskites.

The phase transition in $SrTiO_3$ crystals at 105 K is recognized by a symmetry change from cubic to tetragonal, featuring unit-cell doubling in the lattice. Consequently, three orthogonal domains in tetragonal symmetry are generally formed at temperatures below T_c. In the tetragonal phase, the unit-cell size is doubled, indicating an alternate arrangement of $TiO_6{}^{2-}$ complexes that are rotated in opposite directions, as sketched in Figs. 10.1((a) and (b)). From Fe^{3+} ions substituted for Ti^{4+}, von Waldkirch et al. [82] identified the order variable as related to a rotational angle φ of the $TiO_6{}^{2-}$ complex around a C_4 axis as indicated in Fig. 10.1(a), which illustrates the tetragonal structure around the c axis as occurred below T_c.

Actually, Fe^{3+} spectra observed from perovskite crystals were complicated by the fact that an active octahedral complex modified by doping is accompanied by an oxygen vacancy that is designated by V_0 for the following discussion. Such a paramagnetic $Fe^{3+}-V_0$ complex is deformed axially, exhibiting spectra characterized by a large **D** tensor that is practically uniaxial along the direction connecting Fe^{3+} and V_0. From such spectra observed at the microwave frequency 24 GHz it was too complicated to deduce a quan-

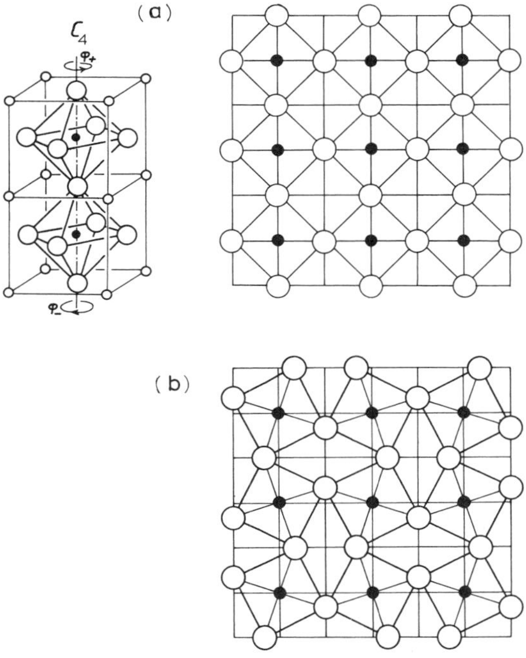

Figure 10.1. Molecular arrangements in $SrTiO_3$ crystals. (a) Librational angles $\pm\phi$ and the normal phase. (b) The phase below 105 K.

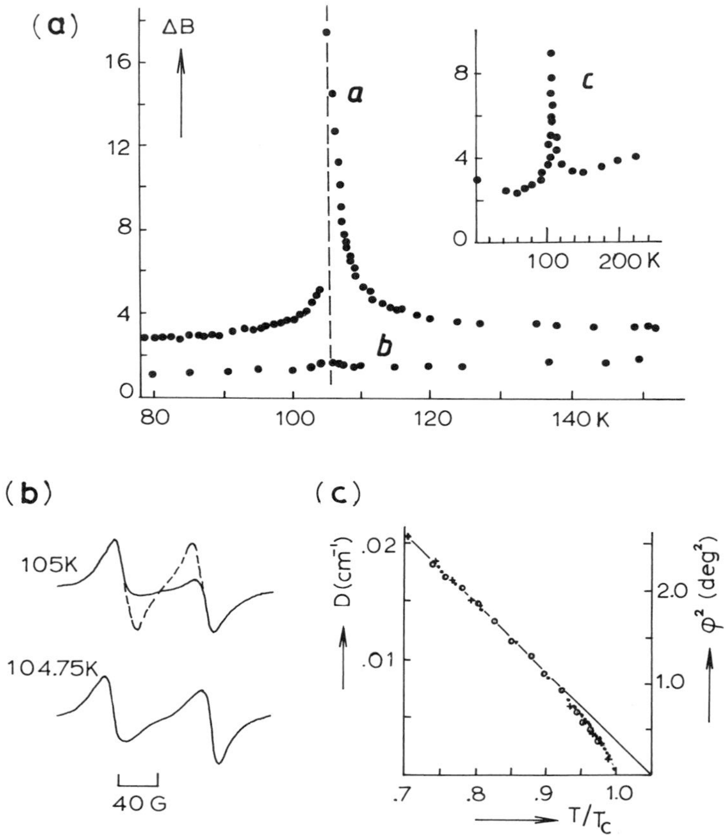

Figure 10.2. Transition anomalies in the Fe^{3+} resonance in $SrTiO_3$. (a) Linewidths (b) An anomalous splitting at 105 K. (c) Anomalous temperature-dependence of $D \propto \phi^2$.

tity related to the structural order, however, Müller et al. [20] managed to obtain the temperature-dependence of the angle φ in various directions of the applied field. In Figs. 10.2((a), (b), and (c)) the transition anomalies observed from the $Fe^{3+}-V_0$ spectra are summarized, where considerable line broadening signifies the nature of the structural change when T_c was approached from above. As seen in Fig. 10.2(a), such broadening is anisotropic: in tetragonal crystals along the b axis, linewidths for $B \parallel a$ and $B \parallel c$ increased nearly four times broader than those at 140 K, while showing virtually no additional broadening for $B \parallel b$.

In the low-temperature phase below T_c, the $Fe^{3+}-V_0$ lines split into two in most directions of the applied field, but the lineshape was most anomalous for $B \parallel [110]$ at temperatures very close to 105 K, as shown in Fig. 10.2(b).

Here, the dotted curve is drawn for a doublet that would be expected from a normal crystal. Nevertheless, the anomaly diminished on lowering temperature, resuming normal lines at temperatures well below 105 K. In the condensate model, such an anomaly in the Fe^{3+} line should be due to fluctuations in the phase mode, and correspond to cell-doubling in $SrTiO_3$ crystals. In the ferroelectric phase transition of TSCC, the anomaly sampled by Mn^{2+} probes exhibited a very similar lineshape, although attributed to fluctuations between opposite domains instead of sublattices. Although macroscopically characterized by entirely different features, the critical regions in these systems were not distinguishable by magnetic resonance anomalies showing a similar lineshape that originates from the binary mechanism.

Considering angular displacements $\pm\varphi$ around the b axis, the order variable σ for the structural change in $SrTiO_3$ can be expressed by a pseudospin vector σ, whose components along $\pm b$ are related by inversion. Short-range correlations among these σ's can then be assumed as expressed by (3.3a). Following Subsection 3.5.1, a specific wavevector $\boldsymbol{q}$ (q_a, q_b, q_c) minimizing the spatial correlation parameter $J(\boldsymbol{q})$ for tetragonal distortion of the lattice in the b direction can be obtained by the condition

$$\sin(q_b b) = 0,$$

which gives $q_b = (\pi/b)m = \frac{1}{2}b^*m$, where $m = 0$ or odd integers, expressing a commensurate arrangement along the b direction. In this case, $\cos(q_b b) = 1$, and so the other components q_a and q_c can be determined from the relations

$$J + 2J'\{\cos(q_a a) + \cos(q_b b)\} = 0 \qquad \text{and}$$

$$J + 2J'\{\cos(q_b b) + \cos(q_c c)\} = 0,$$

namely,

$$\cos(q_a a) = \cos(q_c c) = -\left(1 + \frac{J}{2J'}\right).$$

Therefore, if $0 > J/2J' > -1$ for given values of J and J', the components q_a and q_c can be irrational in reciprocal units $\boldsymbol{a}^*$ and $\boldsymbol{c}^*$, respectively. Consequently, for the cubic-to-tetragonal transition along the unique b axis, the pseudospin arrangement in the critical region can be incommensurate in the ac plane, which is consistent to the anomalies observed by Fe^{3+} probes. When $J' \to \frac{1}{2}J$, $\cos(q_a a) = \cos(q_c c) = 0$, and so $q_a = \frac{1}{2}a^*$ and $q_c = \frac{1}{2}c^*$, indicating that $2a$ and $2c$ are the smallest units of the lattice in this commensurate limit. It is also noted that $q_b = \frac{1}{2}b^*$, therefore the phase transition takes place at any point on the Brillouin zone boundary in the a^*c^* plane, at which the soft modes were indeed identified, as discussed in Chapter 4 (see Fig. 4.6(a)). In this case, the wavevector for the phase transition can be written as

$$q_a = \delta_a a^*, \qquad q_b = \tfrac{1}{2}b^*, \qquad \text{and} \qquad q_c = \delta_c c^*,$$

where δ_a and δ_c are the incommensurability parameters along the a and c directions, respectively.

The observed anomalous broadening near 105 K is clearly due to fluctuations proportional to the pseudospin $\sigma_A = \sigma_0 \cos\phi$, indicating slow sinusoidal fluctuations between $\pm\sigma_0$, although angular variations of Fe^{3+} spectra in $SrTiO_3$ have not been fully determined. Characterized by inversion symmetry, such fluctuations should take place between two opposite sublattices, which are sampled by the fine structure of Fe^{3+} probes as expressed by

$$\Delta D_n(\pm) = \pm c_{1n}\sigma + c_{2n}\sigma^2.$$

In the particular direction $B \parallel [110]$, where $D_n(+) = D_n(-)$, for fluctuations between two sublattices $\pm\sigma$ are not independent and out-of-phase, so that the net fluctuation is given by

$$\Delta D_n = \Delta D_n(+) - \Delta D_n(-) = 2c_{1n}\sigma_0 \cos\phi,$$

which is responsible for the anomalous splitting shown in Fig. 10.2(b). Although not fully substantiated, the critical fluctuation is thus in the phase mode $\sigma = \sigma_P$. For full justification, it is necessary to determine angular dependences of the edge frequency $\Delta\nu_n = 2c_{1n}\sigma_0$.

As seen from Fig. 10.1, active groups are sharing oxygen ions with surrounding groups at the nearest-neighbor positions. These groups are linked with each other, so that their motion cannot be independent, being involved in substantial lattice strains due to structural deformation during the phase transition. In this context, it is conceivable that strain energies for the symmetry change should be dispersed by distorting the lattice in a wide range of temperatures. At least qualitatively, such a mechanism may be considered as responsible for the anomalous broadening observed in a very wide temperature range ($\sim$50 K) in Fig. 10.2(a).

Von Waldkirch et al. [82] also reported transition anomalies observed in the Fe^{3+} spectra from $SrTiO_3$ and $LaAlO_3$ crystals in the tetragonal phase below T_c. They found that the fine-structure splitting D_a on the tetragonal axis exhibited a significant deviation from the mean-field average in the region close to T_c, as shown in Fig. 10.2(c). The observed anomaly in a fine structure splitting can generally be expressed by

$$\langle\Delta D_n\rangle_t = c_{1n}\langle\sigma\rangle_t + c_{2n}\langle\sigma^2\rangle_t.$$

In the noncritical region above T_c, referring to the timescale of measurement t_0, $\langle\sigma\rangle_t = 0$ for a fast fluctuation, while

$$\langle\sigma^2\rangle_t = \langle\varphi^2\rangle_t \propto T_c - T$$

in the mean-field approximation. Therefore, a plot of $\Delta D_a = \langle\Delta D_{n=a}\rangle_t = \langle\varphi^2\rangle_t$ against the temperature T should be primarily a straight line, from which any deviation can be attributed to anomalous unvanished averages $\langle\sigma\rangle_t$ and $\langle\sigma^2\rangle_t$ originating from the condition $t_0 \leq 2\pi/\varpi$. Actually, the plot in Fig. 10.2(c) shows a considerable deviation from the straight line in the critical region from T_c to about $0.9T_c$, indicating that the mean-field approximation is not applicable.

10.2. The Incommensurate Phase in Beta Thorium Tetrabromide

Crystals of β-thorium tetrachloride, $ThBr_4$, exhibit a structural change at $T_i \sim 95$ K from the normal to incommensurate phase, which was first discovered by Raman experiments, and followed by neutron diffraction studies [83]. Bernard et al. [40] performed neutron inelastic scattering experiments on β-$ThBr_4$, and found that the soft mode associated with the phase transition was well defined by a dip at $q_c = 0.310c^*$ in the phonon dispersion curve in $\boldsymbol{q} \| \boldsymbol{c}$ scattering geometry, which was resolved into two branches.

Emery et al. [84] carried out EPR studies on the phase transition in β-$ThBr_4$ crystals using Gd^{3+} probes substituted for Th^{4+} ions, and identified the order variable as related to a rotational angle φ around the z axis of $\bar{4}$ symmetry. They reported various types of anomalous broadening exhibited by Gd^{3+} spectra in the incommensurate phase, depending on directions of the applied field. More recently, Zwanenburg and de Boer [85] used Pa^{4+} probes substituted for Th^{4+}, and observed an anomalous g factor and hyperfine splittings. On the basis of these EPR results, it is evident that the complex Br^- structure, where Th^{4+} is at the center as illustrated in Fig. 10.3(a), is active in the phase transition, exhibiting fluctuations in twist deformation around the z axis in the incommensurate phase. While the central Th^{4+} ion is unchanged, the deformed structure of the surrounding Br^- complex below T_i is characterized by displacements of eight Br^- ions around the z axis of $\bar{4}$ symmetry.

In neutron inelastic scattering results obtained by Bernard et al. [40], it is evident that the two branches in the dispersion curve at 81 K in Fig. 4.9(a) represent phonon fluctuations in amplitude and phase modes. For such a phase transition at irrational $q_c = 0.310c^*$, the active groups signified by deformed structures must be correlated, and cannot behave like rigid bodies individually. We can simply consider that the order variable is associated with a six-ion complex composed of two inverted Br^- tetrahedra above and below each Th^{4+} ion, as highlighted by solid lines in Fig. 10.3(a). Considering the lattice of Th^{4+} ions and alternately arranged Br^--polyhedra in a chain, we notice that these Br^--complexes in the undeformed phase are related by $90°$ rotation around the $\bar{4}$ symmetry axis in alternate succession. In the deformed structure, the order variables associated with these polyhedra are angular distortion that can be expressed in the sinusoidal limit as $\sigma_1 = e_q \exp(i\phi)$ and $\sigma_2 = e_q \exp\{i(\phi + \frac{1}{2}\pi)\}$, where the phase angle ϕ represents a long-wave variation along the $\underline{4}$ symmetry axis.

Considering chains of these order variables for a $ThBr_4$ crystal as illustrated in Fig. 10.3(b), the short-range correlation energy in the chain is expressed as

$$E = -J_c\sigma_1(0)\{\sigma_1(2c) + \sigma_1(-2c)\} - J_c'\sigma_1(0)\{\sigma_2(c) + \sigma_2(-c)\} = -J(q)e_q e_{-q},$$

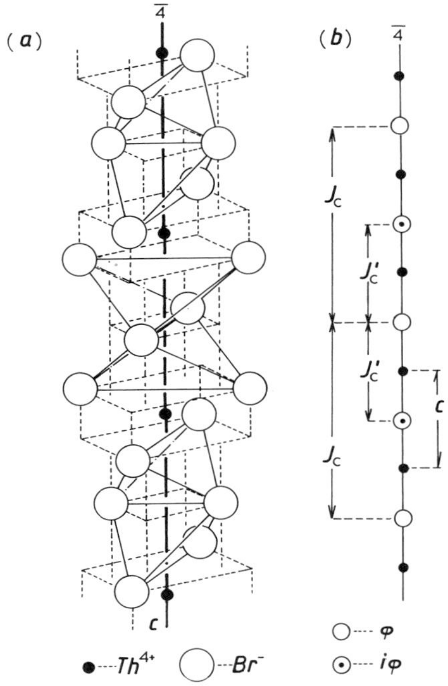

Figure 10.3. (a) Molecular arrangement in $ThBr_4$ crystals along the c axis. (b) The arrangement of pseudospins. ○, ⊙ represent pseudospins that are mutually perpendicular, and ● positions of Th^{4+} ions.

where

$$J(q) = 4J_c \cos(2q_c c) + 4J_c' \cos(q_c c + \tfrac{1}{2}\pi).$$

Here, we have omitted correlations along the a and b axes, where the order-variable distribution is assumed as commensurate with the lattice period. Differentiating $J(q_c)$ with respect to q_c, we obtain the equation,

$$-2J_c \sin 2q_c c - J_c' \cos q_c c = 0,$$

which determines the specific value of the incommensurate wavevector q_c for incommensurate variation along the c axis. Namely,

$$\sin q_c c = \frac{J_c'}{4J_c}$$

gives such an irrational solution, provided that $|J_c'/4J_c| \leq 1$. If commensurate, on the other hand, $q_c = n\pi/c$ where n is an integer, for which J_c' must be zero. Therefore, a nonzero correlation J_c' between "unlike" order variables σ_1 and σ_2 is essential for the modulated structure in $ThBr_4$. In such a modulated lattice, order variables fluctuate in two sinusoidal modes as $\sigma_+ = e_q \exp\{i(q_c - q)z\}$ and $\sigma_- = e_q \exp\{i(-q_c + q)z\}$, owing to the coupling with soft phonons. It is noted that $\sigma_+ = \sigma_-$ at $q = q_c$, however, such degeneracy is lifted by a periodic lattice potential $V(q - q_c) = V_0 \cos 2(q - q_c)z$ that originates from a quartic lattice potential induced by pseudospin correlations, resulting in two fluctuation modes $\sigma_P = e_q \cos \phi$ and $\sigma_A = e_q \sin \phi$, where $\phi = (q - q_c)z$.

The well-defined smooth dip at $q = q_c$ in the phonon dispersion curve (Fig. 4.9(a)) suggests that, at the transition temperature T_i there are spatial fluctuations $\pm q$ in the vicinity of wavevector q_c, signifying a near in-phase coupling between the order-variable mode σ and soft phonons as consistent with the above-modulated structure. In the incommensurate phase at temperatures close to T_i, these two independent lattice modes in the scattering spectra correspond to order-variable fluctuations responsible for EPR anomalies.

In the EPR experiment on $ThBr_4$ by Emery et al. [84], Gd^{3+} impurities were used as the probes substituting Th^{4+} ions. Electrons in the f-shell of a $Gd^{3+}(4f)^7$ ion are described approximately by an idealized L–S coupling scheme, and are much less perturbed by the crystal field than in iron-group ions. As a consequence, the magnetic properties are not much different from the free ionic state given by $^8S_{7/2}$. However, due to an unbalanced impurity charge, the ionic state is perturbed by the field of a vacant Br^- site located in its vicinity. Although the crystal field in undoped crystals is predonimantly cubic and responsible for weak spin-related fields at a Gd^{3+} ion, the quadratic field arising from Br^- vacancy is considered to be more significant for Gd^{3+} than the cubic field at the impurity site.

The **g** tensor of a Gd^{3+} ion is nearly isotropic ($g \sim 1.991$), and the cubic field is of the order of 150 MHz ~ 35 MHz in many other host crystals. Accordingly, we can assume that transition anomalies in $ThBr_4$ crystals originate predominantly from the fluctuating quadratic defect field in modulated crystals. Despite the inaccuracy involved in such approximation, the defect mechanism dominates the Gd^{3+} spectra in β-$ThBr_4$ crystals, judging from relatively large anomalies (~200 G) as reported by Emery et al. [84]. Figure 10.4 shows anomalous Gd^{3+} lines observed for transitions $M_S = \frac{1}{2}5 \leftrightarrow \frac{1}{2}7$ for representative directions of the static field B. Although fluctuations in the cubic field were emphasized by Emery et al., the axial defect field of lower symmetry is regarded as sufficient to account for the anomalies in Gd^{3+} spectra with rather broad linewidths. The defect field can be expressed by a spin-associated fine-structure energy $\mathscr{H}_F = \langle S|\mathbf{D}|S\rangle$ as for ions of the iron-group, and in modulated $ThBr_4$ crystals the spin $\langle S|$ can be considered as if modulated as $\langle S'| = \langle S|\mathbf{a}$, where $\mathbf{a} = \mathbf{1} + \sigma\mathbf{e}$, as discussed in Chapter 9. In a strong-field approximation we have derived the expression for such a variation in the fine-structure energy, i.e., $\Delta E_F = (a_n\sigma + b_n\sigma^2)M_S{}^2$, to account for the EPR anomaly. Using this result, the anomalous broadening in

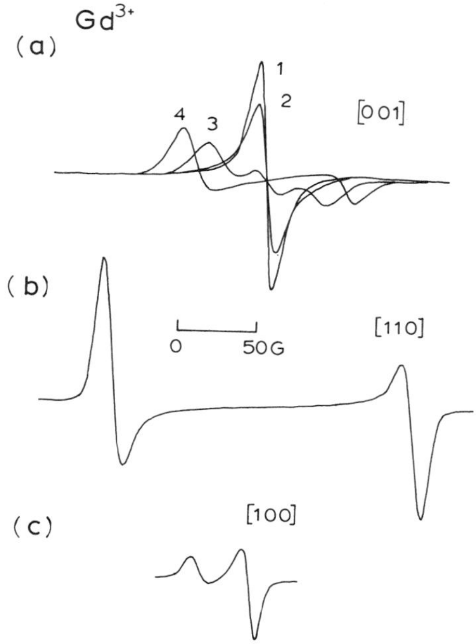

Figure 10.4. Gd^{3+} EPR anomalies in the incommensurate phase of $ThBr_4$ crystals. (a) $B_0 \| [001]$; (b) $B_0 \| [110]$; and (c) $B_0 \| [100]$. (From J. Emery, S. Hubert, and J. C. Fayet, *J. Physique* **46**, 2099 (1985).)

the Gd^{3+} line consists basically of two contributions proportional to σ and σ^2, i.e.,

$$\Delta\nu_n = \nu_1(n)\sigma + \nu_2(n)\sigma^2,$$

which is applicable to both σ_P and σ_A modes. Provided that the coefficients $\nu_1(n)$ and $\nu_2(n)$ are expressed in relation with the defect orientation in the lattice, the above formula should be in identical form to those employed by Emery et al. [84] for numerical simulation of observed lineshapes.

Figure 10.4(a) shows the anomalies observed for $B \| (100)$ at temperatures close to $T_i = 95$ K, where the presence of fluctuations in two modes σ_P and σ_A are evident. It is further noticed that the region where the σ_A mode was observed is wider in the incommensurate phase of $ThBr_4$ than the critical region in TSCC. As remarked, such a difference should be attributed to the presence of significant long-range correlations in the latter. In Figs. 10.4((a), (b), and (c)), the edge intensities in the phase mode are unequal, indicating that the anomalies are predominantly due to fluctuations σ_P, but presumably a contribution from the ${\sigma_P}^2$ term is not negligible. On the other hand, the anomaly shown in Fig. 10.4(b) is a typical case for σ_P, for which however it is unknown if the (110) plane is a significant reflection plane or the coefficient $\nu_2(110)$ happens to be exactly zero.

In contrast to trivalent Gd^{3+}, tetravalent Pa^{4+} ($5f^1$, $I(^{231}Pa) = 3/2$) ions substituted for Th^{4+} exhibited spectra with no influence of a charge defect. Having only one electron in the $5f$ orbit, the spectra are characterized by the Zeeman and hyperfine interactions as well as the nuclear quadrupole interaction. Experiments on Pa^{4+}-doped $ThBr_4$ crystals [85] were performed only at 4.2 K, since the spectral lines became too broad at temperatures above 20 K. At 4.2 K observed Pa^{4+} spectra were characterized by anomalous lineshapes as shown by the spectrum in Fig. 10.5. The results show that the incommensurate phase below 95 K is signified by a pair of four hyperfine edges for distributed resonances in all directions of the static field B, while for $B \parallel c$ these two sets are practically identical, indicating that the Pa^{4+} complex is "conformal" by symmetry to the polyhedral Br^- complexes. The two "sites" for Pa^{4+} spectra were assumed as having emerged at T_i from a single spectrum in the normal phase, when the librational fluctuations are slowed down in the critical region.

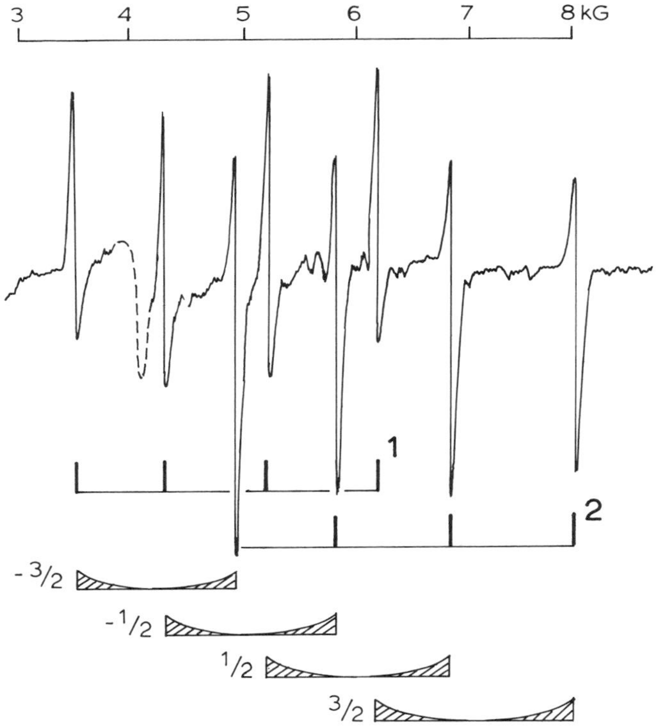

Figure 10.5. An anomaly in Pa^{4+} spectra in $ThBr_4$ at 4.2 K. Anomalous lineshape in each hyperfine line can be explained in terms of Δg_n and ΔK_n similar to VO^{2+} in BCCD.

The Pa^{4+} spectra were analyzed with the spin Hamiltonian

$$\mathscr{H} = \beta\langle S|\mathbf{g}|B\rangle + \langle S|\mathbf{A}|I\rangle + \langle I|\mathbf{P}|I\rangle - \gamma\langle I|B\rangle,$$

where $\mathbf{P} = \mathbf{QT}$, which is similar to the equation used for the analysis of VO^{2+} spectra in BCCD in Subsection 9.3.2, except for the nuclear quadrupole energy, and the nuclear Zeeman energy. Experimentally, **g** and **A** tensors were found as coaxial and cylindrical at both sites, where their unique tensor axis denoted by index 1 and the crystallographic c axis are mutually orthogonal. The principal values of these tensors were reported as

$$g_1 = 1.76, \qquad g_2 \approx g_3 = 1.05;$$
$$A_1 = 695 \times 10^{-4}, \qquad A_2 \approx A_2 \approx 430 \times 10^{-4}\ (\mathrm{cm}^{-1}).$$

Hence, in the ab plane the large g and A anisotropies are considered as responsible for the anomalous broadenings of the transition lines. Figure 10.5 shows a representative Pa^{4+} spectrum in the incommensurate phase of $ThBr_4$ at 4.2 K. Considering fluctuations in amplitude mode, such broadenings can be interpreted with (9.21), as illustrated by the stick diagram in Fig. 10.5. Since the temperature is well below 95 K, there is no visible line of the phase mode, besides, the fluctuations are far from sinusoidal, showing a "soliton-like" distribution.

10.3. Phase Transitions in Deuterated Biphenyl Crystals

Molecular crystals of biphenyl ($C_{12}H_{10}$) were studied primarily for excited naphthalene and phenanthrene molecules in triplet states, when they were accommodated as guest molecules. For a general reference to this subject, interested readers are referred to the book on the triplet spectroscopy by McGlynn et al. [86]. Huchison et al. [87] performed extensive EPR studies on triplet states of small aromatic molecules in various organic host crystals. Among these crystals, Cullick and Gerkin [88] reported phase transitions in deuterated biphenyl crystals ($C_{12}D_{10}$) at 40 K and 15 K, which were first discovered by EPR spectra from phosphorescent triplet states of phenanthrene and naphtalene molecules accommodated as impurities.

According to Hirota and Hutchison [89], the phosphorescent time of ultraviolet excited diphenyl crystals is of the order of 5 s $\sim$ 10 s at 77 K and even longer at lower temperatures, which is sufficiently long to observe the magnetic resonance of excited molecules in triplet states. The triplet state ^{3}S of an aromatic molecule is due to excited π electrons, hence characterized for magnetic resonance observations by an average dipole–dipole interaction of two electrons with parallel spins [90], which is expressed by the fine-structure interaction $\langle S|\mathbf{D}|S\rangle$ for the spin $S = 1$, where $\mathbf{D}$ is a traceless tensor. Hutchison and Mangum [91] showed that these excited guest molecules are well oriented in host crystals in exactly the same way as the host molecule, and signified by two independent principal parameters of the **D** tensor. Such a fine-structure energy can be expressed in the principal form

$$\mathscr{H}_F = D_x S_x^2 + D_y S_y^2 + D_z S_z^2 \qquad \text{where} \quad D_x + D_y + D_z = 0,$$

where x, y, and z are the principal coordinate systems consistent with the molecular symmetry, as shown in Fig. 10.6(b). Owing to the traceless feature of the **D** tensor, $\mathscr{H}_F$ can be expressed in terms of two independent parameters D and E, i.e.,

$$\mathscr{H}_F = DS_z^2 + E(S_x^2 - S_y^2) \qquad \text{where} \quad D = \tfrac{1}{3}3D_z \qquad \text{and}$$

$$E = \tfrac{1}{2}(D_x - D_y),$$

if omitting the constant term. This form of $\mathscr{H}_F$ is normally used for the analysis of triplet spectra.

Because of the fine-structure energy, the triplet state is generally split into two energies, where the energy gap is called the *zero-field splitting*. In this

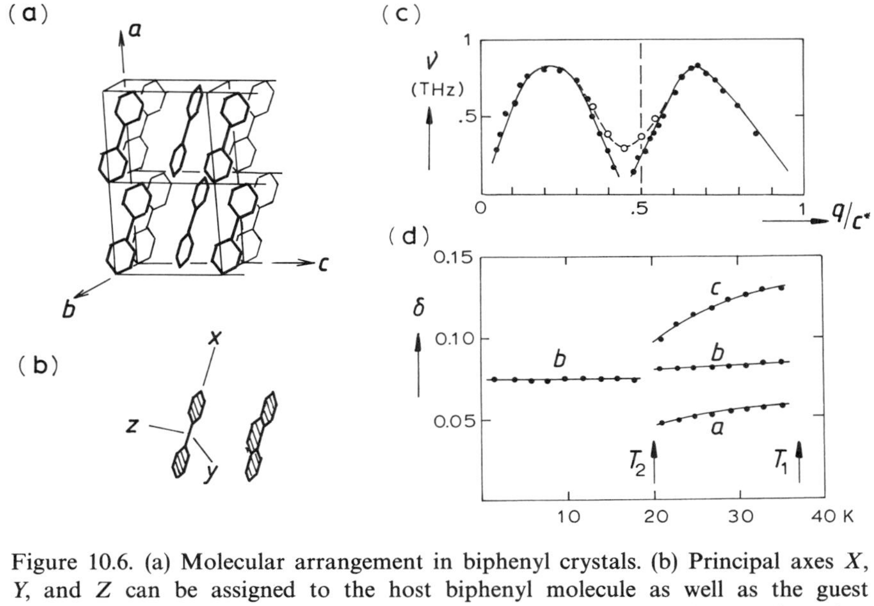

Figure 10.6. (a) Molecular arrangement in biphenyl crystals. (b) Principal axes X, Y, and Z can be assigned to the host biphenyl molecule as well as the guest phenanthrene molecules that are accommodated coaxially. (c) Phonon dispersion curve obtained by neutron inelastic scattering experiments. (d) Incommensuration parameters determined for the phases II and III of the host crystals. (From H. Cailleau: In: *Incommensurate Phases in Dielectrics*, Vol. 2, pp. 72–100, edited by R. Blinc and A. P. Levanyuk (North-Holland, Amsterdam, 1986).)

case, the total spin Hamiltonian is written as

$$\mathscr{H} = \beta\langle S|\mathbf{g}|B\rangle + \mathscr{H}_{\text{F}} + \mathscr{H}_{\text{HF}}(\text{proton}),$$

where the last term $\mathscr{H}_{\text{HF}}$ for the proton hyperfine energy can be omitted for deuterated molecules. The parameters D and E are of the order of 0.1 cm^{-1} and 0.01 cm^{-1}, respectively, and so it is not quite adequate to consider $\mathscr{H}_{\text{F}}$ as a perturbation to the Zeeman energy in a practical magnetic field B. It is usual practice to consider the Zeeman energy together with the diagonal terms of $\mathscr{H}_{\text{F}}$ in the principal directions as the unperturbed state, which is then perturbed by off-diagonal elements of $\mathscr{H}_{\text{F}}$. In the zero field $B = 0$, the two separated energy states by the zero-field splitting are specified by magnetic quantum numbers $M_S = 0$ and ± 1, and the degeneracy for $M_S = \pm 1$ can be lifted by an applied field B.

Following [91], the three energy levels for B applied parallel to one of the principal directions are expressed here as follows:

for $B \parallel \text{z}$,

$$e_{1,2} = D + (\tan \alpha_z)E \pm g_z\beta B \qquad \text{and} \qquad e_3 = 0 \qquad \text{where}$$

$$\tan 2\alpha_z = \frac{E}{g_z\beta B},$$

for $B \parallel \text{x}$,

$$e_{1,2} = \tfrac{1}{2}(D + E)(1 - \tan \alpha_x) \pm g_x\beta B \qquad \text{and} \qquad e_3 = D - E \qquad \text{where}$$

$$\tan 2\alpha_x = -\frac{\frac{1}{2}(D + E)}{g_x\beta B},$$

and for $B \parallel \text{y}$,

$$e_{1,2} = \tfrac{1}{2}(D - E)(1 \pm \tan 2\alpha_y) \pm g_y\beta B \qquad \text{and} \qquad e_3 = D + E \qquad \text{where}$$

$$\tan 2\alpha_y = \frac{\frac{1}{2}(D - E)}{g_y\beta B}.$$

These energy levels are sketched in Fig. 10.7(a) for varying B, where two magnetic resonance transitions at a fixed microwave frequency are indicated by vertical lines, which are generally allowed for magnetic dipole transitions. The parameters D, E, and principal g values can be determined from the observed sets of these magnetic resonance transitions.

Cullick and Gerkin [88] examined triplet spectra from phenanthrene-d_{10} and naphthalene-d_8 molecules in host crystals of biphenyl-d_{10} in the temperature range between 5 K and 80 K, and found that these parameters and the calculated zero-field splitting were temperature-dependent, showing anomalous splitting and discontinuity particularly at about 42 K and 15 K, respectively. While the temperature-dependence is generally attributed to librational fluctuations around the molecular x axis, these authors suggested that the discontinuous change in the fine-structure parameters should be

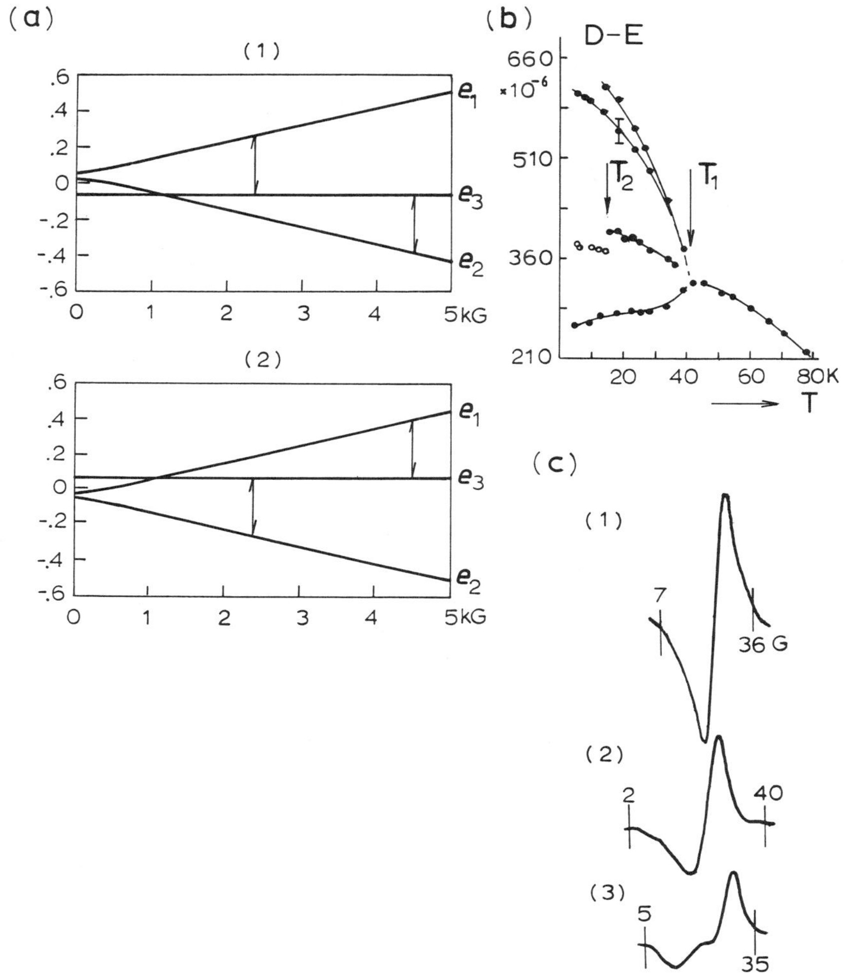

Figure 10.7. (a) Energy diagrams of an excited triplet molecule in a magnetic field. Arrows indicate magnetic resonance transitions. (1) and (2) show two distinct cases arising from different signs of the "zero-field splitting." (b) Temperature dependence of the parameter $D - E$ observed from triplet phenanthrene-d_{10} in phase II of biphenyl crystals. (c) A magnetic resonance line observed with microwaves 4419 MHz at: (1) 42.3 K; (2) 41.5 K; and (3) 40.5 K. (From A. S. Cullick and R. E. Gerkin, *Chem. Phys.* **23**, 217 (1977).)

attributed to structural changes in biphenyl crystals. Cailleau et al. [92] carried out neutron inelastic scattering experiments on biphenyl, and found soft phonons in amplitude and phase modes near the transition temperature $T_1 = 36$ K. They also reported that the inelastic neutron diffraction accompanied a strong first satellite reflection at 20 K, indicating the presence of an intrinsic mode of excitation. The soft modes in two branches are clear evidence for order-variable condensates in undeuterated crystals, which should be responsible for the magnetic resonance anomaly from guest triplet molecules that are sampling condensates in deuterated biphenyl crystals, where $T_1 = 42$ K. Line broadening of the spectra observed near T_1 by Cullick and Gerkin appeared to be "anomalous" as seen in Fig. 10.7(c) which could be attributed to the fluctuating order variable mode σ. While the anomaly in the zero-field splitting in biphenyl crystals at 42 K can be due to slow modulation of condensates during the structural change, four resolved peaks between 42 K and 15 K (Fig. 10.7(b)) could not be interpreted in this model.

Judging from the crystal structure shown in Fig. 10.6(a), the librational motion of phenyl molecules should be highly correlated. As for the short-range correlations, it is conceivable that those between similarly oriented molecules and between dissimilar ones are competitive, becoming explicit as the correlation rate is slowed down to the order of the observing microwave frequency. The incommensurability parameters were determined from neutron results as

$$\boldsymbol{q}_{\mathrm{II}} = \pm(\delta_{\mathrm{a}}\boldsymbol{a}^* - \delta_{\mathrm{c}}\boldsymbol{c}^*) + \tfrac{1}{2}(1 - \delta_{\mathrm{b}})\boldsymbol{b}^* \quad \text{in phase II (36 K} \sim \text{20 K)}$$

and

$$\boldsymbol{q}_{\mathrm{III}} = \tfrac{1}{2}(1 - \delta_{\mathrm{b}})\boldsymbol{b}^* \quad \text{in phase III } (T < 20\ \text{K}),$$

which could be calculated from short-range correlations, if known, and assigned to the phase of the order-variable condensate for structural transitions in phenyl crystals.

10.4. Successive Phase Transitions in A_2BX_4 Family Crystals

Crystals designated by the formula unit A_2BX_4 exhibit successive phase transitions from normal to incommensurate and commensurate phases, ending at ordered phases that are known as ferroelectric in some of these crystals. A variety of inorganic systems belong to this category, including crystals where the ion A^+ can be K^+, Rb^+, ${NH_4}^+$, ${N(CH_3)_4}^+$, etc., and the tetrahedral group ${BX_4}^{2-}$ is ${SO_4}^{2-}$, ${SeO_4}^{2-}$, ${ZnCl_4}^{2-}$, ${BeF_4}^{2-}$, etc. Crystallographically these crystals are isomorphous, exhibiting cascade phase transitions, and "twinning" below the transition temperature T_i [93]. In the phase below T_i, these orthorhombic crystals are spontaneously strained, resulting

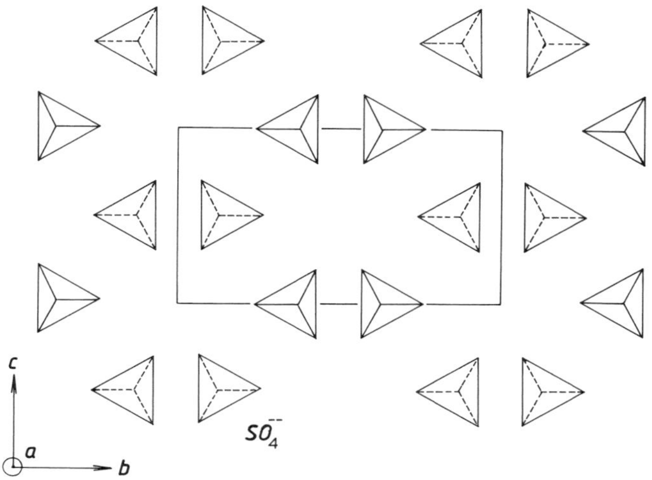

Figure 10.8. Molecular arrangement in $(NH_4)_2SO_4$ in the *bc* plane at room temperature. (From A. Sawada, Y. Makita, and Y. Takagi, *J. Phys. Soc. Japan* **42**, 1918 (1977).)

in the structure composed of three differently oriented *ferroelastic* domains. These domains are signified by temperature-dependent pseudohexagonal symmetry along the *a* axis, as illustrated in Fig. 10.8 for $(NH_4)_2SO_4$ crystals. In each domain, tetrahedral SO_4^{2-} groups are related by $\bar{3}$ screw symmetry along the sliding direction parallel to the *a* axis as illustrated in Fig. 10.9, signifying the presence of a pseudosymmetry in the normal phase, which is however diminishing with increasing temperature. It is noted that such a screw symmetry plays a significant role for successive structural changes in these strained crystals, being responsible for the modulated structure in the incommensurate phase.

Structural changes in the K_2SeO_4 crystals were studied extensively by Iizumi et al. [94] with neutron inelastic scattering experiments. According to their work, the normal-to-incommensurate phase transition is not characterized by the loss of a macroscopic symmetry element, while the mirror symmetry on the *ac* plane is locally violated at temperatures below the transition temperature T_i, when positive K^+ ions are displaced out of the mirror plane, and SeO_4^{2-} tetrahedra are rotated around the *c* axis. While in K_2ZnCl_4 the rotation axis of $ZnCl_4^{2-}$ is slightly tilted from the *c* axis in the mirror plane, soft modes in normal phases were observed near $q \sim \frac{1}{3}a^*$ in K_2SeO_4 crystals. Iizumi et al. [94] showed that in the K_2SeO_4 lattice, the normal modes for one phonon scattering are characterized by the librational fluctuations of SeO_4^{2-} tetrahedra around the *b* axis, accompanying linear displacements of K^+_α and K^+_β along the *c* direction. Clearly the librational behavior of BX_4^{2-} tetrahedra should be described in terms of a pseudospin σ as the primary-

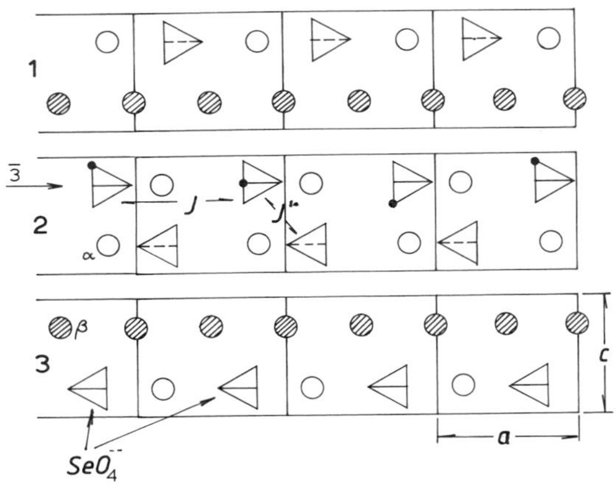

Figure 10.9. Molecular arrangement in K_2SeO_4 in the *ac* plane, where the unit cell are divided into three layers 1, 2, and 3 perpendicular to the *b* axis, and the projections to the *ac* plane are shown for each layer. Open and shaded circles indicate K_α and K_β, respectively. Three unit cells along the *a* axis are approximately equal to the wavelength of modulation. (From M. Iizumi, J. D. Axe, and G. Shirane, *Phys. Rev.* **B15**, 4392 (1977).)

order variable, in general, while displacements of A_α and A_β induced at the transition point T_i are regarded as secondary variables for the ferroelectric phase.

In the observed phonon dispersion curve shown in Fig. 4.5, a dip due to the soft mode occurs at temperatures close to $T_i = 129$ K and in the vicinity of $\boldsymbol{G}_i = 0.7\,\boldsymbol{a}^*$, where the incommensurate wavevector can be written as

$$\boldsymbol{q}_a = \frac{(1-\delta)\boldsymbol{a}^*}{3}.$$

The inelastic neutron scattering is due primarily to the collective mode of heavy anions SeO_4^{2-}. In the phase below T_i of K_2SeO_4 crystals that are characterized by the $\bar{3}$ screw symmetry, SeO_4^{2-} ions are periodically deformed and are responsible for strains in the structure. When representing the deformed SeO_4^{2-} at a lattice site x along the a direction by the order variable $\sigma(x, t)$, the collective mode expressed by a symmetric combination of the Fourier transforms σ_q and σ_{-q} can be pinned by the periodic pseudopotential $V_3 \propto \cos(3q_a x)$ at $q = \pm q_a$. The symmetry of V_3 is consistent with three ferroelastic domains below T_i, which are related quasi-trigonally with respect to the a axis. Therefore, the order variables σ cannot have a binary character, but the secondary-order variables σ_α and σ_β should be responsible for the binary transition at T_c, as inferred from the ferrielectric phase transition of $(NH_4)_2SO_4$ crystals.

In the normal-to-incommensurate transition in K_2SeO_4 crystals, the soft-mode spectrum represented symmetrical fluctuations around q_a, as seen from

Fig. 4.5. Presumably the phase mode fluctuates at a fast rate, so that the modulated structure appears to be related only to symmetrical fluctuations that are pinned by V_3. Needless to say, for such pinning of σ, the fluctuation energy $\Delta\varepsilon$ should be sufficiently lower than $\sigma_0 V_3$. If σ represents a binary mode, the wavevector fluctuations should be given by $\Delta q = \delta a^*$. Iizumi et al. [94] described the mechanism with the Landau formalism, which is however essentially the same as described above.

The incommensurability parameter δ for fluctuations at $q = a^*/3$ can be attributed to two competing correlations J and J' as indicated in Fig. 10.9, which is similar to the model for $ThBr_4$ illustrated in Fig. 10.3. For the incommensurate phase in A_2BX_4, interactions J' between two $BX_4{}^{2-}$ tetrahedra deformed in opposite directions are essential, and a process for $J' \to 0$ can be considered as responsible for the pinning transition to a commensurate phase.

For K_2SeO_4 crystals, Fukui and Abe [95] used paramagnetic VO^{2+} probes to study the nature of phase transitions. These authors observed a temperature change in one of the hyperfine lines shown in Fig. 10.10, where the transition temperatures T_i and T_c are clearly recognizable by changing lineshapes. They observed that a single hyperfine line in the normal phase changed to an anomalous shape described by $\Delta\nu = \nu_1 \cos\phi + \nu_2 \cos^2\phi^2$ for continuously distributed ϕ typical for an incommensurate phase between T_1 and T_c, as illustrated in Fig. 9.1(b). Below T_c the lineshape changed further to a three-line pattern by discrete phases $\phi = 0$, $\pi/3$, and $2\pi/3$, as given by (5.27a), whereby the lineshape was well simulated using the shift parameter $\delta a^* \sim 25°$. The results clearly indicate that incommensurate fluctuations are pinned at T_i and then locked in the pseudoperiodic potential V_3 at T_c. The

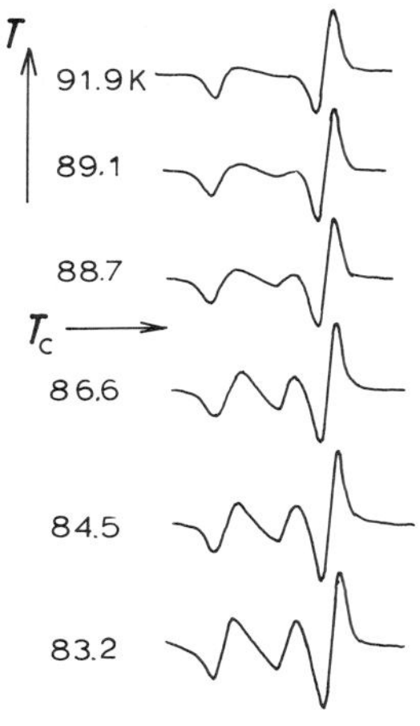

Figure 10.10. Temperature change of an anomalous VO^{2+} resonance line in K_2SeO_3 crystals. (From M. Fukui and R. Abe, *J. Phys. Soc. Japan* **12**, 3942 (1982).)

symmetry change at T_c was not reflected on their VO^{2+} line, but may have been detected with the static field applied in the mirror plane, which however was not discussed in their reports.

The behavior of $BX_4{}^{2-}$ groups in Rb_2ZnCl_4 and related crystals has been studied with paramagnetic probes Mn^{2+} and $SeO_3{}^-$ that are substituted for B^{2+} and $BX_4{}^{2-}$, respectively. Using Mn^{2+} probes, Fayet et al. [96]–[99] studied the modulated phase in Rb_2ZnCl_4 crystals, and similar experiments were performed on the $[N(CH_3)_4]_2ZnCl_4$ system by Fukui and Abe [100] and Kobayashi et al. [101]. These authors have reported that their results are similar to Mn^{2+} anomalies from the incommensurate phase of BCCD crystals. Pezeril et al. [96] identified the incommensurate phase from forbidden lines associated with $(-\frac{1}{2} \rightarrow \frac{1}{2})$ transitions, while giving their interpretation of the anomalies in $(3/2 \rightarrow 5/2)$ transition lines as due to frequency fluctuations described by (9.13).

The transition from the incommensurate phase to the commensurate phase at lower temperature was analyzed as a phase-locking phenomenon, where discommensuration lines should be found in magnetic resonance lines, for which however no positive evidence has yet been presented. The ordered phase in the lowest-temperature range should be ferroelectric, exhibiting polarized domains related by inversion, while no convincing magnetic resonance results have been reported on these systems other than those from ferrielectric $(NH_4)_2SO_4$.

Blinc et al. [102] measured the electric field gradient tensor at ^{87}Rb nuclei in incommensurate phases of Rb_2ZnCl_4 and Rb_2ZnBr_4, showing that the lineshape of the $(\frac{1}{2} \rightarrow -\frac{1}{2})$ ^{87}Rb quadrupole transition modulated via the field gradient could be analyzed generally as $\nu_{1n}\sigma + \nu_{2n}\sigma^2$, depending on the direction of the applied field $|n\rangle$. Blinc et al. [76], [77] have also reported about a line due to what they called "floating" condensates beside the pinned modulation in the vicinity of the paraelectric-to-incommensurate transition temperature of Rb_2ZnBr_4. While observable in the NMR timescale, their findings should be further justified for a possible phase mode, since it was absent in neutron scattering experiments at the incommensurate transition of K_2SeO_4.

10.5. Incommensurate Phases in $RbH_3(SeO_3)_2$ and Related Crystals

Orthorhombic crystals of $RbH_3(SeO_3)_2$ undergo a ferroelectric phase transition at about $T_c = 153$ K, where the symmetry changes to monoclinic with the ferroelectric axis along the *b* direction. Many experimental studies [103] have been performed on $RbH_3(SeO_3)_2$ crystals, since Levanyuk and Sannikov [104] predicted the presence of an incommensurate phase theoretically. Indeed, the incommensurate phase was later verified by many experiments at temperatures just above T_c, although reported transition temperature T_i depended seemingly on the sample quality. As in the A_2BO_4 system, such sequential transitions at T_i and T_c are believed to be due to primary and secondary-order variables for such incommensurate and ordered ferroelectric phases, respectively. In fact, in these experiments the primary transition has been convincingly investigated, while the secondary transition is evident from the observed dielectric anomaly at T_c [103], exhibiting extremely small spontaneous polarization. For deuterated crystals, $RbD_3(SeO_3)_2$, the incommensurate phase was also confirmed by X-ray diffraction studies [105], while the soft-mode vectors were analyzed by Grimm and Fitzgerald [106] for the structural study in the low-temperature phase.

Structures of $RbH_3(SeO_3)_2$ crystals at room temperature and in the ferroelectric phase were studied by a number of groups [107], showing that in the formula unit two SeO_3^{2-} and three hydrogen bonds are crystallographically different. One of the SeO_3^{2-} groups denoted by SeO_3(I) are linked together with one of these hydrogen bonds to form a chain along the *c* axis, while the other SeO_3(II) with the second hydrogen bond is forming another chain along the *a* direction. These parallel chains of SeO_3(I) and SeO_3(II) are, respectively, in separate planes perpendicular to the *b* axis, being linked via the third hydrogen bond. Each Rb^+ ion in the lattice is coordinated by four SeO_3(I) ions in the *ac* plane, and a SeO_3(II) ion along the *b* direction. In Fig. 10.11(a), the crystal structure in the low-temperature phase is sketched after

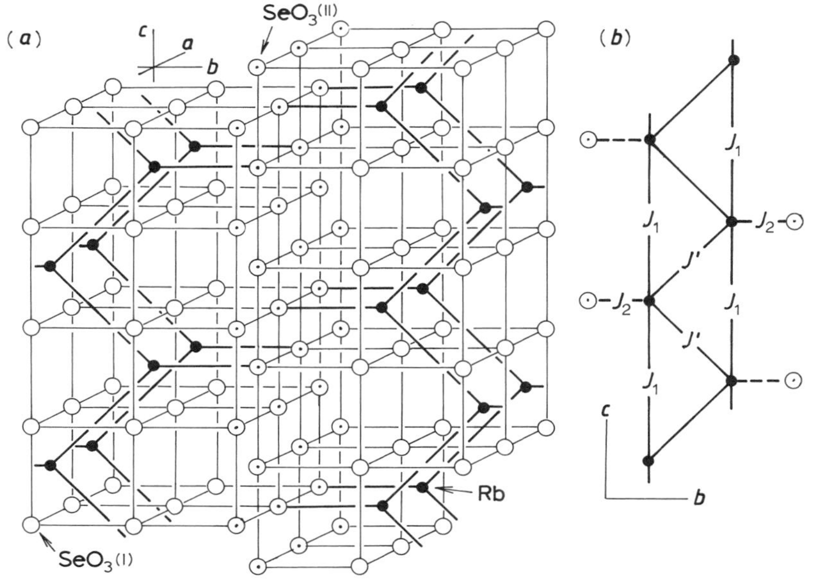

Figure 10.11. (a) Molecular arrangement in the low-temperature phase of $RbH_3(SeO_3)_2$ crystals. (From H. Grimm and W. J. Fitzgerald, *Acta Cryst.* **A34**, 268 (1978).) (b) The pseudospin chain in $RbH_3(SeO_3)_2$.

the neutron study by Grimm and Fitzgerald [106], where a layer of SeO_3(II) ions is sandwiched by layers of SeO_3(I) including Rb^+ ions. Apparently, structural changes in $RbH_3(SeO_3)_2$ crystals originate from reorientation and deformation of these SeO_3 groups, where protons play no significant role in the phase transitions [108], in contrast to other hydrogen-bonding crystals.

According to the crystal structure [106], shown in Fig. 10.11(a), each Rb^+ is at the center of a rectangular complex of four SeO_3(I) ions, and one SeO_3(II) on the pyramidal axis. Such $Rb(SeO_3)_5$ complexes are arranged in zig-zag chains in the *bc* plane, being linked with the SeO_3(II) layer. Waplak et al. [109] and Fukui et al. [110] investigated the behavior of the pyramidal complex with paramagnetic probes Cr^{3+} substituted for Rb^+ ions, and showed that the primary-order variable for the incommensurate phase transition is probably represented by such a Rb complex. Although not fully analyzed for all directions of the applied field, Cr^{3+} spectra shown in Fig. 10.12(a) indicate clearly the transition to the incommensurate phase at T_i, whereas in the spectra of Fig. 10.12(b) the subsequent transition at T_c is evident.

Such an Rb^+ chain abstracted from the crystal structure is shown in Fig. 10.11(b), which can be used as outlined in Chapter 3 to evaluate the short-range correlation scheme in $RbH_3(SeO_3)_2$ crystals, while the pseudospin

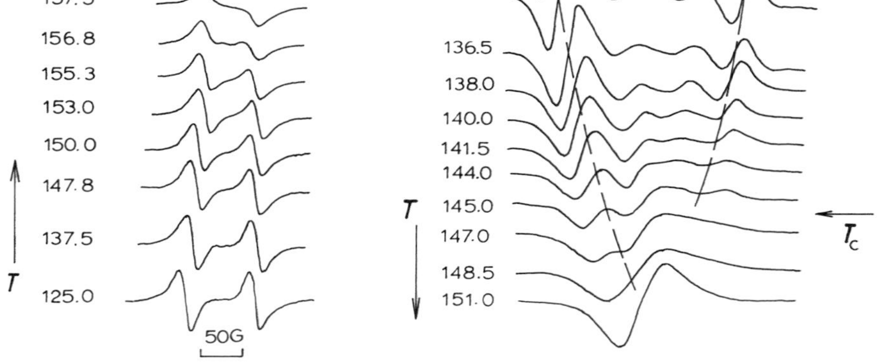

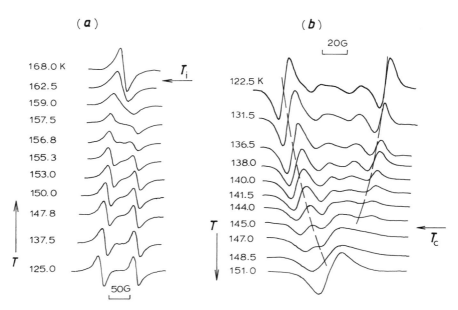

Figure 10.12. Temperature changes of $Cr_3{}^+$ resonance lines in $RbH_3(SeO_3)_2$ crystals. (a) From M. Fukui, C. Takahashi, and R. Abe, *Ferroelectrics* **36**, 315 (1981). (b) From S. Waplak, S. Jerzak, J. Stankowski, and L. A. Shuvalov, *Physica* **106B**, 251 (1981).

vector could be determined from the observed **D** tensor in the Cr^{3+} spectra. Originally predicted by Levanyuk and Sannikov however, the incommensurability in order-variable arrangement can be seen from the following simple analysis.

Considering that two pseudospins σ_α and σ_β at the nearest-neighbor positions α and β are parallel, the initial correlation energy can be calculated from

$$2E = E_\alpha + E_\beta = -\sigma^2 \boldsymbol{e}_q \cdot \boldsymbol{e}_{-q}\{J_\alpha(\boldsymbol{q}) + J_\beta(\boldsymbol{q})\},$$

where these Fourier transforms $\boldsymbol{e}_q$ and $\boldsymbol{e}_{-q}$ are virtually independent of α and β in the long-wave approximation. Therefore, E can be obtained by extremizing $J_\alpha(\boldsymbol{q}) + J_\beta(\boldsymbol{q})$ with respect to $\boldsymbol{q}$. Writing explicitly for the model in Fig. 10.11(b),

$$J_\alpha(\boldsymbol{q}) + J_\beta(\boldsymbol{q}) = 4J_1 \cos(q_c c) + 4J' \cos(q_b b)\cos(\tfrac{1}{2}q_c c) + 4J_2 \cos(q_b b).$$

Since ordering is confined to the chain, we can set $q_b = 0$, and obtain

$$2E \propto J_1 \cos(q_c c) + J' \cos(\tfrac{1}{2}q_c c) + J_2,$$

which is the same equation as (3.21) for two competing interactions J and J'. Extremizing E with respect to q_c, the relation

$$\cos(\tfrac{1}{2}q_c c) = -\frac{J'}{4J_1}$$

can be derived for the incommensurate wavevector q_c, if $|J'/4J| < 1$. Here, it is noted that $q_c = \frac{1}{2}c^*$, when $J' = 0$. Thus, for $|J'| < 4|J_1|$, $\sigma_c \propto \exp(i\phi)$, where $0 \le \phi \le 2\pi$, while the parameter J' can be temperature-dependent.

Pending experimental verification, we can assume that the order variable is a classical vector σ whose component along the c axis is $\sigma_c = \sigma \cos\phi$, and confined to the bc plane; the other component $\sigma_b = \sigma \sin\phi$ may be considered as the secondary-order variable, since quasi-reflection symmetry of the plane of SeO_3(II) ions is locally violated by a finite amplitude of σ. If $J' \to 0$ signifies the approach to T_c, we can consider that $\sigma_b \to \sigma_0 \tanh(\phi/2^{1/2})$ and $\sigma_c = 0$ below T_c. In this limit, σ_b represents the SeO_3(II) network only partially through ions involved in Rb^+ complexes, and not fully compatible to the soft layer mode predicted by Grimm and Fitzgerald [106]. It is also noted that the low-temperature phase is not necessarily ferroelectric, since in the lattice there is no mirror symmetry to be violated by such a mechanism.

10.6. Phase Transitions in $(NH_4)_2SO_4$ and NH_4AlF_4

Ammonium sulfate crystals belong to the A_2BX_4 type, but exhibiting a somewhat unusual phase transition at 223 K, as compared with K_2SeO_4, Rb_2ZnCl_4, and others. Originating from the $\bar{3}$ screw symmetry, the high-temperature phase of $(NH_4)_2SO_4$ crystals is ferroelastic [93], but the transition temperature to the hypothetical normal phase is considered to be above the melting point, and hence no soft mode could be studied for crystals under normal atmospheric pressure. In a single domain sample, the polar phase below 223 K exhibits a *ferrielectric* character in most observations. This transition to the low-temperature phase is first order, and considered as secondary ordering induced by the primary order for spontaneous strains in the high-temperature phase.

So far magnetic resonance results from $(NH_4)_2SO_4$ are from VO_2^+-doped crystals [111] and irradiated crystals [112], [113]. In these works, the behavior of NH_4^+ ions were sampled by VO^{2+} and NH_3^+ radicals, while SO_3^- and SeO_3^- radicals were utilized to study the role played by heavier ions SO_4^{2-}. However, the pseudosymmetry arising from the orientation of SO_4^{2-} groups was not much reflected on the spectra of SO_3^- and SeO_3^- radicals [111], which were found in at least three different orientations, violating the local symmetry of the lattice [114]. On the other hand, NH_3^+ spectra were found to represent NH_4^+ ions at crystallographically different sites α and β. According to the normal mode analysis for K_2SeO_4 [94], the librational motion of SeO_4^{2-} around the c axis accompanies displacements of K^+ ions out of the *ab* plane. Considering a similar mechanism in $(NH_4)_2SO_4$ crystals, such displacements of positive ions may induce a polar distortion in tetrahedral NH_4^+ ions, which is reasonably represented by an NH_3^+ radical. The spectra of NH_3^+ radicals are characterized by a unique ^{14}N-hyperfine tensor axis that is identifiable from its angular dependence. It was found that the unique axes of $(NH_3^+)_\alpha$ and $(NH_3^+)_\beta$ are both almost lying in the *ab* plane above T_c, but slightly deviating out of the plane at temperatures below T_c as illustrated in Fig. 10.13. These two NH_3^+ spectra are independent, whose temperature variations distinguished clearly two NH_4^+ ions at α and β sites, representing two sublattices. Based on the EPR results from irradiated $(NH_4)_2SO_4$ crystals, Fujimoto et al. [112] explained that the spontaneous polarization of a ferrielectric type was originated from these ordered sublattices signified by different spontaneous polarizations in opposite directions. Evidenced by such a discontinuous distortion in NH_4^+ ions, the phase transition in $(NH_4)_2SO_4$ at 223 K is first order.

The phase transition in NH_4AlF_4 crystals at 155 K belongs to a category different from $(NH_4)_2SO_4$, but there is a common feature that NH_4^+ ordering is induced by the collective mode of heavier groups under a critical condition. Fayet et al. [115] carried out magnetic resonance studies on NH_4AlF_4 crystals, and found that the phase transition has dual characters: i.e., order–disorder with respect to NH_4^+ orientation, while the phase transition is involved in continuous displacements of AlF_4 groups.

The crystal structure consists of two-dimensional separate layers of NH_4^+ ions and of a linked AlF_6^{3-} network, which are alternately stacked along the crystallographic c direction, as illustrated in Figs. 10.14(a) and (b), showing the structure in the c plane and a view from a direction perpendicular to the c axis, respectively. Here, ammonium ions are indicated by triangles, whose

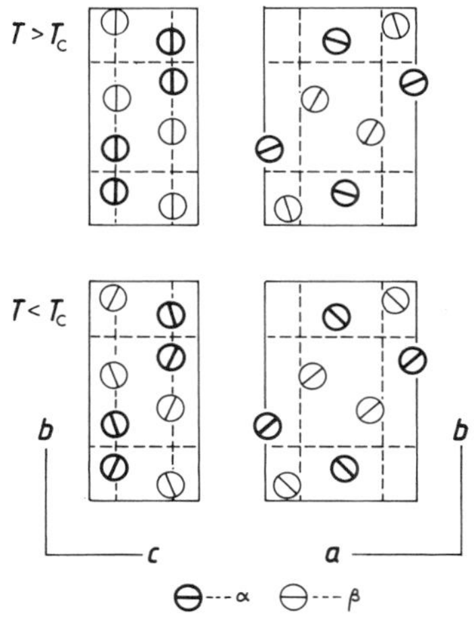

Figure 10.13. Arrangements of NH_4^+ ions above and below T_c in $(NH_4)_2SO_4$ crystals, where SO_4^{2-} ions are omitted for clarity. Inequivalent $(NH_4)_\alpha$ and $(NH_4)_\beta$ ions are indicated by thick and thin circles, respectively. Lines inside these circles indicate directions of the ^{14}N tensor axis in NH_3^+ radicals. (From M. Fujimoto, L. A. Dressel, and T. J. Yu, *J. Phys. Chem. Solids* **38**, 97 (1977).)

orientations are signified by two opposite directions parallel to the c axis. We can specify these two inversion-related orientations of an NH_4^+ ion by a pseudospin σ_a with ± 1 states. Figure 10.14(b) shows that each AlF_6^{3-} octahedron is coordinated by eight ordered σ_a's at the closest positions, either all in $+1$ states or all in -1 states in the ordered arrangement of NH_4^+. According to Fayet et al., these arrangements of ordered NH_4^+ ions have a close relation to small rotational angles $\pm\varphi$ of AlF_6^{3-} groups around the c axis, as indicated in layers 1 and 2 in the figure.

In the magnetic resonance experiments on NH_4AlF_4, Fayet et al. observed magnetic resonance spectra from Fe^{3+} probes that are substituted for Al^{3+}. The spectra in an applied field $B|n\rangle$ are characterized primarily by a uniaxial fine structure energy $\langle n|\mathbf{D}|n\rangle$, representing the AlF_6^{3-} octahedron. In the normal phase, there are two component spectra due to two magnetically inequivalent complexes in all identical unit cells, whereas at temperatures below the transition point resonance lines of each component exhibit anomalous splittings as shown in Fig. 10.15. Despite very broad linewidths of the Fe^{3+} spectra due to hyperfine splittings of six ^9F nuclei ($I = \frac{1}{2}$) in the FeF_6^{3+} complex, these authors reported on the anomalous Fe^{3+} lineshape for $B \| [110]$ that resembled Mn^{2+} spectra in the critical region of TSCC crystals.

Supposing the phase transition is first order, the single line in the anomalous Fe^{3+} spectrum can be interpreted as representing the normal phase coexisting with the low-temperature phase down to about 150 K. If, on the

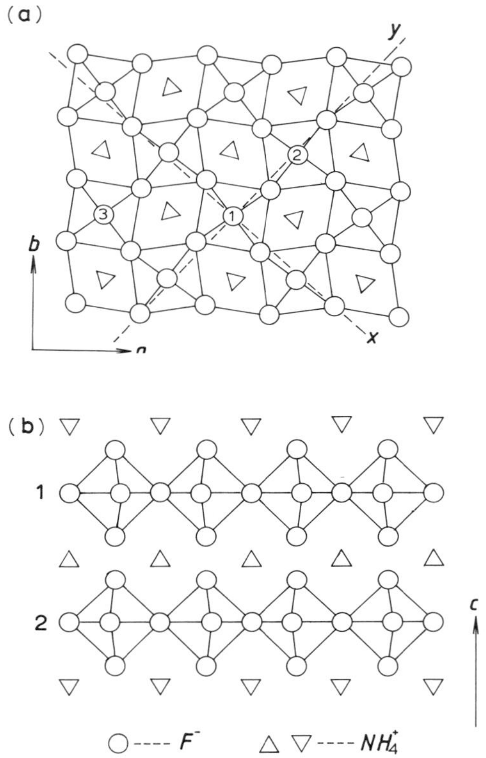

Figure 10.14. Molecular arrangements in NH_4AlF_4 crystals. (a) The layer structure in the *ab* plane. (b) Alternate stack of layers of NH_4^+ and AlF_6^{3-} perpendicular to the *c* axis. (From J. C. Fayer, *Helv. Physica Acta* **58**, 76 (1985).)

other hand, the transition is continuous, the single line may be due to fluctuations in amplitude mode. If the latter is the case, the incommensurability is likely at least in the critical region, although no convincing evidence was obtained from broad Fe^{3+} spectra.

If we consider that angular fluctuations of octahedral AlF_6^{3-} in the network shown in Fig. 10.14(a) can be described by a two-dimensional order variable $\sigma(x, y)$, the dominant short-range correlations should be between adjacent complexes 1 and 2 along the x and y directions and between 1 and 3 along the a and b axes. Denoting these interactions by parameters J and J', respectively, as discussed for perovskite crystals in Section 10.1, we can write exactly the same expression for the short-range energy, from which a specific wavevector for the minimum can be obtained for the transition threshold. As a result,

$$\cos(q_x d) = \cos(q_y d) = -\left(1 + \frac{J}{2J'}\right),$$

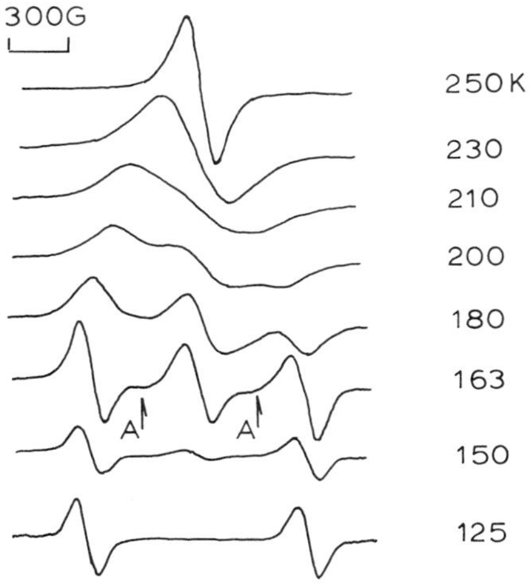

Figure 10.15. EPR anomaly observed in the Fe^{3+} spectra from NH_4AlF_4 for $B_0 \| [110]$ at and below 250 K. (From J. C. Fayer, *Helv. Physica Acta* **58**, 76 (1985).)

where d is the distance between adjacent complexes in the x and y directions. Thus, the wavevector components, $q_x = q_y$, can be irrational, provided that $0 > J/2J' > -1$, resulting in an incommensurate arrangement of $AlF_6{}^{3-}$ octahedra in the low-temperature phase. According to this argument, completely ordered librational states in the $AlF_6{}^{3-}$ network can be achieved in the limit of $J' \to \frac{1}{2}J$, where $q_x = q_y = \frac{1}{2}x^*$. In such an incommensurate phase, fluctuating order variables are given by two components, $\sigma_0 \cos \phi$ and $\sigma_0 \sin \phi$, where $\phi = q_x x - \omega t$. In the Fe^{3+} spectra observed for $B \| x$, the component spectra from complexes 1 and 2 coincide, where the fluctuations σ and $-\sigma$ are sampled by the fine structure term simultaneously, and hence the line broadening is dictated by fluctuations proportional to $\cos \phi$ and $\sin \phi$ in the sinusoidal approximation, while those arising from σ^2 are cancelled out.

In contrast to heavy $AlF_6{}^{3-}$ ions, much lighter $NH_4{}^+$ ions are in rapid rotational motion in the normal phase. In order for such fast $NH_4{}^+$ ions to couple with slow $AlF_6{}^{3-}$, the $NH_4{}^+$ mode should be softened as T_i is approached, so that the orientational ordering can be described in terms of order variables σ_a in double-well potentials. The symmetry of the double-well potential should be determined by the orientation of the $AlF_6{}^{3-}$ octahedron that is surrounded by eight oriented $NH_4{}^+$ ions. Such an arrangement as sketched in Figs. 10.14(a) and (b) can be speculated for the structural strains. Fayet et al. have proposed that somewhat disordered arrangements of eight $NH_4{}^+$ ions for weak extra lines marked A in the spectra of Fig. 10.15, which may arise from interlayer correlations of $AlF_6{}^{3-}$ complexes in the c direction. The first-order phase transition in NH_4AlF_4 crystals appears to be order–

disorder in experiments performed exclusively on NH_4^+ ions, where a displacive role of AlF_6^{3-} groups may likely be left unnoticed.

10.7. Remarks on Proton Ordering in Hydrogen-Bonded Crystals

For classical pseudospins, the ordering threshold is signified by the minimum correlation energy, at which nonzero order variables are considered to emerge. According to the one-dimensional chain model of double-well potentials, the initial collective mode of pseudospins is characterized by zero wavevector and zero frequency, as determined by (5.10). This principle appears to be acceptable for many practical systems showing an ordered arrangement in the low-temperature phase. On the other hand, in such a quantum pseudospin system as in hydrogen bonding crystals the ground state may be degenerate, resulting in a disordered low-temperature phase [116]. In most cases however, proton groups in crystals are not entirely isolated from each other, so that order variables can be treated as classical under the condition of slow correlations. Hence, degenerate ground states have not been found except in quantum liquids.

Traditionally, so-called order–disorder transitions were considered to occur independent of the lattice structure, and hence characterized by the absence of a soft mode at the transition temperature. However, whenever lattice strains are involved, ordering should be related with a specific lattice mode at least in the critical region. On the basis of the one-dimensional chain model in Section 5.3, the initial sinusoidal mode of pseudospins at $k_0 \cong 0$ may couple with the soft lattice mode to form a propagating condensate. On the contrary, in a first-order phase transition the initial pseudospin mode may be characterized by a finite amplitude in a double-well potential with a finite depth, and cannot interact in "resonance" with the sinusoidal soft phonons. Assuming that $k < k_0$ for the initial mode, the corresponding frequency can be complex, i.e., $\omega = \omega_1 + i\omega_2$, where the imaginary part ω_2 represents an energy loss when tunneling through the potential barrier. In this case, the coupling with the soft phonons $\boldsymbol{u}(k, \omega_1)$, as expressed by (4.1), gives rise to an exponential factor $\exp(-\omega_2 t)$, indicating an energy dissipation of the tunneling pseudospin mode. Naturally, such a condensate behaves as relaxational, exhibiting temperature-dependent damping near T_c.

Potassium dihydrogen phosphate KH_2PO_4 crystals, known as KDP, exhibit a ferroelectric phase transition of order–disorder at 122 K. Kaminow and Damen [117] found in Raman studies of KDP an intense overdamped temperature-dependent response in the low-frequency spectra, as shown in Fig. 10.16. These authors showed that their observed lineshapes could fit to the damped harmonic oscillator model with the characteristic frequency $\varpi \propto (T - T_0)^{1/2}$ and a constant damping Γ, behaving very similar to that of a Debye relation.

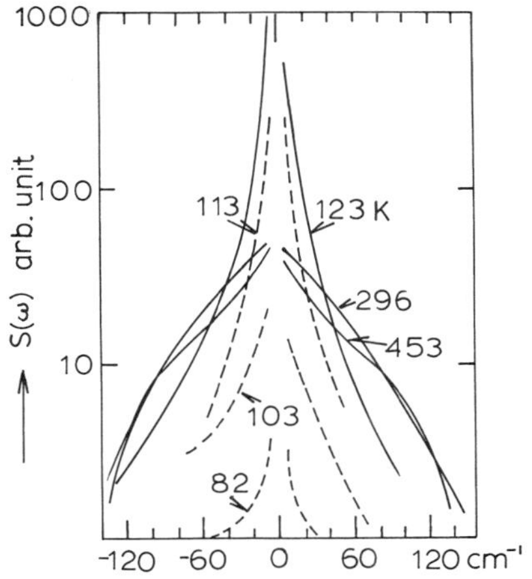

Figure 10.16. Overdamped phonon spectra observed in KDP crystals. (From I. P. Kaminow and T. C. Damen, *Phys. Rev. Lett.* **20**, 1105 (1968).)

Under the circumstances, condensates in the first-order phase transition are localized and cannot propagate. Accordingly, unlike in the critical region of a continuous phase transition, the crystal cannot be modulated, and active groups are directly subjected to thermal interactions with the lattice. In Fig. 10.17(a) the active group in KDP crystals is sketched, where the thermal statistics for proton arrangements is known as the Slater–Takagi model [118]. The group consists of a $PO_4{}^{3-}$ group which is linked with four neighboring

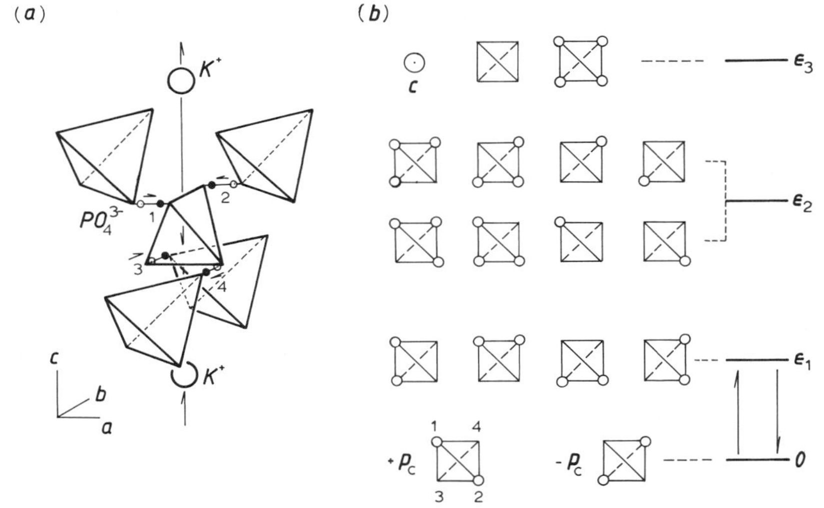

Figure 10.17. (a) Proton configuration around a $PO_4{}^{3-}$ ion in KDP crystals. (b) The Slater–Takagi model.

phosphate ions via four hydrogen bonds. In the ground state, two protons in the hydrogen bonds 1 and 2 in the figure are close to oxygens of the central PO_4^{3-}, while the other protons in bonds 3 and 4 are close to the neighboring PO_4^{3-} ions. In addition, there are 14 excited states depending on the distribution of four protons among these "close" positions to the PO_4^{3-} tetrahedron, as illustrated in Fig. 10.17(b). Among these excited states, four configurations in the first excited energy ε_1 are characterized by two near protons either in (1, 3) or (2, 4) bonds, eight configurations at the second excited energy ε_2 are associated with either one or three protons, and the two configurations with four or no protons in the proximity are the highest excited energy ε_3, as shown in the diagram.

In this model, the transition to the ordered ferroelectric phase occurs at the temperature T_c, which is determined by the relation

$$k_B T_c = \varepsilon_1 \ln 2,$$

suggesting that only those configurations at ε_1 are significant for the phase transition. Furthermore, the ground-state configurations are related by inversion, where the displaced charge distribution in $H_2^{2+}PO_4^{3-}$ constitutes an electric dipole moment either p_c or $-p_c$ along the c axis. Hence, ignoring higher states ε_2 and ε_3, we can consider an approximate model for the phase transition in crystals of KDP-type, where a tunneling inversion of the moment p_c across the potential barrier is represented by the energy ε_1.

Magnetic resonance studies on KH_2AsO_4 crystals, a representative system of KDP type, were carried out independently by two research groups led by Blinc [119] and McDowell et al. [120]. These workers found that irradiated crystals showed no significant change in thermodynamical properties of KH_2AsO_4 by generating radicals AsO_4^{4-}, which were used as paramagnetic probes to study the nature of the ferroelectric phase transition at 97 K. In this case, while the ^{75}As spin ($I = \frac{3}{2}$) exhibited large hyperfine splittings, two protons associated with AsO_4^{4-} give a small hyperfine structure sensitive for a change in the proton arrangement. To study the Slater–Takagi model, McDowell and his associates performed ENDOR (electron–nuclear double resonance) experiments at 4.2 K and determined the interaction tensors from two near protons and two far protons with great accuracy and with clear distinction, while in the EPR spectra hyperfine splittings from the far protons were too small to detect. Figures 10.18(a) and (b) show some of the results of magnetic resonance experiments. A representative spectrum from AsO_4^{4-} radicals in Fig. 10.18(a) exhibits a proton structure of 1:2:1 intensity ratio, implying that two near protons are magnetically equivalent. Figure 10.18(b) shows the results obtained by Blinc's group, showing temperature changes of these proton structures.

In these results, it is significant to notice that the proton structure changes near ~220 K to the intensity ratio 1:4:6:4:1, signifying the complete disorder of four protons, which was however not consistent with the phase transition at 97 K. Such inconsistency as shown in Fig. 10.18(b) can be attributed

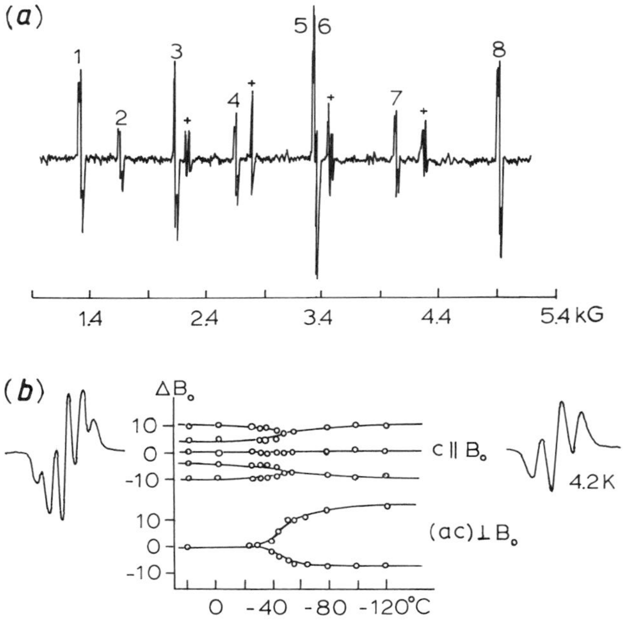

Figure 10.18. (a) A representative EPR spectra of AsO_4^{4-} radicals in irradiated KH_2AsO_4 crystals at 4.2 K. The lines marked 1, 2, ..., 8 are spectra of AsO_4^{4-} at two inequivalent sites, and those marked + represent secondary unidentified species. (From N. S. Dalal, J. A. Hebden, D. E. Kennedy, and C. A. McDowell, *J. Chem. Phys.* **66**, 4425 (1977).) (b) Temperature dependence of the proton hyperfine structure in the EPR spectra of AsO_4^{4-} radicals in KH_2AsO_4. (From R. Blinc, P. Cevc, and M. Schara, *Phys. Rev.* **159**, 411 (1967).)

to a long timescale for the proton arrangement, which was longer than the EPR timescale, so that the electron probe was looking at proton spins as if they were in quasi-static states. As involved in the three energies ε_1, ε_2, and ε_3, the process to disordered protons should be more complex than disorder in a random process among equal energies, and the observed discrepancy in AsO_4^{4-} spectra may not be quite surprising.

References

[1] C. J. Adkins, *Equilibrium Thermodynamics* (McGraw-Hill, London, 1968).
[2] M. W. Zemansky, *Heat and Thermodynamics*, 5th edn. (McGraw-Hill, New York, 1957).
[3] A. B. Pippard, *Classical Thermodynamics* (Cambridge University Press, London, 1964).
[4] C. Kittel and H. Kroemer, *Thermal Physics*, 2nd edn. (W. F. Freeman, San Francisco, 1980).
[5] H. D. Megaw, *Crystal Structures: A Working Approach* (W. B. Saunders, Philadelphia, 1973).
[6] See Chapter 9 in [3].
[7] L. D. Landau and E. M. Lifshitz, *Statistical Physics*, transl. by E. Peierls and R. F. Peierls (Pergamon Press, London, 1958).
[8] C. Kittel, *Introduction to Solid State Physics*, 6th edn., pp. 633–635 (Wiley, New York, 1956).
[9] J. H. E. Stanley, *Introduction to Phase Transition and Critical Phenomena* (Oxford University Press, New York, 1971).
[10] R. A. Cowley and A. D. Bruce, *Structural Phase Transitions* (Taylor and Francis, London, 1981).
[11] F. C. Nix and W. Shockley, *Rev. Mod. Phys.* **10**, 1 (1938); T. Muto and Y. Takagi, *Solid State Physics* **1**, 194 (1955).
[12] R. Blinc and B. Zeks, *Soft Modes in Ferroelectrics and Antiferroelectrics*, Chapter 5 (North-Holland, Amsterdam, 1974).
[13] L. Onsager, *Phys. Rev.* **65**, 117 (1944).
[14] R. Becker, *Z. Angew. Phys.* **6**, 23 (1954); *Theory of Heat*, transl. by G. Leibfried, 2nd edn. (Springer-Verlag, New York, 1967).
[15] M. Born and K. Huang, *Dynamical Theory of Crystal Lattices* (Oxford University Press, Oxford, 1954).
[16] C. Kittel, *Introduction to Solid State Physics*, 6th edn., Chapter 15 (Wiley, New York, 1956).
[17] H. Gränicher and K. A. Müller, *Mat. Res. Bull.* **6**, 977 (1971).
[18] R. Becker, *Theory of Heat*, transl. by G. Leilfried, 2nd edn., page 274 (Springer-Verlag, New York, 1967).

[19] R. Comès, R. Currat, F. Denoyer, M. Lambert, and A. M. Quittet, *Ferroelectrics* **12**, 3 (1976).
[20] K. A. Müller, W. Berlinger, and F. Waldner, *Phys. Rev. Lett.* **21**, 814 (1968).
[21] M. Fujimoto, S. Jerzak, and W. Windsch, *Phys. Rev.* **B34**, 1668 (1986).
[22] A. Yoshimori, *J. Phys. Soc. Japan* **14**, 807 (1959).
[23] M. Fujimoto, *Ferroelectrics* **47**, 177 (1983).
[24] M. Kakudo, S. Bando, and T. Ashida, *Acta Cryst.* **B28**, 1131 (1972).
[25] E. Nakamura, K. Itoh, K. Deguchi, and N. Mishima, *Japan J. Appl. Phys.* Supp. **24-2**, 393 (1985).
[26] S. Jerzak and M. Fujimoto, *Canad. J. Phys.* **63**, 377 (1985).
[27] P. A. Lee, T. M. Rice, and P. W. Anderson, *Solid State Commun.* **14**, 703 (1974).
[28] J. M. Ziman, *Models of Disorder*, page 23 (Cambridge University Press, Cambridge, 1979).
[29] S. Chkazumi, *Physics of Magnetism* (Wiley, New York, 1964).
A. H. Morrish, *Physical Principles of Magnetism* (Wiley, New York, 1965).
Magnetism, edited by G. T. Rado and H. Suhl, in several volumes (Academic Press, New York, 1963).
[30] W. Cochran, *Adv. Phys.* **9**, 387 (1960); **10**, 401 (1961).
[31] P. W. Anderson, In *Fizika Dielectrikov*, edited by G. I. Skanavi (Moskow, 1960).
[32] J. Pryzystava, *Physics of Modern Materials*, vol. 2 (IAEA, Vienna, 1980).
[33] R. E. Peierls, *Quantum Theory of Solids*, page 108 (Oxford University Press, London, 1955).
[34] R. J. Elliott and A. F. Gibson. *An Introduction to Solid State Physics and its Applications*, Chapter 3 (Macmillan, London, 1976).
[35] R. A. Cowley, *Rep. Prog. Phys.* **31**, 123 (1968).
[36] S. M. Shapiro, J. D. Axe, G. Shirane, and T. Riste, *Phys. Rev.* **B6**, 4332 (1972).
[37] K. A. Müller, *Lecture Notes in Physics*, vol. 104, page 209, edited by C. P. Enz (Springer-Verlag, Heidelberg, 1971).
[38] A. Sawada and M. Horioka, *Japan J. Appl. Phys.* Suppl. **24-2**, 390 (1985).
[39] H. Böttger, *Principles of the Theory of Lattice Dynamics* (Physik-Verlag, Weiheim, 1983).
[40] L. Bernard, R. Corrat, P. Delamoye, C. M. Zeyen, S. Hubert, and R. de Kouchkovsky, *J. Phys.* **C16**, 433 (1983).
[41] R. Currat and T. Janssen, *Excitations in Incommensurate Crystal Phases*, page 201 in *Solid State Physics*, vol. 41, edited by H. Ehrenreich and D. Turnbull (Acadmic Press, San Diego, 1988).
[42] M. Wada, H. Uwe, A. Sawada, Y. Ishibashi, Y. Takagi, and T. Sakudo, *J. Phys. Soc. Japan* **43**, 544 (1977).
[43] M. J. Rice, in *Solitons and Condensed Matter Physics*, vol. 8, page 246, edited by A. R. Bishop and T. Schneider (Springer-Verlag, Berlin, 1978).
[44] Cz. Pawlaczyk, H.-G. Unruh, and J. Petzelt, *Phys. Stat. Sol.* **136(b)**, 435 (1986).
[45] M. Fujimoto, Cz. Pawlaczyk, and H.-G. Unruh, *Phil. Mag.* **69**, 919 (1989).
[46] J. A. Krumhansl and J. R. Schrieffer, *Phys. Rev.* **B11**, 3535 (1975).
[47] S. Aubry, *J. Chem. Phys.* **64**, 3392 (1976).
[48] G. L. Lamb Jr., *Elements of Soliton Theory* (Wiley, New York, 1980).
[49] F. C. Frank and J. H. van der Merwe, *Proc. Roy Soc. London* **A198**, 205 (1949).
[50] Xiaoquig Pan and H.-G. Unruh, *J. Phys. Cond. Matter* **2**, 323 (1990).
[51] P. M. de Wolff, *Acta Cryst.* **A30**, 777 (1974); *ibid.* **A33**, 493, 1977.
A. Janner and T. Janssen, *Phys. Rev.* **B15**, 643, 1977.
T. Janssen and A. Janner, *Physics* **126A**, 163 (1984).
T. Janssen, *Phys. Rep.* **168**, 55 (1988).
[52] M. P. Schulhof, P. Heller, R. Nathans, and A. Linz, *Phys. Rev.* **B1**, 2403 (1970).
[53] R. Pinn and B. E. F. Fender, *Physics Today* **38**, 47 (1985).

[54] P. S. Peercy, J. F. Scott, and P. M. Bridenbaugh, *Bull. Amer. Phys. Soc.* **21**, 337 (1976).
J. C. Toledano, G. Errandonea, and J. P. Jaguin, *Solid State Commun.* **20**, 905 (1976).
[55] E. B. Wilson, J. C. Decius, and P. C. Cross, *Molecular Vibrations* (McGraw-Hill, New York, 1955).
[56] J. F. Scott, *Rev. Mod. Phys.* **46**, 83 (1974).
J. F. Scott, *Raman Spectroscopy of Structural Phase Transitions in Light Scattering Near Phase Transitions*, edited by H. Z. Cummins and A. P. Levanyuk (North Holland, Amsterdam, 1983).
[57] G. V. Kozlov, A. A. Volkov, J. F. Scott, G. E. Feldkamp, and J. Petzelt, *Phys. Rev.* **B28**, 225 (1983).
[58] K. Deguchi, N. Aramaki, E. Nakamura, and K. Tanaka, *J. Phys. Soc. Japan*, **52**, 1897 (1983).
[59] J. Petzelt, G. V. Kozlov, and A. A. Volkov, *Ferroelectrics* **73**, 101 (1987).
[60] Cz. Pawlacyk, H.-G. Unruh, and J. Petzelt, *Phys. Stat. Sol.* **(b)136**, 435 (1986).
[61] J. Petersson, *Z. Naturforsch.* **(a)34**, 538 (1979).
[62] A. Abragam and B. Bleaney, *Electron Paramagnetic Resonance of Transition Ions* (Clarendon Press, Oxford, 1970).
[63] F. Bloch, *Phys. Rev.* **70**, 460 (1946).
[64] A. Abragam, *The Principles of Nuclear Magnetism*, page 42 (Oxford University Press, London, 1961).
[65] H. Bethe, *Ann. Physik* **3**, 133 (1929).
[66] A. Abragam and M. H. L. Price, *Proc. Phys. Soc.* **A63**, 409 (1950); *Proc. Roy. Soc.* **A205**, 135; *ibid.* **A206**, 135, 173 (1951).
[67] S. Jerzak and M. Fujimoto, *Canad. J. Phys.* **63**, 377 (1981).
[68] H. J. Rother, J. Albers, and A. Klöpperpieper, *Ferroelectrics* **54**, 107 (1984).
[69] W. Brill and K. H. Ehses, *Japan J. Appl. Phys.* **24**, Suppl. 24-2, 826 (1985).
[70] A. A. Volkov, Yu G. Goncharov, G. V. Kozlov, J. Albers, and J. Petzelt, *JETP* **44**, 606 (1986).
[71] R. Ao and G. Schaack, *Ind. J. Pure Appl. Phys.* Raman Diamond Jubilee, 1988.
[72] W. Brill, W. Schildkamp, and J. Spilker, *Z. Kristallogr.* **172**, 281 (1985).
[73] M. Fujimoto and Y. Kotake, *J. Chem. Phys.* **90**, 532 (1989).
[74] M. Fujimoto and Y. Kotake, *J. Chem. Phys.* **91**, 6671 (1989).
[75] R. N. Rogers and G. E. Pake, *J. Chem. Phys.* **33**, 1107 (1960).
[76] B. W. van Beest, A. Janner, and R. Blinc, *J. Phys.* **C16**, 5409 (1983).
[77] R. Blinc, D. C. Ailion, P. Prelovsek, and V. Rutar, *Phys. Rev. Lett.* **50**, 67 (1983).
[78] R. Blinc, F. Milia, B. Topic, and S. Zumer, *Phys. Rev.* **B29**, 4173 (1984).
[79] R. Blinc, S. Juznic, V. Rutar, J. Seliger, and S. Zumer, *Phys. Rev. Lett.* **44**, 609 (1980).
[80] R. Blinc, *Phys. Rep.* **79**, 331 (1981).
[81] P. Segransan, A. Jánossy, C Berther, J. Mercus, and P. Butaud, *Phys. Rev. Lett.* **56**, 1854 (1986).
[82] Th. von Waldkirch, K. A. Müller, and W. Berlinger, *Phys. Rev.* **B5**, 4324 (1972); *ibid.* 1052 (1973).
[83] S. Hubert, P. Dalamoye, S. Lefrant, M. Lepostollec, and M. Hussonios, *J. Solid State Chem.* **36**, 36 (1981).
[84] J. Emery, S. Hubert, and J. C. Fayet, *J. Physique Lett.* **45**, 693 (1983); *J. Physique* **46**, 2099 (1985).
[85] G. Zwanenburg, Thesis (Catholic University, Nijmegen, 1990).
[86] S. P. McGlynn, T. Azumi, and M. Kinoshita, *Molecular Spectroscopy of Triplet States* (Prentice Hall, Englewood Cliffs, NJ, 1969).
[87] C. A. Hutchison, Jr. and B. W. Mangum, *J. Chem. Phys.* **29**, 952 (1958); *ibid.*, **34**, 908 (1961).

R. W. Brandon, R. E. Gerkin, and C. A. Hutchison, Jr., *J. Chem. Phys.* **41**, 3717 (1964).
R. W. Brandon, G. L. Cross, C. E. Davoust, C. A. Hutchison, Jr., B. E. Kohler, and R. Silbey, *J. Chem. Phys.* **43**, 2006 (1965).
[88] A. S. Cullick and R. E. Gerkin, *Chem. Phys.* **23**, 217 (1977).
[89] N. Hirota and C. A. Hutchison Jr., *J. Chem. Phys.* **42**, 2869 (1965).
[90] A. W. Honig and J. S. Hyde, *Mol. Phys.* **6**, 33 (1963).
[91] C. A. Hutchison Jr. and B. W. Mangum, *J. Chem. Phys.* **34**, 908 (1961).
[92] H. Cailleau, J. C. Messager, F. Moussa, F. Bugant, C. M. E. Zeyen, and C. Vettier, *Ferroelectrics* **67**, 3 (1986).
H. Cailleau, F. Moussa, C. M. E. Zeyen, and J. Bouillot, *J. Physique Coll.* **42**, 704 (1981).
[93] Y. Makita, A. Sawada, and Y. Takagi, *J. Phys. Soc. Japan*, **41**, 167 (1976).
[94] M. Iizumi, J. D. Axe, and G. Shirane, *Phys. Rev.* **B15**, 4392 (1977).
[95] M. Fukui and R. Abe, *J. Phys. Soc. Japan*, **51**, 3942 (1982).
[96] M. Pezeril, J. Emery, and J. C. Fayet, *J. Physique Lett.* **41**, 499 (1980).
[97] M. Pezeril and J. C. Fayet, *J. Physique Lett.* **43**, 267 (1982).
[98] A. Kaziba, M. Pezeril, J. Emery, and J. C. Fayet, *J. Physique Lett.* **46**, 387 (1985).
[99] A. Kaziba and J. C. Fayet, *J. Physique* **47**, 239 (1986).
[100] M. Fukui and R. Abe, *Japan J. Appl. Phys.* **20**, L-533 (1981).
I. Suzuki, K. Ysuchida, M. Fukui, and R. Abe, *ibid.* **20**, L-840 (1981).
[101] T. Kobayashi, M. Suhara, and M. Machida, *Phase Trans.* **4**, 281 (1984).
[102] R. Blinc, D. C. Ailion, J. Dolinsek, and S. Zumer, *Phys. Rev. Lett.* **50**, 67 (1983).
[103] L. A. Shuvalov, N. R. Ivanov, N. V. Gordeyeva, and L. F. Kirpichnikova, *Soviet Phys. Crystallogr.* **14**, 554 (1970).
K. Gesi, K. Ozawa, and Y. Makita, *Japan J. Appl. Phys.* **12**, 1963 (1973).
[104] A. P. Levanyuk and D. G. Sannikov, *Soviet Phys. Solid State*, **12**, 1418 (1971).
[105] Y. Makita and S. Suzuki, *J. Phys. Soc. Japan*, **36**, 1215 (1974).
[106] H. Grimm and W. J. Fitzgerald, *Acta Cryst.* **A34**, 268 (1978).
[107] A. B. Tovbis, T. S. Davydova, and V. I. Simonov, *Soviet Phys. Crystallogr.* **17**, 81 (1972).
R. Tellgren, D. Armed, and R. Luminga, *J. Solid State Chem.* **6**, 250 (1975).
[108] L. A. Shuvalov, N. R. Ivanov, N. V. Gordeyeva, and L. F. Kirpichnikova, *Phys. Lett.* **A33**, 490 (1970).
[109] S. Waplak, S. Jerzak, J. Stankowski, and L. A. Shuvalov, *Physica* **106B**, 251 (1981).
[110] M. Fukui, C. Takahash, and R. Abe, *Ferroelectrics* **36**, 315 (1981).
[111] M. Fujimoto, T. J. Yu, and K. Furukawa, *J. Phys. Chem. Solids* **39**, 345 (1978).
[112] M. Fujimoton, K. Furukawa, and T. J. Yu, *J. Phys. Chem. Solids* **40**, 101 (1979).
[113] N. Shibata, R. Abe, and I. Suzuki, *J. Phys. Soc. Japan*, **41**, 2011 (1976).
R. Abe and N. Shibata, *J. Phys. Soc. Japan*, **43**, 1308 (1977).
[114] S. Jerzak, private communication
[115] J. C. Fayet, *Helv. Physica Acta* **58**, 76 (1985).
[116] R. M. Stratt, *J. Chem. Phys.* **84**, 2315 (1985).
[117] I. P. Kaminow and T. C. Damen, *Phys. Rev. Lett.* **20**, 1105 (1968).
[118] J. C. Slater, *J. Chem. Phys.* **9**, 16 (1941).
Y. Takagi, *J. Phys. Soc. Japan* **3**, 271 (1948).
[119] R. Blinc, P. Cevc, and M. Schara, *Phys. Rev.* **159**, 411 (1967).
[120] N. S. Dalal, C. A. McDowell, and R. Srinivasan, *Chem. Phys. Lett.* **4**, 97 (1969); *Phys. Rev. Lett.* **25**, 823 (1970); *Mol. Phys.* **24**, 1051 (1972).
N. S. Dalal and C. A. McDowell, *Phys. Rev.* **B5**, 1074 (1972).
N. S. Dalal, J. A. Hebden, and C. A. McDowell, *J. Chem. Phys.* **62**, 4404 (1975).

Index